Surviving the Climate Crisis

This is the first textbook to adopt an integrated perspective of climate change in Australia, drawing on research detailed in the latest Intergovernmental Panel on Climate Change (2021, 2022) Sixth Assessment Reports to make it the most up-to-date resource available. It fills a knowledge gap in an ever-increasing hot topic for the country, its people, economy and environment.

Australia has been identified by a number of respected sources as a 'climate change hotspot', with all major sectors of the economy considered vulnerable or highly vulnerable to the anticipated adverse impacts of climate change. The chief industry sectors examined in this book include energy, transportation, manufacturing, agriculture, fisheries, forestry, tourism and mining. Other chapters focus on other key thematic areas, such as protected areas and world heritage sites (including their natural and cultural values), coastal and island environments, biosecurity, biodiversity and ecosystem services, human health, water resources, cities and settlements, rural and regional areas, and Indigenous communities.

Ideal for advanced undergraduate and graduate students with limited science backgrounds, this book will inform those undertaking business, management, sustainability, education, environmental, development or heritage studies and other social science programmes.

Praise for the Book

"This book is a climate change tour-de-force. Covering climate science and the history of climate knowledge, impacts past, present and future, and the solutions that we can, and must employ, it is an invaluable one-stop-shop for anyone engaged in the most challenging issue of our time."

- **Lesley Hughes,** Distinguished Professor of Biology, Macquarie University, Australia, and board member of Australia's Climate Change Authority

"Australia is on the frontlines of the climate crisis. Read *Surviving the Climate Crisis* by Stephen M. Turton to understand the science, the impacts, and the solutions to this defining threat."

- **Michael Mann,** Presidential Distinguished Professor, University of Pennsylvania, and author of *The New Climate War*

Surviving the Climate Crisis
Australian Perspectives and Solutions

Stephen M. Turton

CRC Press
Taylor & Francis Group
Boca Raton London New York

CRC Press is an imprint of the
Taylor & Francis Group, an **informa** business

First edition published 2023
by CRC Press
6000 Broken Sound Parkway NW, Suite 300, Boca Raton, FL 33487-2742

and by CRC Press
4 Park Square, Milton Park, Abingdon, Oxon, OX14 4RN

CRC Press is an imprint of Taylor & Francis Group, LLC

© 2023 Stephen M. Turton

ISBN: 978-1-032-03953-4 (hbk)
ISBN: 978-1-032-03947-3 (pbk)
ISBN: 978-1-003-18990-9 (ebk)

DOI: 10.1201/9781003189909

Typeset in Minion
by codeMantra

Contents

Preface   xi

Acknowledgements   xiii

Author   xv

List of Acronyms   xvii

1   Introduction   1
2   The Science of Climate Change   9
     Introduction   9
     Fundamentals of the Greenhouse Effect   9
          Radiation balance   9
          Understanding greenhouse gases and their sources and sinks   11
     Drivers of Climate Variability and Change   15
          Orbital forcing (Milankovitch cycles)   15
          Solar forcing   17
          Volcanic forcing   18
          Internal forcing   18
          Anthropogenic forcing   19
     Past Climates   20
          Paleo-reference periods   21
          Past 800 ka   21
          Past 2 ka   24
     Changes in the Modern Era   24
          Dealing with certainty and uncertainty   24
          Past 140 years   26
               Greenhouse Gases   26
               Global Mean Surface Temperature   27
               Precipitation   27
               Cryosphere Changes   28
               Oceanic Changes   28
               Sea-Level Changes   29
               Hot and Cold Extreme Events   29
               Poleward and Upward Shifts in Climate Zones and the Terrestrial Biosphere   29
               Changes in Tropical Cyclones   30
               Compound Extreme Events   30
     Future Changes in Climate   30
          Climate models   30
          Shared socio-economic pathways   31

Projections of future climates 32
Global Mean Surface Air Temperature 32
Carbon Cycle Changes 33
Precipitation and Water Cycle Changes 33
Oceanic Changes 34
Sea-Level Changes 34
Hot and Cold Extreme Events 34
Cryosphere Changes 35
Changes in Tropical and Extra-Tropical Cyclones 35
Compound Extreme Events 35
3 The Climate of Australia: Past, Present and Future 37
Introduction 37
Past Climates 37
Present Climates 40
Climate drivers 40
Pacific Decadal Oscillation 40
El Niño Southern Oscillation 41
Indian Ocean Dipole 43
Southern Annular Mode 44
Madden-Julian Oscillation 44
Climate types 45
Historical Changes in Climate 47
Changes in key regional climate indicators 48
Temperature Averages and Extremes 48
Precipitation and Hydrology 48
Fire Weather 49
Oceanic Conditions 49
Sea Level 50
Tropical Cyclones 50
Compound Extreme Events 50
Future Climates 51
Projected changes in key regional climate indicators 51
Large-Scale Climate Drivers 51
Changes in Climate Types and Zones 52
Changes in Other Climate Indicators 52
Australia at 3°C Plus 54
4 Responding to Climate Change 57
Introduction 57
Conceptual Framework and Key Definitions 58
Natural, economic and social systems 58
Climate change impacts, exposure and sensitivity 59
Adaptive capacity 59
Vulnerability, coping capacity, resilience and risks 60
Planned adaptation and maladaptation 60
Mitigation 61
Climate-Resilient Pathways and Opportunities 62
Scenarios and pathways 62
Mitigation and adaptation pathways 63
International Climate Change Conventions and Agreements 68
United Nations Framework Convention for Climate Change 68
Kyoto Protocol 69

Paris Agreement	71
Intergovernmental Science-Policy Platform on Biodiversity and Ecosystem Services	75
United Nations 2030 Agenda for Sustainable Development	75
Sustainable Development Goals	75
Roles and Responsibilities for Climate Change Action in Australia	77
5 Natural Systems	83
Introduction	83
Terrestrial and Freshwater Ecosystems	88
Climate impacts and risks	88
Coastal and Marine Ecosystems	93
Climate impacts and risks	93
Climate-Resilient Development Pathways for Ecosystems, Ecosystems Services and Biodiversity	98
Mitigation and adaptation strategies	98
Integration with the sustainable development goals	103
6 Economic Systems	105
Introduction	105
Energy and Transportation	107
Socio-economic context	107
Managing risk	110
Climate impacts and risks	111
Climate-resilient development pathways and opportunities	112
Greenhouse Gas Emissions	112
Mitigation Pathways	113
Adaptation Pathways	117
Integration with the Sustainable Development Goals	118
Agriculture and Land	119
Socio-economic context	119
Managing risk	121
Climate impacts and risks	122
Natural Capital	123
Quality and Quantity of Crops and Pastures	123
Biosecurity	126
Heat Stress on Livestock	126
Distribution of Crops and Livestock	127
Climate-resilient development pathways and opportunities	127
Greenhouse Gas Emissions	127
Mitigation Pathways	130
Adaptation Pathways	133
Integration with the Sustainable Development Goals	135
Mining	135
Socio-economic context	135
Managing risk	139
Climate impacts and risks	139
Disturbance to Mine Infrastructure and Operations	140
Changing Access to Supply Chains and Distribution Routes	140
Challenges to Worker Health and Safety Conditions and Community Relations	141
Challenges to Environmental Management	141
Climate-resilient development pathways and opportunities	142
Greenhouse Gas Emissions	142
Mitigation Pathways	142

Adaptation Pathways   144
Integration with the Sustainable Development Goals   147
Manufacturing and Construction   147
Socio-economic context   147
Managing risk   147
Climate impacts and risks   149
Climate-resilient development pathways and opportunities   150
Greenhouse Gas Emissions   150
Mitigation Pathways   150
Adaptation Pathways   152
Integration with the Sustainable Development Goals   153
Tourism   154
Socio-economic context   154
Managing risk   156
Climate impacts and risks   157
Impacts on Tourist Travel Decisions   157
Impacts on Tourism Destinations   158
Climate Change Threats to World Heritage Sites   159
Climate-resilient development pathways and opportunities   159
Greenhouse Gas Emissions   159
Mitigation Pathways   162
Adaptation Pathways   163
Integration with the Sustainable Development Goals   164

7   Social Systems   169
Introduction   169
Cities, Settlements and Built Environments   172
Climate impacts and risks   172
Rising Average Temperatures and Extreme Heatwaves   174
Changes in Rainfall   175
Water Scarcity   175
Bushfire Risk   176
Rising Sea Levels   177
Human Health and Well-Being   178
Climate impacts and risks   178
Heatwaves   179
Bushfires   179
Water Scarcity   180
Floods and Storms   180
Indigenous Peoples   181
Background context   181
Climate impacts and risks   182
Climate mitigation and adaptation   184
Cross-Cutting Social Issues   184
Compound extreme events   184
Breakdown of institutions and governance systems   185
Climate-Resilient Development Pathways and Opportunities   186
Mitigation pathways   186
Adaptation pathways   188
A Leadership Role for Local Government   188

Adaptation Strategies for Local Government Areas 190

Integration with the sustainable development goals 192

8   Synthesis 201

References **209**

Index **229**

Preface

Climate change – primarily driven by human activities – has been identified by numerous scientific publications as among the greatest existential threats to humanity and other lifeforms on Earth. The latest Intergovernmental Panel for Climate Change (IPCC, 2021) Synthesis Report states, 'it is unequivocal that human influence has warmed the atmosphere, ocean and land.' It warns that the Paris Agreement goals of 1.5°C and 2°C above pre-industrial levels will be exceeded this century without dramatic reductions in greenhouse gas emissions (IPCC, 2021).

Australia has been identified as a 'climate change hotspot', with its natural, economic and social systems considered vulnerable or highly vulnerable to climate change (Lawrence et al., 2022). This vulnerability is due to the geographical location of Australia and dominant sub-tropical high-pressure weather systems (Turton, 2017). It has long held the global distinction of being the driest continent with permanent habitation, with most of its population located in the more temperate coastal southeast, south and southwest of the country (see map Chapter 3, Figure 3.1). Sixty-six per cent of Australia is arid and semi-arid (see Chapter 3, Figure 3.3). Over the past 50 years, there has been a detectable drying and warming of continental south-eastern and particularly south-western parts of the country in response to climate change (IPCC, 2021). Meanwhile, the sparsely populated northwest of the continent has witnessed increased rainfall and warming temperatures.

Australia's future climates will likely follow similar patterns to those observed since the 1950s. Minimum and maximum temperatures will increase, as will the number of hot days and warm spells. Strong drying will continue in the southwest of Western Australia and parts of South Australia and western Victoria. Indications are that moderate drying will continue across the highly populated continental southeast. However, rainfall may continue to increase across the far northern fringes of the Northern Territory and the Kimberly region of Western Australia and parts of Cape York Peninsula in Queensland. Rising sea levels and extreme storm surge events will affect all coastal areas of the country, but to varying degrees. Future changes in Australia's climate depend strongly on global greenhouse gas emissions over the coming decades (IPCC, 2021). Even with lower emissions, Australia is facing many decades of warming, changes in rainfall, rising sea levels and more extreme climate-driven events.

Australia has received considerable global media attention due to catastrophic bushfires over large areas of the southeast of the country. An astonishing 21% of Australia's forests burned during the 2019–2020 bushfire season, an amount considered globally unprecedented. This extreme event was a consequence of years-long droughts linked to warming and drying associated with climate change (Boer et al., 2020). International media attention has also featured Australia's world heritage-listed Great Barrier Reef (GBR), a global icon considered one of the world's seven natural wonders. Over the past 7 years, four mass coral bleaching events have affected the GBR in varying degrees (2016, 2017, 2020 and 2022). The first three events resulted in a decision by the International Union for the Conservation of Nature (IUCN, 2020) to re-classify the conservation outlook status of the GBR from 'significant concern' to 'critical' due to a deteriorating trend in the condition of its outstanding universal values. This decline in its conservation status is due to oceanic warming and associated marine heatwaves driving mass coral bleaching events.

Record-breaking east coast floods in 2022 have also propelled Australia into the international arena as a country at the frontier of climate change. Climate change is set to triple the cost of natural disasters in Australia without urgent government spending on resilience, according to a recent report by the country's Commonwealth Scientific and Industrial Research Organisation (CSIRO) (Naughtin et al. 2022).

Australia has been captivated by an existential climate crisis, with demand for action on climate change considered one of the main reasons for the defeat of the sitting Liberal-National Party Coalition Government at the May 21 2022 federal election. The latest State of the Environment Report (Cresswell et al., 2021), the release of which (in July 2022) was delayed by the previous Australian Government, paints a dire picture of the recent impacts of climate change on the country's environment, ecosystems services and biodiversity. This report is alarming as crucial sectors of Australia's economy rely on the health of its natural capital (e.g. agriculture and tourism). This timely book will examine climate change impacts, mitigation and adaptation in Australia. It will also explore climate-resilient development pathways for the country's natural, economic and social systems. There is a lack of up-to-date resources and university-level textbooks that adopt an integrated perspective of climate change in Australia. This book aims to fill the knowledge gap in a rapidly growing challenge for the country, its people, economy and the environment.

This book is for advanced undergraduate and graduate students with limited science backgrounds. This audience includes those undertaking business, management, sustainability, education, environmental studies, development studies, heritage studies and other social science programmes. Those with science or technology backgrounds will also benefit from the content, as they will be exposed to societal and economic aspects of climate change in Australia. The content may also interest other readers, including individuals in the government and private sectors. There is increasing demand for knowledge and understanding of climate change in both sectors' day-to-day operational and strategic planning activities. There is an urgent need for governments and businesses to find ways to achieve climate-resilient development. Finally, academic readers outside of Australia may find the content helpful for comparing climate change risks, mitigation and adaptation strategies in their regions.

Acknowledgements

I begin by acknowledging the *Gubbi Gubbi* and *Jinibara Peoples*, the traditional custodians of the land on which I wrote this book, paying my respects to their elders past, present and emerging.

The genesis for this book arose when I was teaching a graduate-level (master's) course on climate change impacts, mitigation and adaptation in Australia. This popular course regularly comprised a class of international and Australian students with diverse undergraduate backgrounds, including geography, environmental science, engineering, health sciences, education, business, management, social sciences and humanities. At the time, no suitable textbook was available to guide students through such a complex topic in an integrated way. Hence, my decision to write this book. I especially thank the many students for being the inspiration for this book. Your feedback has helped shape the book in a manner that I hope is useful to others undertaking similar courses.

My interest in weather and climate began at a young age, and I was fortunate to have science and geography teachers at high school who encouraged this interest. When I read geography at the University of Canterbury in the late 1970s, I was privileged to have excellent lecturers and mentors. I particularly acknowledge my master's thesis advisor, Emeritus Professor Andy Sturman, for encouraging my interests in climatology and, more recently, anthropogenic climate change impacts and adaptation in Australia. Over many years, I had engaging climatology conversations with my PhD advisor at James Cook University, the late Professor Mike Bonell. His mentorship and encouragement were much appreciated, even after leaving Australia for a post with UNESCO in Paris. More recently, I thank Professor Grant Stanley, Deputy Vice-Chancellor of Research at Central Queensland University, for supporting my adjunct professorial appointment and affording access to online resources. His encouragement and kindness are much appreciated.

I am very grateful for the many offers to read earlier versions of chapters and for the constructive feedback and encouragement from Emeritus Professor Nigel Stork, Adjunct Professor Peter Valentine, Dr Miriam Goosem, Dr David Turton and Ian Stannard. I thank the excellent editorial and production teams at CRC Press (Taylor & Francis Group, UK) for supporting this book project and acknowledge Alice Oven for her assistance along the journey.

Lastly, I want to thank my family, colleagues and friends for their support and encouragement over the past 18 months. My wife, Wendy Turton, has proofread and provided constructive comments on every chapter of this book. Wendy's attention to detail has been nothing short of outstanding!

Author

Stephen M. Turton, born in 1959, grew up in Southern Africa and New Zealand and has lived in Australia for 40 years. He graduated from the University of Canterbury in 1982 (BSc and MSc in geography) and James Cook University in 1992 (PhD in geography and environmental science). He is now retired and an Adjunct Professor in Environmental Geography in the Research Division at Central Queensland University. From 2005 to 2016, he held several senior roles as Director and Professor of Geography at James Cook University in Cairns. From 2003 to 2005, he was an Associate Professor and Director of Research for the Rainforest Cooperative Research Centre. From 1984 to 2003, he was a Lecturer, Senior Lecturer and Associate Professor in Geography at James Cook University. He is a former Councillor of the Royal Geographical Society of Queensland, a former Councillor of the Institute of Australian Geographers, a former Councillor of the Association for Tropical Biology and Conservation and a former member of the Wet Tropics Management Authority's Scientific Advisory Committee. He is a Past President of the Australian Council of Environmental Deans and Directors, a Past President of the Institute of Australian Geographers and Immediate Past Chair of the National Committee for Geographical Sciences, Australian Academy of Science. He was also an expert reviewer for the Intergovernmental Panel for Climate Change's (IPCC) Fifth and Sixth Assessment Reports, Working Group 2 (Impacts and Adaptation). He is a Distinguished Fellow of the Institute of Australian Geographers and a recipient of the Royal Geographical Society of Queensland's highest honour, the J.P. Thomson Medal. He has published over 130 scientific reports, book chapters and journal articles in rainforest ecology, environmental geography and climate change adaptation in tourism and natural resource management. Steve and his wife, Wendy, a fellow graduate of the University of Canterbury and James Cook University, have three sons, all living in Australia.

List of Acronyms

AAS	Australian Academy of Science	**GMSL**	Global Mean Sea Level
ABARES	Australian Bureau of Agricultural and Resource Economics and Sciences	**GMST**	Global Mean Surface Temperature
		GSAT	Global Mean Surface Air Temperature
ABS	Australian Bureau of Statistics	**GWP**	Global Warming Potential
ABS	Australian Bureau of Statistics	**IOD**	Indian Ocean Dipole
ACCU	Australian Carbon Credit Unit	**IPBES**	Intergovernmental Science Platform on Biodiversity and Ecosystem Services
ACT	Australian Capital Territory		
AEMO	Australian Energy Market Operator		
AISC	Australian Industry and Skills Committee	**IPCC**	Intergovernmental Panel for Climate Change
ARENA	Australian Renewable Energy Agency	**IUCN**	International Union for the Conservation of Nature
BCA	Business Council of Australia		
BOM	Bureau of Meteorology	**LGA**	Local Government Area
BP	Before Present	**LGM**	Last Glacial Maximum
CCS	Carbon Capture and Storage	**LIG**	Last Interglacial
CCUS	Carbon Capture, Use and Storage	**LNG**	Liquid Natural Gas
CE	Common Era	**MBA**	Market-Based Mechanism
CEPF	Critical Ecosystem Partnership Fund	**MCA**	Minerals Council of Australia
CER	Clean Energy Regulator	**MDB**	Murray Darling Basin
CMIP	Coupled Model Inter-comparison Project	**MJO**	Madden-Julian Oscillation
		NAP	National Adaptation Plan
COA	Commonwealth of Australia	**NCCARF**	National Climate Change Adaptation Facility
COP	Conference of Parties		
CSIRO	Commonwealth Scientific and Industrial Organisation	**NCP**	Nature's Contributions to People
		NDC	Nationally Determined Contribution
DCCEE	Department of Climate Change and Energy Efficiency	**NEM**	National Electricity Market
		NFF	National Farmers Federation
DISER	Department of Industry, Science, Energy and Resources	**NSW**	New South Wales
		NT	Northern Territory
ENSO	El Niño-Southern Oscillation	**OECD**	Organisation for Economic Co-operation and Development
ERF	Effective Radiative Forcing		
ERF	Emissions Reduction Fund	**PDO**	Pacific Decadal Oscillation
EV	Electric Vehicle	**PIA**	Planning Institute of Australia
FFDI	Forest Fire Danger Index	**PV**	Photovoltaic
GBR	Great Barrier Reef	**RET**	Renewable Energy Target
GCM	Global Circulation Model	**RF**	Radiative Forcing
GDP	Gross Domestic Product	**SAM**	Southern Annular Mode
GHG	Greenhouse Gases	**SDG**	Sustainable Development Goal

SSP	Shared Socio-economic Pathway	**UNFCCC**	United Nations Framework Convention on Climate Change
TCI	Tourism Climatic Index		
UN	United Nations	**WA**	Western Australia
UNEP	United Nations Environment Programme	**WHA**	World Heritage Area
		WMO	World Meteorological Organization

Introduction

Global climate change is one of the greatest challenges facing humanity in the twenty-first century.

Angela Merkel BVO (2010), former Chancellor of Germany

Over the past 12,000 years, human societies, economies and nature have benefited from a remarkably stable but variable climate during the Holocene interglacial epoch (Figure 1.1). Climate variability over this multi-millennial period was driven by natural processes, as carbon dioxide (CO_2) levels in the atmosphere ranged from only 265 to 285 parts per million (see Chapter 2, Box 2.1). Natural climate drivers included orbital forcing associated with Milankovitch cycles, small changes in solar output, volcanic eruptions and the effects of the El Niño-Southern Oscillation (ENSO) phenomenon in the Pacific Ocean (see Chapter 2). These natural drivers of the climate system have affected surface temperatures and rainfall patterns across the Earth at a range of spatial and temporal scales. However, natural, economic and social systems have largely remained within their 'coping ranges' during the Holocene, with only occasional short departures from the overall climatic stability provided by relatively constant CO_2 levels (Figure 1.1). Holocene climatic stability enabled the growth of human civilisations worldwide and included the domestication of crops and animals and the construction of the world's first cities.

With the advent of the industrial revolution in the late 18th century, levels of greenhouse gases (GHGs) from human activities (e.g. burning fossil fuels, agriculture and deforestation) began to rise rapidly, and the Age of Humans or Anthropocene[1] was ushered in during the mid-20th century. In the Anthropocene, natural drivers of climate variability continue (e.g. solar, volcanic, ENSO). However, rates of climate change are steadily increasing each decade as GHGs from human activities continue to accumulate in the atmosphere (see Chapter 2, Figure 2.5). Rather than a stable and variable climate – afforded by consistent levels of natural GHGs – the climate system is now changing or exhibiting discernible trends, notably atmospheric and oceanic warming (Figure 1.1). Changing 'baseline' climate conditions force natural, economic and social systems beyond their normal coping ranges. Either autonomous or planned adaptation is then required to reduce the vulnerability of these systems to changes in baseline climate conditions and increased climatic variability (see Chapter 4, Figure 4.1). As GHGs continue to rise, there will be further shifts in Earth's baseline climate system and corresponding increases in the frequency, severity and duration of more extreme

---

[1] The Anthropocene is a proposed geological epoch dating from the commencement of significant human impact on Earth's geology and ecosystems, including – but not limited to – anthropogenic climate change (Waters et al. 2016). It is generally accepted that the Anthropocene started in the 1950s.

DOI: 10.1201/9781003189909-1

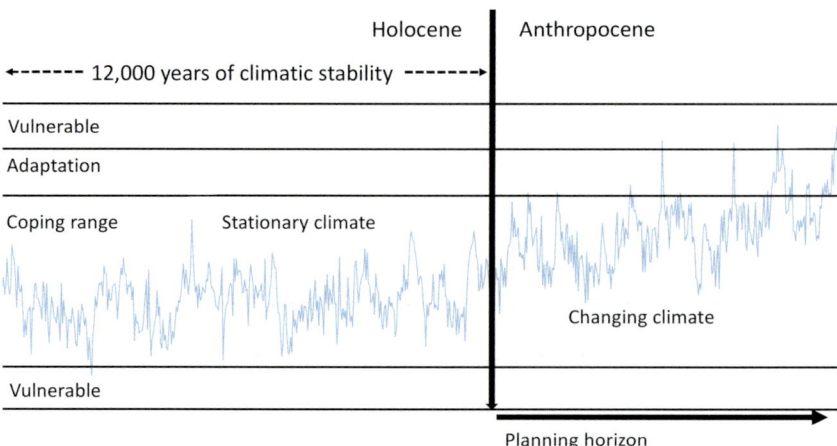

Figure 1.1 Schematic diagram showing the variable but stationary (stable) Holocene climate, compared with the variable but changing Anthropocene climate (i.e. a discernible upward trend). Under the stationary Holocene, climate, natural, economic and social systems largely remained within their coping ranges. Under the changing Anthropocene climate, such systems must adapt to changing conditions to avoid vulnerability. The planning horizon assumes that climate change will continue and various systems will require autonomous or planned adaptation to avoid vulnerability (see Chapter 4, Figure 4.1). Axes are not drawn to scale. (Adapted from Jones and Mearns, 2005.)

climate-driven events, such as floods, droughts, heatwaves, bushfires and storm surges. Australia is highly exposed to climate change and associated increases in more extreme climate-driven events (see Chapter 3).

The science of climate change is not new, and many scientists have been active in sharing their profound knowledge about atmospheric physics and chemistry for 200 years (Table 1.1). Their significant contributions have included compiling 'unequivocal' evidence that human activities have altered Earth's climate system and demonstrating that rates of global change are accelerating in the 21st century (see Chapter 2). Significant theoretical and practical advances in the study of the atmosphere began with Joseph Fourier in 1822, who first discovered that the atmosphere retained heat. While Fourier never used the well-known term 'greenhouse effect' in his publications, his work paved the way for later discoveries of the role of GHGs in maintaining habitable temperatures for complex lifeforms to exist on Earth. Debate continues as to who first demonstrated the absorption of thermal (infrared) radiation by carbon dioxide (CO_2) and water vapour (H_2O). Some claim that Eunice Foote, an amateur American scientist, made this discovery in 1856, 3 years before John Tyndall, an English physicist. Most sources cite

Tyndall as the first person to make this crucial scientific discovery, but both should be recognised for their contributions to atmospheric science.

The pioneering work of Foote and Tyndall was taken a step further in 1896 by Svante Arrhenius, a Swedish physicist, who was the first person to theorise that a doubling of atmospheric CO_2 would result in about 4°C of atmospheric warming. Arrhenius' theory still stands true, even though his research pre-dated the establishment of a near-global climate observing network in 1905. This monitoring network would later validate his theory (Table 1.1). Thirty years later, E.O. Hulburt, an American physicist, would verify the earlier work of Foote, Tyndall and Arrhenius by developing the 'CO$_2$ theory' of the ice ages.

One of the most significant climate researchers of the 20th century was Milutin Milankovitch, a Yugoslavian (now Serbian) mathematician, astronomer and climatologist. His ground-breaking research on 'planetary climatology' is still highly regarded (Table 1.1). Notably, his detailed explanation of Earth's long-term climate changes caused by changes in the position of the Earth in comparison to the Sun, now known as 'orbital forcing' or Milankovitch cycles (see Chapter 2, Table 2.4).

The study of Earth's climate system increased rapidly in the second half of the 20th century as

Table 1.1 Significant theoretical and practical advances, milestones and discoveries in climate science. Adapted from Weart (2008), Mason (2020), IPCC (2013) and IPCC (2021).

Year	Climate science advances, milestones and discoveries
1822	Joseph Fourier's 'greenhouse effect' analogy that the atmosphere retains heat established, but Fourier never actually used the term.
1840	Louis Agassiz glacial theories propose that ice ages had occurred in Earth's past.
1856	Eunice Foote discovers the absorption of thermal radiation by CO_2 and H_2O (i.e. greenhouse gases).
1861	John Tyndall first demonstrates infrared absorption by various gases (some sources credit this to Foote, 3 years earlier).
1896	Svante Arrhenius becomes the first person to investigate the effect that doubling atmospheric CO_2 would have on global climate.
1897	Ocean temperatures measured to a depth of 150 m.
1900	T.C. Chamberlin describes the global geological cycle.
1905	Near-global climate observing network established.
1931	E.O. Hulburt calculates 4°C warming with a doubling of atmospheric CO_2, supporting the CO_2 theory of the ice ages, originally proposed by Tyndall.
1940	Systematic radiosonde (vertical atmospheric layers) monitoring established.
1941	Milutin Milankovitch demonstrated the multi-millennial-scale effects of changes in Earth's position relative to the Sun (orbital forcing) as essential drivers of long-term climate, including initiating the commencement and cessation of periods of glaciation.
1955	Hans Suess identifies isotopic signature of fossil fuels in the atmosphere.
1956	Gilbert Plass calculates warming using a layered model of the atmosphere, i.e. radiative absorption by CO_2.
1957	Roger Revelle and Hans Suess propose that oceans are not absorbing all anthropogenic CO_2.
1958	Bert Bolin and Erik Erikson determine that carbon sequestration in the deep ocean is too slow to keep up with the rate of CO_2 emissions from human activities.
1958	David Keeling measures yearly increases in atmospheric CO_2 for the first time.
1967	Suki Manabe and Richard Wetherald model CO_2 with water vapour feedback.
1969	World's first deep ice core (Greenland).
1978	Continuous satellite temperature measurements commence.
1979	Charney Report, a scientific assessment published on CO_2 and climate.
1990	IPCC First Assessment Report.
1999	Vostok, Antarctica 420,000-year ice core.
2000	Earth systems models with coupled carbon cycles. Ocean data to 2,000 m from Argo floats.
2001	Satellites measure enhanced greenhouse effect at the top of the atmosphere.
2010	Ice sheet mass loss established.
2015	Satellites measure enhanced greenhouse effect at Earth's surface.
2021	IPCC Sixth Assessment Report, energy and sea level budgets closed.

new technologies (e.g. satellites, radiosondes, supercomputers, ocean and atmosphere sensor networks) were deployed (Table 1.1). Scientists like Hans Suess, an Austrian-born American physicist, began to study the effects of combusted fossil fuels on the atmosphere, including identifying their isotopic signatures. By the late 1950s, it was evident from work by several researchers (i.e. Revelle and Suess, and Bolin and Erikson) that CO_2 concentrations were increasing in the atmosphere and oceans. They also verified that the rate of carbon sequestration (uptake) in the shallow and deep oceans was insufficient to keep pace with the atmospheric rise of GHGs from human activities (see Chapter 2, Figure 2.2).

One of the most significant scientific advances in climate science measurements occurred in 1958, when David Keeling, an American atmospheric scientist, established constant monitoring of ambient CO_2 concentrations at the Mauna Loa Observatory in Hawaii (Table 1.1). Keeling charted the steady rise of CO_2 levels in the atmosphere at a fine scale, and the 'Keeling Curve' is named in his honour as one of the most fundamental measures of global change (see Chapter 2, Figure 2.5). In 1969, the world's first deep ice core was extracted from the Greenland Ice Sheet, allowing scientists to measure changes in well-mixed GHGs back over several glacial and interglacial cycles (Table 1.1). Some 30 years later, a deeper ice core was taken at Vostok, a Russian research station high on the east Antarctic plateau. This ice core extended the climate record back to 420,000 years before present (BP). The Vostok ice core record was later extended back 800,000 years BP (see Chapter 2, Figure 2.3).

The deployment of sophisticated weather satellites in the 1970s, including the first continuous satellite temperature measurements, was a crucial step in global-level assessments of anthropogenic climate change (Table 1.1). In the late 1970s, the President of the United States, Jimmy Carter, ordered an official scientific assessment of CO_2 and climate, culminating in the Charney Report in 1979. The report is an exemplar of robust science, and the success of its predictions over the past four decades has underpinned the science of global warming (see Chapter 2), including the crucial work of the Intergovernmental Panel on Climate Change (IPCC). The IPCC published its First Assessment Report in 1990 (Table 1.1) and its most recent Sixth Assessment Report in 2021 (see Chapter 4).

Despite the overwhelming scientific evidence (Table 1.1, see also Chapter 2), it remains common in Australia to see politicians, media commentators or social media users cast doubt on the role of humans in driving climate change (Turton, 2022). However, this denialism is now almost non-existent among academics and scientists, according to a recent study by Lynas et al. (2021). They examined the peer-reviewed literature and found more than 99% of researchers now endorse the evidence for human-induced (anthropogenic) climate change. This revised figure is even higher than the 97% reported by Cook et al. (2013), which has become a widely cited statistic by both climate change deniers and those who accept the evidence.

Why has the needle evidently shifted even more firmly in favour of the evidence-based consensus? What has happened to the 3% of researchers who rejected the consensus of human caused climate change back in 2013? The answer lies in the more recent study. Lynas et al. (2021) re-examined the literature published since 2012, and their analysis is based on the same methods as the Cook et al. (2013) study, with a few minor refinements. Both studies searched the Web of Science database – an independent worldwide repository of scientific paper citations – using the keywords 'global climate change' and 'global warming'. However, Lynas et al. (2021) added 'climate change' to the other two keyword searches, because the researchers found that most climate-contrarian papers would not have been returned with only the two original terms. Cook et al. (2013) examined 11,944 climate research papers. They found almost one-third of them expressed a position on the cause of global warming. Of these 4,014 papers, 97% endorsed the consensus position that humans are the cause, 1% were uncertain, and 2% explicitly rejected it.

Another review paper (Benestad et al., 2015) examined 38 climate-contrarian papers published over the preceding decade and identified a range of methodological flaws and sources of bias. One of the reviewers commented that 'every single one of those analyses had an error – in their assumptions, methodology, or analysis – that, when corrected, brought their results into line with the scientific consensus'. For example, many of the contrarian papers had 'cherrypicked' results that supported their conclusion, while ignoring important context and other data sources that contradicted it. Some of them simply ignored fundamental physics. Benestad et al. (2015) also made the important point that 'science is never settled and that both mainstream and contrarian papers must be subjected to sustained scrutiny'. This is the cornerstone of the 'scientific method', and few if any climate scientists would disagree with this statement.

Action on climate change was a feature of the May 2022 Australian federal election and played a significant role in the rise of conservative 'Teal' independents in some of Australia's most affluent 'Blue Ribbon' electorates in Sydney, Melbourne and Perth, and contributed to wins by several Greens Party candidates in inner-Brisbane electorates. Both disparate groups campaigned for stronger federal leadership and action on climate change. The election of a new federal Labor Party Government is poised to end the internal 'Climate Wars' that plagued previous Liberal-National Party Coalition Governments and thwarted effective climate policy in the country for almost a decade.

This book aims to provide an integrated perspective of the impacts of anthropogenic climate change on Australia's natural, economic and social systems and identify climate-resilient development pathways for these systems now and in the future (Figure 1.2). The book addresses the following compelling questions:

1. What are the main drivers of natural climate variability, and how do they interact with anthropogenic drivers of climate change, such as rising greenhouse gases and land use change?
2. Why is it important to understand past changes in Earth's climate and the significance of changes to the natural 'greenhouse effect'?
3. What have been the main changes in the global climate system since the industrial revolution in the late 18th century?
4. What are the likely projections for global climate attributes this century, such as rises in atmospheric and oceanic temperatures, changes in rainfall patterns, sea level changes, ice sheet mass changes and increases in extreme climate-driven events?
5. How has Australia's climate changed in the deep and recent past? What are the likely future changes in regional climate attributes this century, including land and sea temperatures, rainfall patterns, sea level rise and severe climate-driven events, such as heatwaves, storms, floods, droughts and bushfires?
6. How can Australia cope with a changing climate and increasing climatic variability? What are the options to build climate-resilient development pathways for natural, economic and social systems? How can policy- and

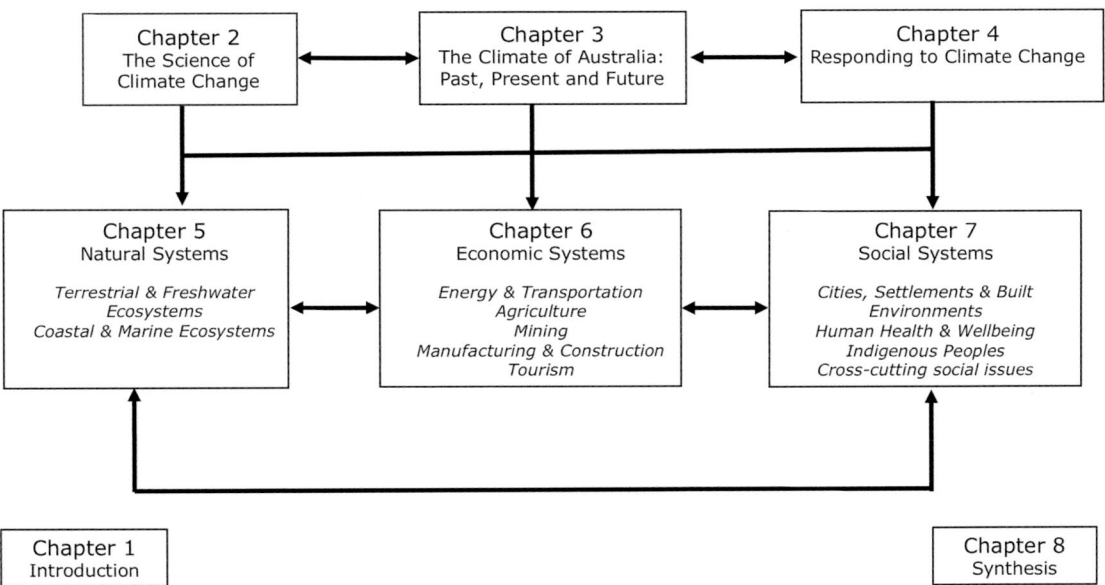

Figure 1.2 The structure of the book showing the six main chapters and how they link together.

decision-makers integrate these pathways with the United Nations' Sustainable Development Goals (SDGs) at a range of planning scales?

The book comprises eight chapters. The six main chapters are written to assist readers with digesting the content in a sequential and integrated way (Figure 1.2). Chapter 2 provides a general introduction to the science of climate change. It describes Earth's climate system's fundamental physics and chemistry, and examines the role of humans in driving recent changes. The first section introduces the fundamentals of the greenhouse effect. This includes an explanation of the radiation balance and the role of natural and anthropogenic GHGs and their sources and sinks in the climate system. Second, climate variability and change drivers or 'forcings' are discussed. Natural drivers include external and internal climate factors, such as orbital forcing, solar forcing, volcanic forcing and ENSO cycles. Anthropogenic forcing of the climate system, primarily driven by GHG emissions, has been evident from about 1750 and has escalated since 2000. Third, past climates or paleoclimates are discussed. This evaluation provides context for comparing amounts and rates of climate change in the past with those experienced today. It also provides insights into how future climates may compare with much warmer periods in the geological record. Fourth, global changes in crucial climate indicators (e.g. GHGs, temperature averages and extremes, rainfall and hydrology, land and sea ice) are evaluated in the modern era (since about 1880). Last, future global changes in key climate indicators are presented for a range of plausible emissions scenarios.

Chapter 3 provides an overview of climate change in Australia. Past climate changes are discussed first. Understanding past climates, particularly the magnitude and rate of change in the geological record, provides context for historical and future changes driven by human activities. Second, the present climate is evaluated, including explaining the main drivers of Australia's climate at different spatial and temporal scales. The main climate types found across the country form part of the discussion. Third, changes in key climate indicators in Australia (e.g. land and ocean temperature averages and extremes, fire weather, rainfall and hydrology) are evaluated in the modern era (since 1910). Last, future changes in key climate

indicators for Australia are presented for a range of plausible emissions scenarios. Climate risks to the country of a 3°C warmer world are discussed. This chapter covers the material needed to give context to later chapters that focus on climate-resilient development pathways for Australia's natural, economic and social systems.

Chapter 4 considers how Australia may respond to a changing climate now and in the future. Responding to climate change requires strategies to reduce emissions of GHGs and enhancement of carbon sinks (climate mitigation), as well as strategies to cope with climate change impacts and risks that are inevitable even with mitigation policies (climate adaptation). First, a conceptual framework evaluates key terms and concepts discussed in later chapters that consider natural, economic and social systems. Key concepts include climate vulnerability, adaptive capacity, climate risk and resilience, and planned adaptation and maladaptation. Second, a discussion of climate-resilient development pathways for natural, economic and social systems provides context for a more detailed focus on climate change impacts, mitigation and adaptation for Australia in Chapters 5–7. Third, there is an overview of international climate change conventions, protocols and agreements for which the Australian Government has international obligations. These address climate mitigation and adaptation, and include the United Nations Framework Convention for Climate Change (UNFCCC), the Intergovernmental Science-Policy Platform on Biodiversity and Ecosystem Services (IPBES), the United Nations 2030 Agenda for Sustainable Development and its associated SDGs. The last section provides a high-level overview of the roles and responsibilities for climate change action in Australia.

Chapter 5 examines the impacts of climate change on Australia's ecosystems, ecosystem services and biodiversity. Australia is one of the world's 17 megadiverse countries. The country also contains two global biodiversity hotspots and many world heritage areas and Ramsar wetland sites recognised for their outstanding universal values. There are two vulnerable groups of ecosystems: (1) terrestrial and freshwater ecosystems and (2) coastal and marine ecosystems. The discussion by group evaluates any observed impacts of climate change, along with any significant issues; potential climate impacts, threats and risks in the context of their vulnerability (i.e. exposure and sensitivity) to climate change and adaptive capacity. It then considers climate-resilient development pathways and opportunities and ways to integrate these with the relevant SDGs. Addressing the SDGs for each group of ecosystems informs incremental and transformative strategies to avoid maladaptation pathways in the future.

Chapter 6 examines the impacts of climate change on aspects of Australia's economy. There are five highly vulnerable groups of economic sectors: (1) energy and transportation, (2) agriculture and land, (3) mining, (4) manufacturing and construction, and (5) tourism. The discussion by sector evaluates their socio-economic context; approaches to managing business and other risks; any observed impacts of climate change, along with any significant issues; potential climate impacts, threats and risks in the context of their vulnerability (i.e. exposure and sensitivity) to climate change and adaptive capacity. It then considers climate-resilient development pathways and opportunities by sector and ways to integrate these with the relevant SDGs. Addressing the SDGs for each economic sector informs incremental, structural and transformative strategies to avoid mal-mitigation and maladaptation pathways in the future.

Chapter 7 examines the impacts of climate change on aspects of Australian society. Four socially orientated themes are examined that vary in their vulnerability to climate change and increasingly extreme climate-driven events: (1) cities, settlements, and built environments; (2) human health and well-being; (3) Indigenous peoples; and (4) cross-cutting social issues. The emphasis for each theme is on the local government area (LGA) scale. The discussion by theme evaluates any observed impacts of climate change, any significant issues, potential climate impacts, threats and risks in the context of their vulnerability (i.e. exposure and sensitivity) to climate change and adaptive capacity. The second section evaluates climate-resilient development pathways for social systems at the LGA scale. This discussion includes identifying climate mitigation and adaptation pathways for vulnerable infrastructure, properties, communities and populations. The last section integrates the four themes with the relevant SDGs. Addressing the SDGs at the LGA scale informs incremental, structural and transformative strategies to avoid mal-mitigation and maladaptation pathways in the future.

2

The Science of Climate Change

David Keeling altered our perspectives about global change with his painstaking observations of atmospheric carbon dioxide. The now famous 'Keeling curve', measuring both the pulse of Nature and a steadily rising human impact on atmospheric composition, is invariably hailed as our most rigorous and fundamental measure of global change.

James E. Hansen (2005), tribute to Charles David Keeling (1928–2005)

INTRODUCTION

The United Nations Framework Convention on Climate Change (UNFCCC, see Chapter 4) – in its Article 1 – defines climate change as 'a change of climate which is attributed directly or indirectly to human activity that alters the composition of the global atmosphere and which is in addition to natural climate variability observed over comparable time periods' (United Nations Framework Convention on Climate Change, 1992). It makes an important distinction between climate change attributable to human (anthropogenic) activities that change the atmospheric composition and climate variability caused by natural drivers. Understanding the differences between natural climate variability and anthropogenic climate change and how they interact is the main aim of this chapter. Covering such a dense topic in one chapter has many challenges. However, this chapter attempts to cover the basic science to inform later chapters that focus on climate change impacts, climate mitigation and climate-resilient adaptation pathways in Australia. The fundamentals of Earth's radiation balance and the role of the carbon cycle and greenhouse gases (GHGs) in the climate system are discussed first. Second, it will examine Earth's climate system's natural and anthropogenic drivers. Third, it will discuss changes in the climate system over the past 56 million years (Ma) but

will highlight significant changes over the past 800 thousand years (ka) before the present (BP). It will then provide a more in-depth analysis of the last 2 ka BP, including fine-scale instrumental data collected from about 1880 Common Era[1] (CE) onwards. The final section evaluates climate models, emissions scenarios (shared socio-economic pathways; SSPs) and projected changes in Earth's climate system this century. This chapter takes a global perspective, while Chapter 3 focuses on past changes in Australia's climate, drivers of its present climate and future climate change projections.

FUNDAMENTALS OF THE GREENHOUSE EFFECT

Radiation balance

The radiation (or energy) balance is a natural phenomenon that warms Earth's climate system to an ambient temperature that allows complex life to exist (Figure 2.1). On average, 343 W/m^2 (Watts per square metre) of solar radiation reaches the top of Earth's atmosphere. It then easily transfers through the atmospheric layers to the surface – much like sunlight

---

[1] Common Era (CE) is the secular equivalent of AD (*anno Domini*), which means 'in the year of the Lord' in Latin (the year 0). Unless, specified otherwise all dates cited in the text (e.g. 1750) refer to the Common Era.

DOI: 10.1201/9781003189909-2

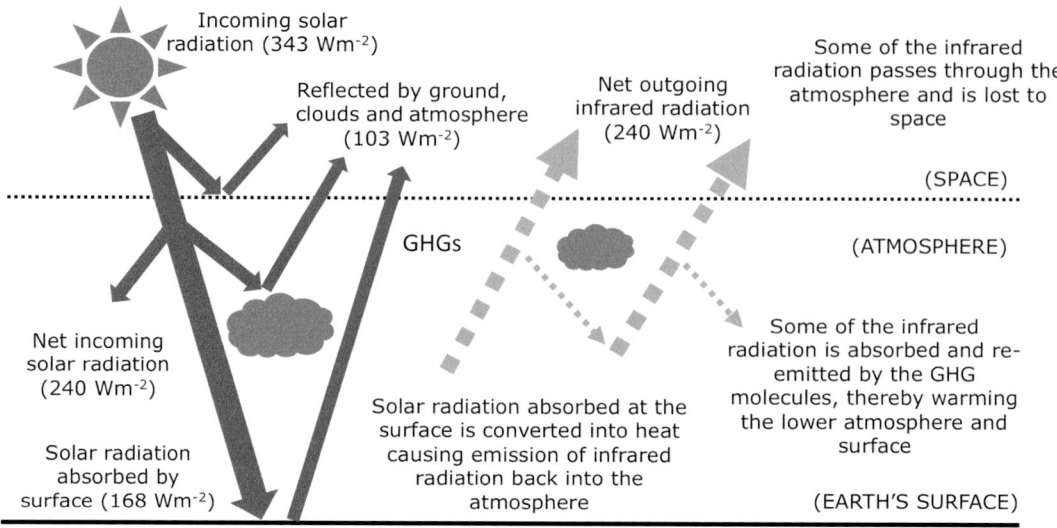

Figure 2.1 The energy (radiation) balance and the natural greenhouse effect, showing the role played by GHGs in trapping heat that would otherwise escape to space. Bold lines refer to solar radiation or energy produced from the Sun, and dashed lines to infrared radiation or heat produced in Earth's climate system. Human activities (burning of fossil fuels, agriculture and deforestation) are increasing the concentrations of GHGs (see Table 2.1), giving rise to the enhanced greenhouse effect that is driving warming of our atmosphere and oceans. (Adapted from United Nations Environment Program/-GRID-Arendal, 2005.)

through a greenhouse. Some 30% of this solar energy (103 W/m²) is reflected to space (7% from Earth's surface, 23% from clouds and atmospheric particles or aerosols), 23% (79 W/m²) is absorbed in the atmosphere by water vapour, dust and ozone and 47% (168 W/m²) is absorbed at the surface (Figure 2.1). Hence, about 70% (240 W/m²) of the total incoming solar energy is absorbed by Earth's climate system (net incoming solar radiation). Satellite measurements show the atmosphere radiates heat to space equivalent to 59% (200 W/m²) of the incoming solar radiation. In comparison, the surface radiates only 12% equivalent energy (40 W/m²) through the atmospheric window.[2] Hence, net outgoing infrared radiation equals 240 W/m² (equivalent to about 70% of the incoming solar radiation). In other words, net incoming solar radiation is balanced by net outgoing infrared radiation, thereby maintaining stable surface temperatures (the natural greenhouse effect).

[2] Atmospheric window refers to wavelengths of the electromagnetic spectrum that can be transmitted through Earth's atmosphere. Atmospheric windows occur in the visible, infrared and radio regions of the spectrum. Reference here is to the infrared wavelengths of the spectrum or simply heat.

The ability of Earth's atmosphere to retain heat was first established by Joseph Fourier in 1822 (see Chapter 1, Table 1.1).

To fully appreciate how Earth's climate system 'balances' the energy budget (Figure 2.1), it is necessary to consider atmospheric processes occurring at three vertical levels: (1) the surface of Earth, where most of the solar heating occurs; (2) the top of Earth's atmosphere, where solar radiation enters the system; and (3) the atmosphere in between, where GHGs play a critical role in absorbing and re-radiating infrared radiation or heat upwards and downwards (see Table 2.1). The quantity of incoming and outgoing radiation (or net radiation) must be equal to maintain stable temperatures at each level. This process means processes at the surface must remove 48% (168 W/m²) of incoming solar radiation that the ocean and land surfaces absorb. Three interacting processes facilitate the removal of this energy (or heat) from Earth's surface into the atmosphere and space: (1) evaporation (25%), convection (5%) and net thermal radiation (17%). However, only 12% (40 W/m²) of the thermal (infrared) radiation from the surface directly escapes through the atmospheric window into space (Figure 2.1) – leaving a net 5%–6% (~20 W/m²)

Table 2.1 Lifetimes, global warming potentials (potency) and abundances (ppbv) of significant greenhouse gases

Gas	Global warming potential (GWP) (100 Years)	Lifetime (Years)	Atmospheric abundance parts per million by volume (ppmv) for CO_2 and parts per billion by volume (ppbv) for the other gases 2019 level (1750 CE: pre-industrial level)
CO_2	1	>300	409.9 (280.0)
CH_4	21	9–15	1,866 (715)
N_2O	310	120	332 (270)
HCF-23	11,700	264	0.032 (0)
HCF-134a	1,300	14.6	0.108 (0)
HCF-152a	140	1.5	0.006 (0)
CF_4	6,500	50,000	0.085 (0.04)
C_2F_6	9,200	10,000	0.005 (0)
SF_6	23,900	3,200	0.009 (0)

Sources: World Meteorological Organization (2019) and Gulev et al. (2021).

Notes: CE, Common Era (years); CO_2, carbon dioxide; CH_4, methane; N_2O, nitrous oxide; HCF, hydrofluorocarbon; CF_4, tetrafluoromethane; C_2F_6, hexafluoroethane; SF_6, sulphur hexafluoride; CF_4 and C_2F_6 are perfluorocarbons (PFCs).

which is absorbed by GHG molecules in the atmosphere (see Table 2.1). GHG molecules radiate heat in all directions, including downwards, where they warm the air in the lower layers and at the surface, thereby facilitating the natural greenhouse effect. This process maintains Earth's temperature at about 33°C warmer than it would otherwise be without these natural GHGs. Specifically, Earth has a moderate global mean surface temperature (GMST) of 15°C, rather than a frigid global average of −18°C that would prevail without GHGs.

This natural greenhouse effect has been remarkably stable during the Holocene interglacial period (see Chapter 1, Figure 1.1 and Box 2.1 below). Only changes in solar radiation input, concentrations of GHGs (Table 2.1), quantities of stratospheric particles (aerosols) from major volcanic eruptions or combinations of all these climate drivers can affect Earth's radiation balance (Figure 2.1). For example, human activities – such as burning fossil fuels, industrial processes, agriculture and deforestation – increase the quantity of various GHGs released into the atmosphere. These rising gases are trapping extra heat leading to planetary global warming (the enhanced greenhouse effect). These natural and anthropogenic 'destabilising' influences on Earth's climate system are referred to as climate forcings (or feedbacks) and are discussed in more detail below.

Understanding greenhouse gases and their sources and sinks

Carbon dioxide (CO_2), nitrous oxide (N_2O) and methane (CH_4) are the primary (well-mixed) GHGs in Earth's atmosphere (Table 2.1). Well-mixed GHGs have lifetimes long enough to be relatively homogeneously mixed in the atmosphere (Gulev et al., 2021). The basic properties of these gases are central to understanding the natural and human drivers (or forcings) of Earth's climate system. These common GHGs are primarily natural but also include human-caused (anthropogenic) additional releases of the same gases, plus more potent human-made gases,[3] such as perfluorocarbons, hydrofluorocarbons

[3] There are several entirely human-made GHGs in the atmosphere, such as the halocarbons and other chlorine- and bromine-containing substances, dealt with under the Montreal Protocol (United Nations Environment Program, 2021a).

and ozone-depleting substances into the atmosphere (Table 2.1).

Natural and human-made GHGs vary widely in their potential to contribute to global warming and hence their ability to force the climate system (Table 2.1). Given that CO_2 is the dominant well-mixed GHG, other gases are often expressed as CO_2 equivalents (CO_2-e), calculated by multiplying the mass of emissions by the global warming potential (GWP) of the gas. For example, methane (CH_4) and sulphur hexafluoride (SF_6) have GWPs 21 and 23,900 times greater than CO_2, respectively. The industrial revolution began in 1750. Since then, concentrations of CO_2, CH_4 and N_2O have increased by 147%, 261% and 123%, respectively (Table 2.1). These GHG levels are unprecedented in the past 800 ka and highly likely to be the highest ambient concentrations in the last 2 Ma (Gulev et al., 2021, see Figure 2.2 and Box 2.1). Over the 12 ka before 1750, CO_2 levels only ranged from 265–285 ppm by volume (ppm), and as of June 2021, the global average concentration was 418.94 ppm (see Figure 2.5). The lifetime (or longevity) of different GHGs in the atmosphere is also important (Table 2.1). CO_2 has the longest atmospheric 'residence' time of all the primary GHGs, while CH_4 has the shortest.

CO_2 is by far the most important anthropogenic GHG – accounting for about 75% of all anthropogenic GHG emissions (Table 2.2). Its long atmospheric lifetime (Table 2.1) means that atmospheric concentrations of CO_2 will remain elevated for many decades, even after mitigation efforts to reduce CO_2 are implemented through multilateral instruments – such as the Kyoto Protocol and Paris Agreement (see Chapter 4). CH_4 is the main component of natural gas and is the second largest contributor to anthropogenic GHG emissions (Table 2.1). It is also produced by growing rice, grazing cattle and fugitive losses from coal mines. Importantly, CH_4 has a GWP of 21 – making it a potent GHG, but unlike CO_2, it has a much shorter atmospheric lifetime. However, CH_4 eventually breaks down to CO_2, thereby contributing to long-term warming (Table 2.1). N_2O is also a potent GHG, with a GWP of 310 and a long atmospheric residence time of about 120 years (Table 2.1). The largely anthropogenic sources of this gas include agriculture (fertiliser use and microbial activity in soil and water), nylon and nitric acid production, fuel-fired power plants and vehicle emissions. Finally, there is a suite of human-made GHGs (e.g. HFCs, PFCs and SF_6) that – except for HCF-152a – have incredibly high GWPs and long atmospheric lifetimes (Table 2.1).

Despite being an abundant GHG, water vapour (H_2O) is not evaluated here because its concentration depends on temperature and other atmospheric conditions and not directly upon human activities (Gulev et al., 2021). However, surface warming caused by human production of other GHGs (Table 2.1) increases atmospheric H_2O because warmer temperatures enhance evaporation, and the resultant water remains in the atmosphere in the form of vapour. This phenomenon creates a positive feedback loop in which warming leads to more warming (see Figure 2.1).

Ozone (O_3) is also a GHG but behaves very differently from the well-mixed GHGs (Gulev et al., 2021). The forcing effects of O_3 depend on its altitude in the atmosphere (see Table 2.5). Most O_3

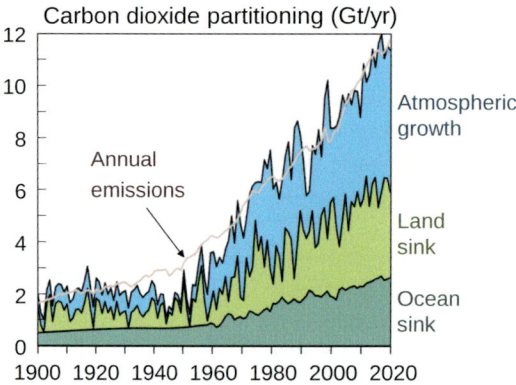

Figure 2.2 Annual CO_2 emissions and CO_2 partitioning (Gt/year) from 1900 to 2020. (Credit: 2020 Global Carbon Budget/Creative Commons/Public Domain.)

Table 2.2 Global anthropogenic greenhouse gas sources and emissions (CO_2 equivalent) in 2018

Gas	Source	Emissions (Mt CO_2-e/ year) 2018	CO_2 equivalent percentage
CO_2	Deforestation, decay of biomass, etc	1,193	2.4
CO_2	Energy (fossil fuels)	33,742	68.9
CO_2	Other	1,507	3.1
CH_4	Agriculture	3,517	7.2
CH_4	Natural gas consumption, coal mining, etc.	4,781	9.8
N_2O	Agriculture, industry, transportation, etc.	3,064	6.3
High GWP gases (HFCs, PFCs and SF_6)	Consumer products, refrigerants, aluminium production, semiconductor manufacturing	1,136	2.3
All GHGs		48,940	100.0

Source: Climate Watch (2021).
Notes: CO_2, carbon dioxide; CH_4, methane; N_2O, nitrous oxide; GWP, global warming potential; Mt, megatonne (= 1 million tonnes); GHG, greenhouse gas; HCF, hydrofluorocarbon; CO_2-e, carbon dioxide equivalent; SF_6, sulphur hexafluoride; HFC, hydrofluorocarbon; PFC, perfluorocarbon.

exists naturally in the stratosphere, ranging from about 10 to 50 km above Earth's surface. Its presence in the stratosphere has a slight net warming effect on the climate system. However, it is critical for life on Earth because it absorbs harmful ultraviolet (UV) radiation from the Sun, preventing UV rays from reaching Earth's surface. In the troposphere – the lowest layer of the atmosphere – O_3 is an air pollutant that is harmful to breathing and the principal agent of urban smog. O_3 is also an important GHG that contributes to global heating. However, O_3 is not homogeneous in the troposphere. It only lasts for days to weeks, so levels of O_3 often vary regionally (e.g. over the south and east Asia) and with the time of year and day (e.g. summer afternoons).

As well as understanding the physical and chemical properties of GHGs (Table 2.1), it is also essential to understand the sources of natural and anthropogenic GHGs and their emission rates into the atmosphere each year (Table 2.2). The burning of fossil fuels for electricity, transportation and heating contributed nearly 69% of all the CO_2-e emissions into the atmosphere as CO_2 in 2018, while agriculture, burning of natural gas and coal mining together contributed 17% of CO_2-e emissions in the form of methane (CH_4). In the same

year, agriculture, industry and transportation contributed a further 6.3% CO_2-e emissions in the form of nitrous oxide (N_2O). Notably, combined CO_2 emissions from all sources contributed about 75% of all CO_2-e emissions in 2018.

At the beginning of the 20th century, natural and anthropogenic sources emitted about 1.8 Gt/ year of CO_2 into the atmosphere, with 0.5 and 1.0 Gt C/year absorbed by natural land and ocean sinks and 0.3 Gt C/year remaining in the atmosphere (Figure 2.2). Sinks include terrestrial and marine ecosystems that remove CO_2 from the atmosphere by sequestering carbon in biomass (terrestrial vegetation, aquatic and marine plants and phytoplankton), dead organic matter and soils. However, since the 1960s, annual emissions of CO_2 (and other well-mixed GHGs) from primarily anthropogenic sources (e.g. fossils fuels) have risen sharply (Figure 2.2 and Table 2.1). For the 1959–2019 period, the total CO_2 emissions (~12 Gt C/year) were partitioned among the atmosphere (5.4 Gt C/year or 45%), ocean (2.9 Gt C/year or 24%) and land (3.8 Gt C/year or 32%). Both ocean and land CO_2 sinks have increased roughly in line with the atmospheric increase, but with significant decadal variability on land and possibly in the ocean (Friedlingstein et al. 2021). During strong

El Niño events (see Figure 2.4), there is often a slight increase in the ocean sink and a significant decrease in the land sink.

In addition to global GHG emissions by gas type (Table 2.2), GHG emissions can also be broken down by the economic activities that lead to their production (Table 2.3). The combustion of coal, natural gas and oil for electricity and heat is the largest single source of global GHGs. However, emissions from this source do vary significantly from country to country. Industry contributes about one-fifth of all GHGs, mainly from emissions due to coal, gas and oil-burning onsite for energy, but not industrial electricity (grouped with electricity and heat in Table 2.3). There are also GHG emissions from chemical, metallurgical and mineral processing and waste management activities. Almost a quarter of global GHG emissions originate from agriculture, forestry and other land uses (Table 2.3). Most emissions in this sector are from agriculture (crop cultivation and livestock), followed by deforestation. Importantly, this value does not include carbon sinks (Figure 2.2), which offset about 20% of the emissions from this sector (Food and Agriculture Organization, 2014). Transportation produces 14% of global GHGs, mainly from fossil fuels used for road, rail, air and marine transportation. About 95% of the world's transportation energy is produced by petroleum-based fuels (petrol and diesel). Buildings contribute 6% of global GHG emissions, mainly from onsite energy generation (but not electricity use in buildings) and burning fuels for heat and cooking food in buildings. The remaining 10% of global GHG emissions refers to all emissions from the energy sector which are not directly related to electricity or heat production, such as fuel extraction, refining, processing and transportation (Table 2.3).

Global GHG emissions vary significantly among countries (Friedlingstein et al., 2021). In 2020, the highest CO_2 emissions were from China (27.9%), the USA (13.5%), the European Union (27 member states; 7.5%) and India (6.8%). The rest of the world contributed ~43%, including 1.4% from Australia. These relative figures are indicative of the past decade. Despite Australia's small contribution to global GHG emissions, it has one of the highest per capita carbon emissions globally, averaging about 15.4 tonnes (t) of CO_2 per person in 2020. This figure compared with the United States and Canada at 14.2 t, the Russian Federation at 10.8 t, China at 7.4 t, the United Kingdom at 4.9 t and India at 1.8 t.

Another way to examine global CO_2 emissions is to consider cumulative emissions into the atmosphere since the beginning of the industrial revolution. Over the period of 1750–2020, the United States has contributed 416.7 billion tonnes of CO_2, compared with 266.2 Bt for the European Union, 235.6 Bt for China, 115.3 Bt for Russia, 78.2 Bt for the United Kingdom, 54.4 Bt for India, 33.6 Bt for Canada, 20.1 Bt for Mexico, 18.6 Bt for Australia and 16.2 Bt for Brazil (Friedlingstein et al., 2021). Given that CO_2 is a long-lived GHG (Table 2.1), it is more equitable to apply this cumulative metric as the basis for international agreements concerning GHG emissions reduction targets, such as those prescribed in the Kyoto Protocol and Paris Agreement (see Chapter 4).

Table 2.3 Global anthropogenic greenhouse gas emissions by economic sector

Economic sector	GHG emissions (%)
Electricity and heat production	25
Industry	21
Agriculture, forestry and other land use	24
Transportation	14
Buildings	6
Other energy	10
Total	100.0

Source: Intergovernmental Panel on Climate Change (2014a).
Note: Emissions and sinks associated with land use are not included in these values.

DRIVERS OF CLIMATE VARIABILITY AND CHANGE

Earth's climate system is driven by a phenomenon called radiative forcing (RF). This process is

> the change in the net, downward minus upward, radiative flux (in W m^{-2}) at the tropopause (= top of troposphere) due to a change in a driver of climate change, such as a change in the concentration of carbon dioxide (CO_2) or the output of the Sun.
>
> *(Intergovernmental Panel on Climate Change, 2021)*

Hence, RF measures the net change in Earth's radiation balance (see Figure 2.1) in response to some external perturbation (Gulev et al., 2021). A positive RF results in warming and negative RF in cooling.

Earth's climate system has always been influenced by natural atmospheric drivers of climate (explained further below), including significant changes in incoming solar radiation – driven by long-term orbital fluctuations (orbital forcing). Solar variability – caused by changes in radiation outputs from the Sun itself – also affects Earth's climate at shorter time scales (solar forcing). Another natural driver of climate variability is episodic major volcanic eruptions (volcanic forcing) that have historically resulted in short intervals of global cooling (through negative RF). In addition to these external drivers, there are also internal drivers within the climate system itself that cause short-term variations in climate, such as the El Niño-Southern Oscillation (ENSO) phenomenon. With the advent of the industrial revolution in 1750, another new driver, anthropogenic forcing of the climate system, has been primarily responsible for enhancing natural climate variability and driving observed changes (or trends) in the climate system (Forster et al., 2021, Eyring et al., 2021, Gulev et al., 2021). This section evaluates natural and anthropogenic drivers of Earth's climate system.

Orbital forcing (Milankovitch cycles)

Orbital RF of Earth's climate system (Table 2.4) was first recognised about 80 years ago by the Serbian astrophysicist Dr Milutin Milankovitch (see Chapter 1, Table 1.1). Milankovitch demonstrated

Table 2.4 Orbital (Milankovitch) cycles and their role in Earth's climate

Orbital cycle	Periodicity (ka)	Climate effects
Eccentricity	96	Currently, Earth's eccentricity is near least elliptical (most circular). Thus, about 7% more incoming solar radiation reaches Earth on January 2 (*perihelion*) than it does on July 4 (*aphelion*). When the Earth's orbit is most elliptical, about 23% more incoming solar radiation reaches Earth at its perihelion compared with its aphelion.
Obliquity	41	The Earth has a tilt on its polar axis, known as obliquity, and this is the main reason for the seasons. It varies between 22.1° and 24.5° perpendicular to Earth's orbital plane (current tilt=23.4°). The greater the obliquity, the more extreme the seasons. Larger tilt angles favour periods of deglaciation.
Precession	26	As Earth rotates, it wobbles slightly upon its axis due to tidal forces caused by gravitational influences from the Sun and Moon. The trend of the direction in this wobble relative to fixed positions of stars is known as the axial precession. This phenomenon makes seasonal contrast more extreme in one hemisphere and less in the other.

Source: Buis (2020).
Notes: ka is thousands of years; *eccentricity* is the shape of Earth's orbit around the Sun; *obliquity* is the angle of Earth's axial tilt with respect to its orbital plane; *precession* refers to the direction Earth's axis of rotation is pointed.

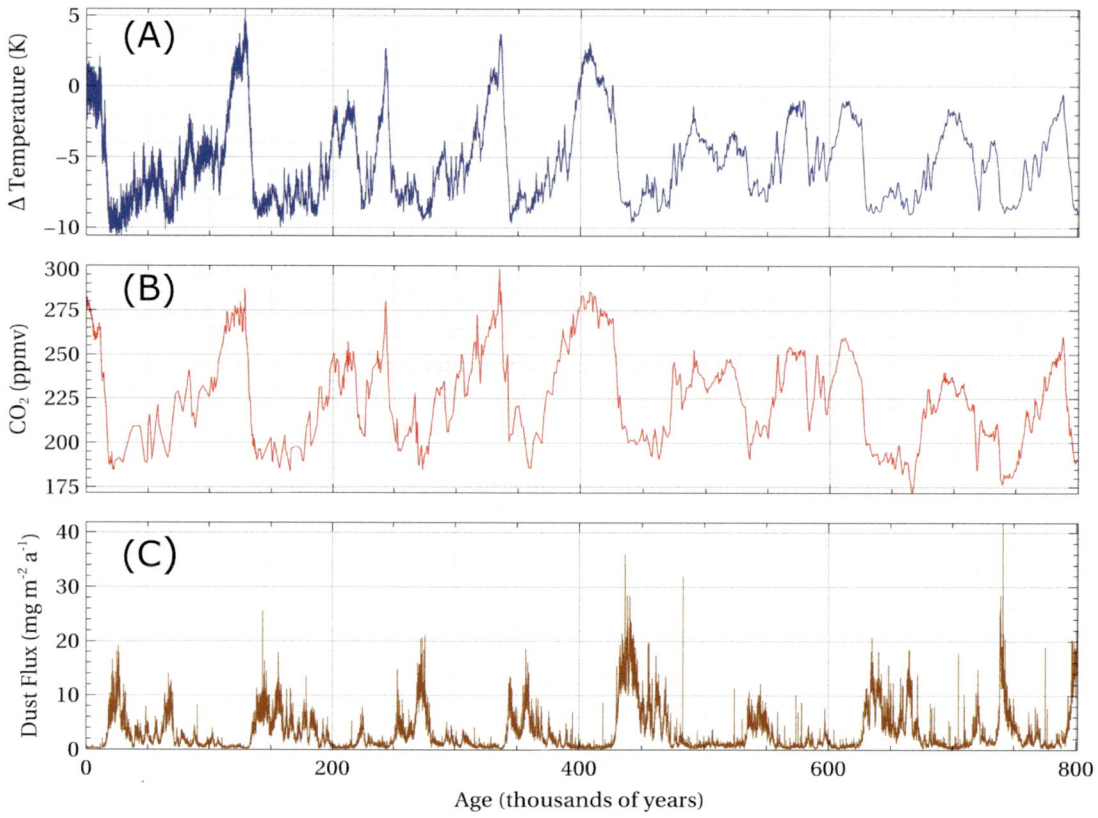

Figure 2.3 (a) Temperature (K), (b) CO_2 (ppmv) and (c) dust fluxes (mg/m²/a¹) from the EPICA Dome C ice core covering the last 800,000 years before the present (Lüthi et al., 2008b). (Credit: Fabrice Lambert/Wikimedia/Public Domain.)

the multi-millennial-scale effects of changes in Earth's position relative to the Sun as essential drivers of long-term climate, including initiating the commencement and cessation of periods of glaciation (see Figure 2.3). Milankovitch identified three types of Earth orbital movements that affect solar radiation reaching the top of Earth's atmosphere (see Figure 2.1). These orbital changes are known as Milankovitch cycles (Table 2.4). They account for variations of up to 25% in the quantity of incoming solar radiation reaching Earth's mid-latitudes over time scales of 26–96 ka (Buis, 2020).

Orbital cycles (Table 2.4) operate separately and in unison to drive Earth's climate over very long-time intervals, resulting in more significant changes in climate over tens of thousands of years (Buis, 2020). Milankovitch combined the orbital cycles to produce a detailed mathematical model for calculating differences in solar radiation reaching various latitudes and corresponding surface temperatures. Importantly, it was assumed changes in solar radiation at some

latitudes, and some seasons were conducive to the growth or retreat of the continental ice sheets in the Northern Hemisphere. Some 10 years after the death of Milankovitch, research efforts began to pay serious attention to these orbital cycles. For example, a seminal publication by Hays et al. (1976) that analysed deep-sea sediment cores found Milankovitch cycles corresponded with periods of significant climate change. Ice ages occurred when Earth was transitioning different stages of orbital variation (Table 2.4, see also Figures 2.3 and 2.6a). They appropriately described these orbital cycles as the 'pacemaker' of the ice ages.

In addition to past climate records derived from marine sediment cores, ice core data also provide scientists with powerful glimpses into Earth's climate history over tens and hundreds of millennia. Cores drilled deep into the East Antarctic and Greenland ice sheets contain ancient compacted snows (much like tree rings), allowing researchers to sample ice accumulation, air temperature and (importantly) air chemistry from the deep past (Jouzel and Masson-Delmotte, 2010). Such records track the glacial/interglacial cycles driven by Milankovitch cycles (Table 2.4, Figures 2.3 and 2.6a) and permit scientists to compare contemporary with past concentrations of primary GHGs (Figure 2.3b). Significant variations are evident when researchers examine past changes in natural well-mixed GHGs (CO_2, CH_4 and N_2O) in ice core records from Antarctica, dating back about 800 ka BP (Petit et al., 1999, Lüthi et al., 2008a). Over this multi-millennial time scale, CO_2 levels – as shown by ice core data – have fluctuated between 170 and 300 ppm (Figure 2.3b) but have never approached current levels (Table 2.1). In addition to significant temperature changes (Figures 2.3a and 2.6a), the ice core records also show evidence of dry/wet cycles on Earth – as depicted by dust fluxes transported via atmospheric winds and then deposited on the ancient snows in Antarctica. Glacial periods were drier and conducive to elevated dust levels in the atmosphere (Figure 2.3c).

Solar forcing

While orbital forcing controls levels of solar radiation reaching Earth at scales of tens of millennia, short-term variations in solar output can also affect Earth's climate system. The Sun's radiation output varies in response to the frequency of dark sunspots, bright spots on the Sun and other solar phenomena. The Sun undergoes a regular 11-year sunspot cycle, where solar radiation levels go up or down – as does the amount of material the Sun ejects into space from to time through solar flares. Solar activity since 1900 was high but not exceptional compared to the past 9 ka (Arias et al., 2021). Total solar irradiance (TSI) very likely increased over the first 70 years of the 20th century and decreased after that (Gulev et al., 2021). However, TSI did not change significantly between 1986 and 2019. Hence, solar RF only played a very minor role in Earth's climate system during that measurement period (see Table 2.5 below).

Table 2.5 Summary of effective radiative forcing (ERF) values (W/m²) for a range of climate system drivers

Driver	Global mean effective radiative forcing (ERF) by emissions and drivers	
	1750–2019 (W/m²)	Level of Confidence
CO_2	2.16 [1.90–2.41]	H
CH_4	0.54 [0.43–0.65]	H
Halo-carbons	0.41 [0.33–0.49]	H
N_2O	0.21 [0.18–0.24]	H
O_3	0.47 [0.24–0.71]	H
H_2O^{str}	0.05 [0.00–0.10]	M
Aerosol-radiation interactions	−0.22 [−0.47 to 0.04]	M
Aerosol-cloud interactions	−0.84 [−1.45 to −0.25]	M
Land use (including irrigation)	−0.20 [−0.30 to −0.10]	M
Surface albedo (black + organic carbon aerosol on snow and ice)	0.08 [0.00–0.18]	L
Combined contrails and aviation-induced cirrus	0.06 [0.02–0.10]	L
Total anthropogenic	**2.72 [1.96–3.48]**	**H**
Solar irradiance	0.01 [−0.06 to 0.08]	M

Source: Adapted from Forster et al. (2021).
Global mean ERF values are shown, with their 5% to 95% ranges in square brackets. Negative ERF drivers and their values are shaded. Levels of confidence for various assessments are shown.
Notes: CO_2, carbon dioxide; CH_4, methane; N_2O, nitrous oxide; O_3, ozone; H_2O^{str}, stratospheric water vapour. An ERF assessment for volcanic aerosols is not provided due to the episodic nature of volcanic eruptions. Level of confidence (see Box 2.2): H, high, M, medium and L, low.

Volcanic forcing

Major volcanic eruptions change Earth's radiation balance because volcanic particulate (aerosol) clouds absorb infrared radiation and scatter a significant amount of incoming solar radiation back to space (see Figure 2.1). Even though large-scale volcanic activity may only last for a few days (or be sporadic over several weeks), large amounts of sulphuric gases and ash may influence the climate system several years post-eruption. These sulphuric gases convert to sulphate aerosols, which are sub-micron droplets containing about 75% of sulphuric acid (Wolfe, 2020). Following major eruptions (i.e. where the gas and ash cloud enters the stratosphere or reaches above 10 km in altitude), these volcanic aerosols can linger for 2–4 years before eventually dissipating. Several major eruptions over the past 200 years have all resulted in short intervals of global cooling (or negative RF). The most recent climate-impacting eruption was Mt Pinatubo in the Philippines in June 1991. The eruption released an estimated 20 million tonnes of sulphur dioxide and ash particles to a height of more than 40 km above the surface. This massive injection of aerosols into the stratosphere caused a 10% decline in solar radiation reaching Earth's surface (see Figure 2.1), resulting in global average temperatures falling by about 0.5°C for at least 3 years. Hence, strong individual volcanic eruptions can cause multi-annual variations in solar radiation reaching Earth's surface. However, patterns over the last 100 plus years are not unusual in the context of the past 2.5 ka (Forster et al., 2021). Paradoxically, the 15 January 2022 Hunga Tonga-Hunga Ha'apai submarine volcanic eruption near Tonga is predicted to impact climate not through surface cooling due to sulphate aerosols, but instead through surface warming due to the positive RF from the excess stratospheric water vapour (Millán et al. 2022).

Internal forcing

The main internal driver of Earth's climate system over short time intervals is the ENSO phenomenon in the Pacific Ocean (Gulev et al., 2021). El Niño and La Niña are opposite phases of a natural – but powerful – climate phenomenon across the tropical Pacific Ocean that swings back and forth with an average periodicity of 3–7 years (Figure 2.4). The immense size of the Pacific Ocean means these opposite phases of the ENSO cycle are responsible for significant changes in global temperatures and precipitation patterns.

During normal conditions (or neutral conditions), the trade winds in the equatorial Pacific blow towards the west, shifting vast amounts of warm surface water and atmospheric moisture towards the western Pacific around northern Australia and Southeast Asia (Figure 2.4a). As the surface water moves into the western equatorial

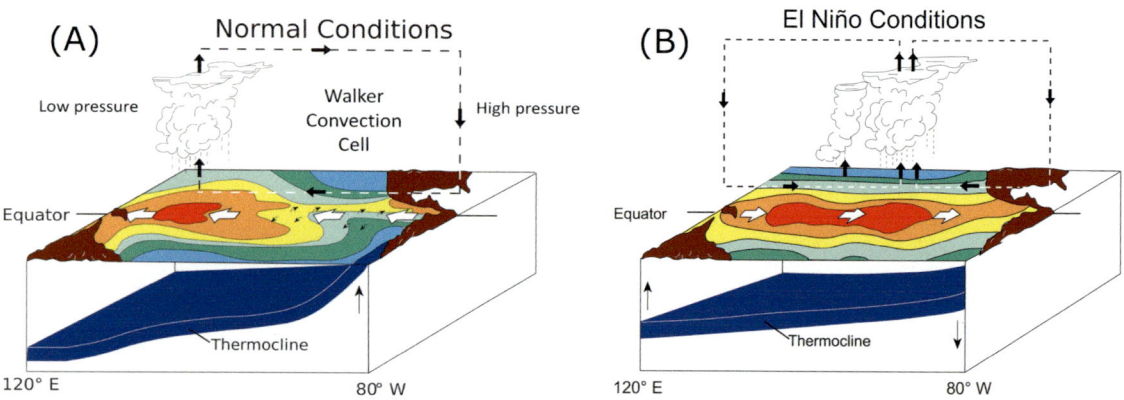

Figure 2.4 Diagram (a) normal conditions in the equatorial Pacific. Low pressure in the western Pacific and high pressure in the eastern Pacific cause the trade winds to move surface water to the west, leading to upwelling and a shallow thermocline near South America (top). Diagram (b) El Niño conditions in the equatorial Pacific. Weakening or reversal of the trade winds transport warm surface water eastward towards South America, disrupting coastal upwelling. (Credit: Wikimedia/Public Domain.)

Pacific, the upwelling of nutrient-rich cold water dominates the surface waters in the eastern Pacific. In response to the coastal upwelling off South America, a very shallow thermocline is present in the eastern Pacific (Figure 2.4a). During normal ENSO conditions, atmospheric indices favour low pressure, convection and rainfall over the western Pacific. Meanwhile, in the eastern Pacific, high pressure dominates with dry conditions.

The strength of the Walker Circulation (or convection cell) is generally moderate under normal conditions (Figure 2.4a). This east to west circulation results from a pressure gradient between the equatorial eastern Pacific (high pressure) and the western Pacific (low pressure). This circulation ensures that a warm-water pool accumulates towards the west, while a cool-water pool develops in the east near South America (Figure 2.4a). During La Niña events, the Walker Circulation (Figure 2.4a) strengthens and pushes warm water even further west, allowing increased upwelling of cold water near South America. Hence, strong La Niña conditions enhance the upwelling of cold water in the eastern Pacific – much like a giant air conditioner – which results in below-average global temperatures due to the negative RF of the climate system. However, with strong El Niño conditions (Figure 2.4b), the Walker Circulation weakens (or even breaks down), causing warm water to approach the South American coastline. Rainfall patterns change in response to these warmer ocean temperatures, with increased rainfall in the central Pacific near and east of the international dateline. The thermocline in the eastern Pacific becomes deeper, effectively preventing any upwelling of cold water (Figure 2.4b). The lack of a cold-water pool in the eastern Pacific causes above-average global temperatures through the positive RF of the climate system. Finally, under prevailing neutral ENSO conditions (Figure 2.4a), global temperatures are generally close to their long-term average. Chapter 3 examines the effects of ENSO cycles and other climate drivers on Australia's climate.

Anthropogenic forcing

Since 1750, but particularly since 1880, human activities have become significant drivers of the climate system for the first time – notably those responsible for emitting GHGs into the atmosphere (Table 2.2) and those associated with aircraft contrails and land surface changes. Natural drivers of the climate system and their differing RFs continue to be part of the climate system. However, effective radiative forcing (ERF) is generally applied in contemporary studies of the atmosphere (Forster et al. 2021). ERF measures the energy imbalance after allowing for atmospheric temperatures, water vapour and clouds to adjust to the forcing agent while keeping surface conditions (specifically temperature) unchanged (Andrews et al., 2021).

In addition to the coarse-scale CO_2 data from ice cores (Figure 2.3b), a crucial fine-scale record of CO_2 measurements has been undertaken at the Mauna Loa Observatory in Hawaii since the late 1950s (Figure 2.5). This site is often referred to because it is very isolated from anthropogenic GHG sources. This record is known as the Keeling Curve, in recognition of Dr David Keeling from the Scripps Institution of Oceanography, who established the monitoring programme. The curve shows seasonal records of CO_2, driven by changes in photosynthesis activity across the vast deciduous and evergreen forests of the Northern Hemisphere and the steady overall increase in CO_2 due to human activities. The Australian Government's Bureau of Meteorology (BOM) and Commonwealth Scientific and Industrial Research Organisation (CSIRO) established a long-term GHG monitoring network at Cape Grim, Tasmania, in 1976. This remote global 'baseline' site samples some of the purest air in the world circulating in the Southern Ocean's westerly wind belt and is unaffected by regional pollution sources (Commonwealth Scientific and Industrial Research Organisation, 2022).

Anthropogenic emissions from various activities (Table 2.2) have driven changes in well-mixed GHGs in the industrial era, resulting in positive ERF values (Tables 2.5). This increase in GHGs has culminated in the climate system's uptake of excess radiation (energy) and hence has driven atmospheric and oceanic warming (see Figure 2.1). The most significant contribution to total ERF has been caused by the increase in the atmospheric concentration of CO_2 since 1750 (Table 2.5). Increases in concentrations of CH_4, halo-carbons, N_2O, O_3 and stratospheric water vapour (H_2O^{str}) have also contributed to a range of positive ERFs of the climate system since 1750 (Table 2.5), but particularly in recent decades (Forster et al., 2021). The January 2022 Tongan submarine volcanic eruption (see

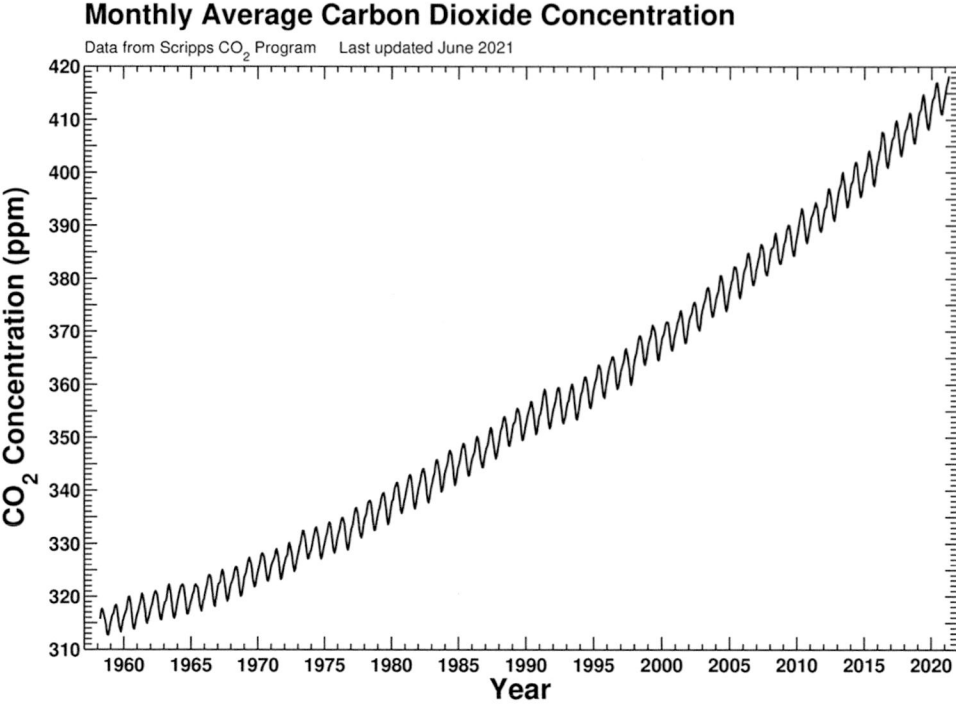

Mauna Loa Observatory, Hawaii
Monthly Average Carbon Dioxide Concentration

Data from Scripps CO_2 Program   Last updated June 2021

Figure 2.5 Monthly average atmospheric CO_2 concentration versus time at Mauna Loa Observatory, Hawaii, USA (20°N, 156°W) where CO_2 concentration is in parts per million in the mole fraction (ppmv). The curve is a spline through the data. (Credit: Data from Scripps CO_2 Program, Scripps Institution of Oceanography/Open Access/Public Domain.)

Volcanic forcing above) is likely to result in a small positive forcing of the climate system due to its high production of H_2O^{str} (Millán et al. 2022).

While rising GHGs are important anthropogenic drivers of climate at the global scale, land use change is another important driver of climate change at regional scales (Table 2.5). When land is cleared of vegetation – such as through deforestation – the reflectivity (albedo) of the surface typically increases. If dense forest with a dark green canopy is replaced with grasses or irrigated crops, the albedo increases, which means less incoming solar radiation gets absorbed by the surface (Figure 2.1). The reflected solar radiation is then lost to space, resulting in a negative ERF of the local or regional climate (Table 2.5). With less absorption of solar radiation by the surface, there is less energy available for conversion to infrared radiation and absorption by GHGs, hence less heating of the troposphere (Arias et al., 2021).

Human activities, notably the burning of biomass and fossil fuels, has resulted in a substantial increase in emissions of aerosols (organic and inorganic particles), their precursors and hence an overall increase in atmospheric aerosol concentrations since 1750 (Forster et al., 2021). Atmospheric aerosols (particulates) interact with incoming solar radiation and water in clouds, resulting in negative ERFs of the climate system. Aerosols have contributed a total ERF of −1.1 [−1.7 to −0.4] W/m² over 1950–2019 (Table 2.5). However, black and organic carbon aerosols have decreased the surface albedo over some areas of ice and snow in the Arctic, resulting in a weak positive ERF (Table 2.5). Similarly, a small positive ERF from contrails and aviation-induced cirrus has been observed but with low confidence (Forster et al., 2021).

PAST CLIMATES

There is an adage that the past is the key to the present. This viewpoint is particularly relevant when comparing past 'baseline' climates and past rates of climate change with those of the Anthropocene epoch (see Chapter 1, Figure 1.1). Various techniques have been applied to reconstruct past climates from the 'deep' geological past through to more recent epochs – such as the Pleistocene and Holocene – and through to the modern instrumentation era. This section begins by examining paleo-reference periods from 56 Ma to 12 ka BP to better understand past concentrations and rates of change of atmospheric CO_2. Second, it considers changes in climate over the past 800 ka, as derived from ice core records (see Figure 2.3). Third, it elaborates on finer scale changes in climate over the past 2 ka, derived from ice core data and other proxy measures of atmospheric CO_2 concentrations. Finally, it describes climate changes derived from instrumental records from 1880 to the present.

Paleo-reference periods

Nine geological reference periods have been proposed as examples of distinct climate states (Gulev et al., 2021). These paleo-reference periods (Box 2.1) represent colder and warmer times than the present, and periods of rapid climate change, thereby providing context and insights into a plausible range of future climates. Climate scientists can also use these reference periods to test the performance of climate models (see the 'Climate models' section below) under different anthropogenic climate forcing pathways (see the 'Shared socio-economic pathways' section below). GMST is considered the key indicator of the changing state of the climate system (Gulev et al., 2021). GMST is calculated using proxy data from a range of sources, including marine sediments, ice core records (see Figure 2.3a), tree rings (dendrochronology) and other well-established isotopic dating methods (see Forster et al., 2021). From the 1880s, instrumental measurements are used to calculate GMST.

There is now high confidence that average CO_2 concentrations in the Early Eocene Climatic Optimum (1150–2500 ppm) and Miocene Climatic Optimum (1150–2500 ppm) CO_2 were higher than those before industrialisation in 1750 at 285 ppm (Box 2.1). While there is some uncertainty with the marine sediment record used to obtain these proxy values, the last time atmospheric CO_2 concentrations were as high as today (see 'Modern' in Box 2.1 and Figure 2.5) was very likely more than 2 Ma ago (Gulev et al., 2021). Importantly, rates

of change in CO_2 concentrations per century over the past 100 years and the modern era are much higher than those experienced over the past 56 Ma (Box 2.1).

Past 800 ka

There is a rich record of past climates extending back 800 ka – carefully reconstructed from Antarctic ice core records and deep-sea marine sediments (Figure 2.3). These records show Earth's climate has cycled between ice ages and warm interglacial periods under the influence of orbital forcing (see Table 2.4), with each cycle taking tens of thousands of years or more (Figure 2.6a). Interglacial peaks correspond with periods of high sea level, while low points refer to glacial peaks when the sea level was lower. During the Last Interglacial (LIG) (125–118 ka BP), GMST was about 1°C higher than the present and like that for its potential value by 2,100 (see Table 2.6: SSP1–2.6). However, throughout the LIG (Box 2.1), CO_2 concentrations were only 280 ppm, compared with more than 419 ppm in July 2022 (Figures 2.3b and 2.5). Moreover, sea levels were about 10 m higher during the LIG, which poses questions about likely rates and amounts of sea-level change in future centuries (Rohling et al., 2019). Nonetheless, the LIG is not a perfect climate analogue[4] for the future because incoming solar radiation was higher than today – driven by Earth's favourable orbital forcing position relative to the Sun during that period (Table 2.4).

The Last Glaciation Maximum (LGM) occurred about 20 ka BP (Figure 2.6a) during the last phase of the Pleistocene epoch (see Box 2.1). CO_2 concentrations were only 180–200 ppm during this time (Figure 2.3b). At the peak of the LGM, GMST was about 4°C–5°C lower than today (Figure 2.6a), and the sea level was more than 134 m lower from 29 to 21 ka ago (Lambeck et al., 2014). The main period of deglaciation occurred from about 18 to 12 ka BP (Box 2.1) when sea levels rose rapidly and CO_2 concentrations increased from 195 to 265 ppm (Figure 2.3b).

---

[4] Climate analogues involve identifying places or areas that experience similar climatic conditions. These places/areas may be separated in space or time (i.e. with past or future climates). Climate analogues may be helpful when starting to consider adaptation strategies to a changing climate.

BOX 2.1: Paleo-reference periods, listed from oldest to youngest

Period	Age (year)	Description of the climate state (relative to 1850–1990) CO_2 concentrations (ppm)	Rate of change (ppm per century)
Palaeocene-Eocene thermal maximum (PETM)	55.9–55.7 Ma	A geologically rapid, large-magnitude warming event with a large pulse of carbon released to the ocean-atmosphere system, decreasing ocean pH and oxygen content. CO_2 concentrations 800–1,000 ppm → 1,400–3,150 ppm. Global mean temperature: 10°C–25°C>than 1850–1900.	4–42
Early Eocene climatic optimum (EECO)	53–49 Ma	Prolonged 'hothouse' period. CO_2 concentrations 1,150–2,500 ppm. Global mean temp.: 10°C–18°C>than 1850–1900. GMSL: 70–76 m>than 1850–1900.	–
Miocene climatic optimum (MCO)	16.9–14.7 Ma	Prolonged warm period. CO_2 concentrations 400–600 ppm.	–
Mid-Pliocene warm period (MPWP)	3.3–3.0 Ma	Warm period when CO_2 concentrations were similar to present values. CO_2 concentrations 360–420 ppm. Global mean temp.: 2.5°C–4.0°C>than 1850–1900. GMSL: 70–76 m>than 1850–1900.	–
Last Interglacial (LIG)	129–116 ka	Most recent interglacial, similar to mid-Holocene, but with a more pronounced seasonal insolation cycle. Monsoon was enhanced. CO_2 concentrations 266.20–285.4 ppm (Dome C). Global mean temp.: 0.5°C–1.5°C>than 1850–1900. GMSL: 5–10 m>than 1850–1900.	–
Last Glacial Maximum (LGM)	23–19 ka	Most recent glaciation when global temperatures were lower, with greater cooling near the poles. Precipitation lower over most regions, and the atmosphere was dustier (see Figure 2.3c). CO_2 concentrations 184.9–193.1 ppm (Dome C). Global mean temp.: 5.0°C–7.0°C<than 1850–1900. GMSL: 134–125 m<than 1850–1900.	–

Period	Age (year)	Description of the climate state (relative to 1850–1990) CO_2 concentrations (ppm)	Rate of change (ppm per century)
Last Deglacial Transition (LDT)	18–12 ka	Warming that followed the LGM, with decreases in ice cover in both Polar Regions. Sea level rose rapidly. CO_2 concentrations 195.2–265.3 ppm (Dome C). GMSL: 120–50 m < than 1850–1900.	7.1 (Dome C)
Mid-Holocene (MH)	6.5–5.5 ka	Middle of the present interglacial when CO_2 concentration was similar to the onset of the pre-industrial era (1750 CE), but orbital configurations (see Table 2.4) led to warming and shifts in the hydrological cycle, especially the Northern Hemisphere monsoons. CO_2 concentrations 260.1–268.1 ppm (Dome C). Global mean temp.: 0.2°C–1.0°C > than 1850–1900. GMSL: –3.5 to +0.5 m than 1850–1900.	–
Last millennium	850–1850 CE	Climate variability during this period is better documented on annual to centennial scales than during earlier reference periods. Climate changes were driven by solar, volcanic, land cover and anthropogenic RFs, including significant increases in GHGs since 1750 CE. CO_2 concentrations 278.0–285.0 ppm. Global mean temp.: –0.14°C ~ 0.24°C than 1850–1900. GMSL: –3.5 to +0.5 m than 1850–1900.	–6.9 ~ 4.7 (Law Dome)
Last 100 years	1919–2019	CO_2 concentrations 302.8–306.0 ppm → 409.5–410.3 ppm.	103.9–107.1
Modern	1995–2014	CO_2 concentrations 359.6–360.4 ppm → 396.7–397.5 ppm. Global mean temp.: 0.66°C–1.00°C > than 1850–1900. GMSL: 0.15–0.25 m > than 1850–1900.	192.3–198.3

Source: Adapted from Gulev et al. (2021), Arias et al. (2021).

The Holocene interglacial began about 12 ka BP (Box 2.1, Figure 2.6a) and reached its orbital forcing (warming) maximum in the mid-Holocene (6.5–5.5 ka BP). Rates of sea-level rise slowed from 8.2 ka to about 2.5 ka BP (Lambeck et al., 2014), after which they remained stable until the renewed sea-level rise at about 100–150 years ago. CO_2 concentrations were remarkably constant during the Holocene, ranging from only 265 to 285 ppm (Box 2.1).

Past 2 ka

Other than normal variability – modulated by natural climate drivers (e.g. volcanic and solar RFs) – centennial-scale GMSTs were generally stable from 2 to 1 ka ago (Figure 2.6b). However, they were in a general cooling trend for about 1,000 years before fossil fuel-based industrialisation in 1750 (Figure 2.6b). Since then, the GMST has increased about 1°C in a period less than 1/3,000th the width shown in Figure 2.6a (or the past 800 ka). There is a prominent role for volcanic activity over the past 2 ka that explains the GMST trends. Büntgen et al. (2020) identified 12 major volcanic eruptions from 100 to 1,200 (1.1 ka), compared with 17 eruptions from 1,200 to 1,900 (0.7 ka). As discussed above, major volcanic eruptions are associated with short intervals of negative RF (and hence cooling) of the climate system. Therefore, heightened volcanic activity over the past 1 ka has been linked to the general cooling trend of GMST before the industrial revolution (Figure 2.6b).

Climate variability from 850 to 1850 is better documented on annual to centennial scales than earlier reference periods (Box 2.1). CO_2 concentrations were relatively stable during this millennium period, ranging from 278 to 285 ppm. During this time, the climate system was driven mainly by variations in solar and volcanic RFs and increases in land cover change. However, after 1750, anthropogenic forcing became a more important climate driver due to increases in GHGs associated with rapid industrial, agricultural and population growth.

CHANGES IN THE MODERN ERA

With the advent of instrumental measurements of surface air temperatures in the modern era (*circa* 1880), it was now possible to measure temperature variations at an annual scale. Similarly, with the commencement of measurements of CO_2 and other GHG concentrations in the late 1950s (Figure 2.5), it became feasible to examine fine-scale seasonal and annual trends in GHGs and their effects on Earth's climate system. The first section will discuss the IPCC's treatment of certainty and uncertainty in their 6th Assessment Reports (2021, 2022a, 2022b). The last section will evaluate changes in key global climate indicators over the past 150 years, focusing on documenting changes in key climate indicators from 2000 to 2019.

Dealing with certainty and uncertainty

A rigorous process is followed for evaluating and communicating the degree of certainty (or uncertainty) in IPCC Assessments (Chen et al., 2021). First, expert teams consider what evidence of climate change exists. Evidence may be observational, experimental, theoretical, statistical or derived from computer models. Second, they need to evaluate any evidence regarding its type, quality, quantity, consistency and scientific agreement. For example, 'there is high agreement that observed changes in the atmosphere, ocean, cryosphere (land and sea ice) and biosphere provide unequivocal evidence of a world that has warmed'.

Third, expert teams need to decide whether there is enough evidence and agreement to evaluate confidence in their findings. In the IPCC process (see Chen et al., 2021), confidence is 'a qualitative measure of the validity of a determination, based on the type, amount, quality and consistency of evidence (e.g., data, mechanistic understanding, theory, models, expert judgement) and the degree of agreement'. Fourthly (if there is enough evidence and consensus), the expert team may evaluate confidence in their findings. Through this process, a statement of fact is established, with a prescribed level of confidence expressed using five qualifiers: *very low, low, medium, high* and *very high*, and described as a typeset in italics. For example, 'together, ice sheet and land glacier mass loss were the dominant contributors to global mean sea-level rise during 2006–2018 (*high confidence*)' (Gulev et al., 2021).

Once the amount of confidence is established as a statement of fact, the expert team may finally determine whether there is enough confidence and

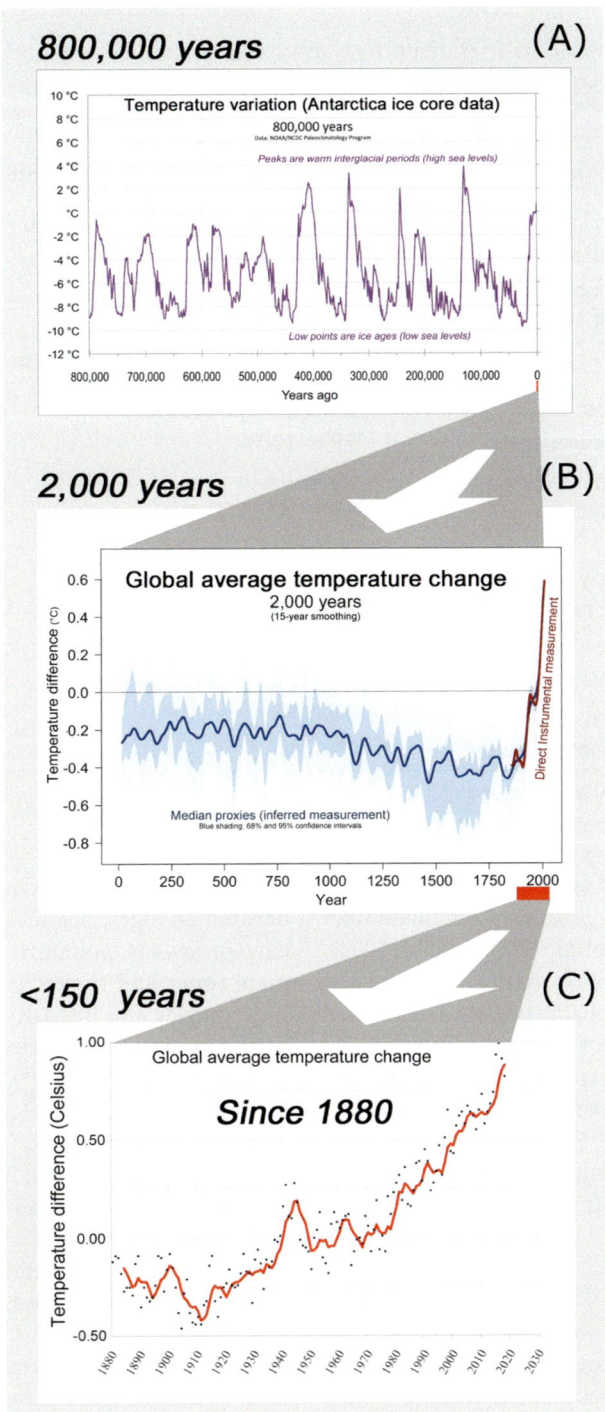

Figure 2.6 Three charts of global average temperature three respective time periods: 800,000 years, 2,000 years and 139 years, showing current global warming in the perspective of geologic time. (a) Top graph: 0.0 line is the average temperature of the last one thousand years; (b) the 0.0 line is the 1951–1980 average, the blue line is median proxies (15-year smoothing), and the red line is direct instrumental measurement; and (c) the 0.0°C line is the 1901–2000 average and annular data is smoothed with a 5-year trailing average (red line). (Credit: RCraig/Wikimedia/Public Domain.)

BOX 2.2: Terms used to indicate the assessed confidence and likelihood of an outcome or a result

Confidence terminology	Degree of confidence in being correct
Very high confidence	At least nine out of ten chance
High confidence	About eight out of ten chance
Medium confidence	About five out of ten chance
Low confidence	About two out of ten chance
Very low confidence	Less than one out of ten chance

Likelihood	Outcome probability
Virtually certain	99%–100%
Extremely likely	95%–100%
Very likely	90%–100%
Likely	66%–100%
More likely than not	>50%–100%
About as likely as not	35%–66%
Unlikely	0%–33%
Very unlikely	0%–10%
Extremely unlikely	0%–5%
Exceptionally unlikely	0%–1%

Source: After Chen et al. (2021).

quantitative or statistical evidence. If the evidence meets these criteria, it is possible to evaluate likelihood (Box 2.2). Chen et al. (2021) define likelihood as 'a quantitative measure of uncertainty in a finding, expressed probabilistically (e.g., based on a statistical analysis of observations or model results, or both, and expert judgement by the author team or from a formal quantitative survey of expert views, or both)'. For example, 'it is *virtually certain* that the average surface warming will continue to be higher over land areas than oceans' (Arias et al., 2021).

Past 140 years

The scale of recent changes (2000–2019) and the present state of many aspects of the climate system (Box 2.1) are unprecedented over many centuries to many thousands of years (Intergovernmental Panel on Climate Change, 2021, Eyring et al., 2021). Changes in key global climate indicators over the past 140 years and regional patterns are now discussed individually. These indicators include changes in GHGs (GHGs), GMST changes,

precipitation changes, changes in the cryosphere, oceanic changes, sea-level changes, hot and cold extreme events, poleward and upward shifts in climate zones and the terrestrial biosphere, changes in frequency and intensity of tropical cyclones and compounding effects of extreme events.

GREENHOUSE GASES

Over the past 100 years, CO_2 concentrations (Box 2.1) have increased by 25% from about 306 to 410 ppm (about 105 ppm per century). However, from 1995 to 2014, CO_2 concentrations rose from 360 to 398 ppm – representing a very high rate of nearly 200 ppm per century. Land and ocean have absorbed a near-constant proportion of CO_2 emissions (~56% per year) from human activities from 1960 to 2020 (Figure 2.2), with some regional differences (*high confidence*) (Gulev et al., 2021). The remaining 44% of CO_2 emissions have been added to the atmosphere over this time.

Average annual GHG emissions during 2010–2019 were higher than in any previous decade, but the growth rate between 2010 and 2019 was lower than that between 2000 and 2009 (*high confidence*) (Intergovernmental Panel on Climate Change, 2022a). Notably, in 2019, atmospheric CO_2 concentrations (Figure 2.5) were higher than at any time in at least 2 Ma (Box 2.1) (*high confidence*). Concentrations of CH_4 and N_2O (Table 2.1) were higher than at any time in at least 800 ka (*very high confidence*) (Arias et al., 2021, Gulev et al., 2021). Since 2011, GHG concentrations have continued to increase in the atmosphere, reaching annual averages of 410 ppm for carbon dioxide (CO_2), 1,866 ppb for methane (CH_4) and 332 ppb for nitrous oxide (N_2O) in 2019 (Table 2.1).

GLOBAL MEAN SURFACE TEMPERATURE

GMST was 1.09°C [range: 0.95°C–1.20°C] higher in 2011–2020 than 1850–1900 (*high confidence*) – a period less than 1/20,000th the width of Figure 2.6a (or the past 800 ka). Importantly, each of the last four decades has been successively warmer than any decade that preceded it since 1850 (*high confidence*) (Figure 2.6c). Well-mixed GHGs (Table 2.5) were *likely* the primary driver of tropospheric warming since 1979 (*high confidence*). GMST has increased faster since 1970 than in any other 50-year period over the last 2 ka (Arias et al., 2021, see Figure 2.6b) (*high confidence*). GMST during the most recent decade (2011–2020) – relative to 1850–1900 – exceeded those of the most recent multi-century warm period, around 6.5 ka BP of 0.2°C–1°C (*medium confidence*) (Box 2.1). Before that, the next most recent warm period was the Last Interglacial Maximum (~125 ka BP) when the multi-century temperature of 0.5°C–1.5°C (relative to 1850–1900) overlaps the observations of the most recent decade (Box 2.1, Figure 2.6a and c) (*medium confidence*).

More significant GMST increases have occurred overland (1.60°C) than over the ocean (0.88°C) between the periods of 2011–2020 and 1850–1900 (*high confidence*) (Arias et al., 2021). Data shows that warming over land is about 80% larger than warming over the ocean (*high confidence*). Of concern, more significant warming over land alters key water cycle attributes, such as evapotranspiration, precipitation and runoff (*medium confidence*) (Intergovernmental Panel on Climate Change, 2021). While there has been a general upward trend in GMST since 1800 (Figure 2.6c), significant changes have also been observed in both Polar Regions (Arias et al., 2021). The Arctic mean surface temperature has increased two to three times

faster than the GMST rate of change (*high confidence*), while the Antarctic Peninsula mean surface temperature has increased 1.5 times faster than that for the Southern Hemisphere as a whole (*high confidence*) (Gulev et al., 2021).

PRECIPITATION

Human influence has likely contributed to the pattern of observed global precipitation (rain, snow, hail, fog) since the 1950s (Eyring et al., 2021). Precipitation over land has *likely* increased since 1950, with a faster rate since 1980 (*medium confidence*). Additionally, the frequency and intensity of heavy precipitation events have increased since the 1950s over most land areas for which observational data are enough for trend analysis (*high confidence*), and human-induced climate change is *likely* the main driver (Gulev et al., 2021).

Significant regional changes in precipitation have been driven by historical and recently declining anthropogenic aerosols and rising GHGs (see Table 2.5 for ERFs). Decreases in global land monsoon[5] precipitation from the 1950s to the 1980s were partly attributed to anthropogenic Northern Hemisphere aerosol emissions (i.e. global dimming). However, precipitation increases since then have resulted from rising GHG concentrations, decadal to multi-decadal internal variability and a decline of anthropogenic aerosol emissions over Europe and North America (*medium confidence*).

Changes in global precipitation patterns have also occurred due to the expansion of the tropics (Turton, 2017, Eyring et al., 2021). Mid-latitude storm tracks have *likely* shifted poleward in both hemispheres since the 1980s, with marked seasonality in trends (*medium confidence*). Human influence *very likely* contributed to the poleward shift of the Southern Hemisphere extra-tropical jet stream during the austral summer (Arias et al., 2021, see also Chapter 3). Finally, human-induced climate change has contributed to increased agricultural[6]

---

[5] A monsoon is a seasonal change in the direction of the prevailing, or strongest, winds of a region. Monsoons cause wet and dry seasons throughout much of the tropics and sub-tropics.

[6] Agricultural drought is "characterized by lack of sufficient moisture in the surface soil layers to support crop and forage growth" (Mannocchi et al., 2004).

and ecological[7] droughts in some world regions due to increased land evapotranspiration (*medium confidence*).

CRYOSPHERE CHANGES

Significant changes in the global cryosphere have resulted from the worldwide retreat of land glaciers, sea ice and ice sheets over recent decades (Gulev et al., 2021). Anthropogenic influence is *likely* the primary driver of glacial retreat since the 1990s and the decrease in Arctic sea ice between 1979–1988 and 2010–2019 (~40% in September and 10% in March). There is *medium confidence* that the worldwide retreat of glaciers since the 1950s is unprecedented in at least the last 2 ka (Arias et al., 2021). Meanwhile, the rate of ice sheet loss increased by a factor of four between 1992–1999 and 2010–2019 (Eyring et al., 2021).

There are some differences in rates of cryosphere change between Earth's Polar Regions. For example, anthropogenic influence has *likely* contributed to the observed melting of the Greenland ice sheet and *medium agreement* for the Antarctic ice sheet mass loss (Gulev et al., 2021). The West Antarctica ice sheet dominated the total Antarctic ice mass losses (Arias et al., 2021), with combined West Antarctica and its peninsula annual loss rates increasing since about 2000 (*very high confidence*). In total, the Antarctic ice sheet has likely lost 2670 ± 530 Gt of ice (Arias et al. 2021), contributing 7.4 ± 1.5 mm to global mean sea-level (GMSL) rise over 1992–2020 (see the 'Sea-Level Changes' section below).

Some components of Earth's cryosphere are approaching critical climate thresholds. Over 2011–2020, the annual average Arctic sea ice area reached its lowest level since at least 1850 (*high confidence*), and its size in September 2020 was smaller than at any time in at least the past 1ka (*medium confidence*) (Intergovernmental Panel on Climate Change, 2021). However, there is no significant trend in Antarctic sea ice from 1979 to 2020 due to opposing regional trends (east and west differences) and sizeable internal variability (Gulev et al., 2021).

---

[7] Ecological drought is an "episodic deficit in water availability that drives ecosystems beyond thresholds of vulnerability, impacts ecosystem services, and triggers feedbacks in natural and/or human systems" (Crausbay et al., 2017).

OCEANIC CHANGES

Oceans, seas and bays cover 71% of Earth's surface area and contain 96.5% of all its water. Therefore, understanding changes in key oceanic climate indicators is crucial to gaining insights into recent and future changes in the climate system. Since the 1970s, it is *virtually certain* that the upper ocean (0–700 m) has warmed and *highly likely* that human influence is the primary driver (Gulev et al., 2021, Eyring et al., 2021). There is *high confidence* that this ocean warming represents about 90% of the global energy budget (see Figure 2.1, Table 2.5) (Arias et al., 2021). In comparison to the paleo-reference periods (Box 2.1), the global ocean has warmed faster over the past century than at any time since the end of the Last Deglacial Transition (~11 ka BP) (*medium confidence*).

Rising ocean temperatures are driving other significant oceanic changes (Eyring et al., 2021). For example, human influence has *very likely* changed near-surface ocean salinity (*high confidence*). Similarly, it is *virtually certain* that anthropogenic CO_2 emissions are the primary driver of the current global acidification (pH) of the open surface ocean (Arias et al., 2021). Acidification is driven by solution of excess CO_2 emissions by seawater and the formation of carbonic acid. To put this in context, open surface ocean pH as low as in recent decades is unusual in at least the last 2 Ma (*medium confidence*) (Box 2.1) (Gulev et al., 2021). Finally, there is *high confidence* that dissolved oxygen (O_2) levels have dropped in many upper ocean regions since the mid-20th century. There is *medium confidence* that human influence contributed to this fall (Intergovernmental Panel on Climate Change, 2021). Deoxygenation of oceans has varied regionally due to differences in ocean heating. Some of the most significant dissolved O_2 declines have occurred in tropical waters, by as much as 40% from 1960 to 2019 (Gulev et al., 2021).

Significant changes in oceans' temperature and chemistry have impacted many marine organisms (Intergovernmental Panel on Climate Change, 2021). Over at least the last 20 years, the geographical range of many marine organisms has shifted towards the poles and towards greater depths (*high confidence*), indicative of shifts towards cooler waters (Arias et al., 2021). Meanwhile, there is *high confidence* that the range of a smaller subset of organisms has shifted equatorward and to shallower depths due to ocean warming (Arias et al., 2021). Finally, there is *high confidence* that phenological metrics associated with the life cycles of many organisms have also changed over the last two decades or longer (Gulev et al., 2021). If such changes continue, significant disruptions to marine ecosystems and marine fisheries are *likely*.

SEA-LEVEL CHANGES

Since 1750, the sea level has increased steadily in response to global heating. GMSL increased by 0.20 m [range: 0.15–0.25 m] between 1901 and 2018 (Gulev et al., 2021). This rising rate is greater than over any preceding century in at least the last 3 ka (*high confidence*) (Box 2.1). Notably, the average rate of sea-level rise was 3.7 mm/year [range: 3.2–4.2 mm/year] between 2006 and 2018 (*high confidence*) (Arias et al., 2021). The anthropogenic influence was *very likely* the main driver of these increases since at least 1971 (Arias et al., 2021). Contributors to sea level rise have come from several sources. Overall, thermal expansion of seawater explained 50% of sea-level rise during 1971–2018, while ice loss from glaciers contributed 22%, ice sheets 20% and changes in land water storage 8% (Gulev et al., 2021). However, the balance of contributions from terrestrial sources has changed over the past 20 years. Ice sheet and land glacier mass loss were the dominant contributors to GMSL rise during 2006–2018 (*high confidence*).

Sea level does not change equally across the planet. Other processes come into play at regional scales that modify the local or regional sea-level change relative to GMSL. These include vertical land motion, ocean circulation and density changes, and gravitational, rotational and deformational effects arising from water and ice mass redistribution between land and the ocean (Gulev et al., 2021). Consequently, there is a tendency for these processes to increase sea levels in the tropics and reduce those in the mid-to-high latitudes, thereby producing differing regional trends. For example, the strongest regional trends over the period from January 1993 to January 2020 were experienced in the Southern Hemisphere, from east of Madagascar in the Indian Ocean, east of New Zealand in the Pacific Ocean and east of Rio de la Plata/South America in the south Atlantic (Gulev et al., 2021).

HOT AND COLD EXTREME EVENTS

As the baseline climate shifts due to anthropogenic forcing of the climate system (see Table 2.5), so does the distribution of extreme climate events around the altered baseline. Because of global heating, it is *virtually certain* that hot extremes (including heatwaves) have become more frequent and more intense across most land regions since the 1950s. In contrast, cold extremes (including cold waves) have become less frequent and less severe (Arias et al., 2021). There is *high confidence* that anthropogenic climate change is the primary driver of these changes (Intergovernmental Panel on Climate Change, 2021). Of note, some recent hot extremes observed over the past decade would have been *extremely unlikely* to occur without human influence on the climate system (Gulev et al., 2021). Similarly, marine heatwaves have approximately doubled in frequency since the 1980s (*high confidence*), and human influence has *very likely* contributed to most of them since at least 2006 (Arias et al., 2021).

POLEWARD AND UPWARD SHIFTS IN CLIMATE ZONES AND THE TERRESTRIAL BIOSPHERE

As the tropical zone warms, it expands vertically and horizontally. This tropical expansion largely explains why Earth's climate zones have shifted poleward in both hemispheres since the 1970s (Turton, 2017, Eyring et al., 2021), and there is *high confidence* in these findings (Intergovernmental Panel on Climate Change, 2021). Notably, extra-tropical cyclones and tropical cyclones have *likely* shifted poleward since the 1980s (Arias et al., 2021). Such climate changes have affected the terrestrial biosphere. For example, over the last century, there has been a poleward and upslope shift in the distribution of many land species (*very high confidence*) as well as increases in species turnover within many ecosystems (*high confidence*) (Gulev et al., 2021). Significant regional differences have been due to poleward shifts in climate zones (Arias et al., 2021). First, the growing season has, on average, lengthened by up to 2 days per decade since the 1950s in the Northern Hemisphere extra-tropics (*high confidence*). Second, there is *high confidence* that Southern Hemisphere storm tracks and associated precipitation have migrated poleward over recent decades, especially in the austral summer and autumn (see Chapter 3).

CHANGES IN TROPICAL CYCLONES

As stated above, tropical cyclones (including hurricanes and typhoons) are shifting poleward in response to anthropogenic heating of the planet. However, other aspects of changes in tropical cyclone attributes are much *less certain*. For example, there is *low confidence* in long-term (multi-decadal to centennial) trends in the frequency of all-category tropical cyclones (Arias et al., 2021). However, event attribution studies and physical understanding indicate that human-induced climate change increases heavy precipitation associated with tropical cyclones (*high confidence*). However, data limitations inhibit the precise detection of past trends on a global scale (Gulev et al., 2021). In terms of intensity change, it is *likely* that the worldwide proportion of severe (Categories 3–5) tropical cyclone occurrence has increased over the last four decades, and the latitude where tropical cyclones in the western North Pacific reach their peak intensity has shifted northward (Intergovernmental Panel on Climate Change, 2021). Studies have shown that these observed changes cannot be explained by internal variability alone (*medium confidence*).

COMPOUND EXTREME EVENTS

Individual extreme weather and climate events may interact and worsen the effects of other extreme events. This interaction results in a greater risk for compound extreme events. Human influence has *likely* increased the chance of compound extreme events since the 1950s (Intergovernmental Panel on Climate Change, 2021). For example, increases in the frequency of concurrent heatwaves and droughts on the global scale (*high confidence*), fire weather in some regions of all inhabited continents (*medium confidence*) and compound flooding in some locations (*medium confidence*) (Arias et al., 2021).

FUTURE CHANGES IN CLIMATE

Climate models

Climate models – also known as 'general circulation models' (GCMs) – use differential equations based on the fundamental laws of physics, fluid motion and chemistry to characterise how energy and matter interact in different parts of the ocean, atmosphere and land (Arias et al., 2021). GCMs

separate Earth's surface into a three-dimensional grid of cells. The physical processes modelled in each cell are transferred to neighbouring cells to simulate the exchange of matter and energy over time. Grid cell size defines the resolution of the GCM. Higher-resolution GCMs have smaller grid sizes and require supercomputers to run many simulations for the entire planet. GCMs may also be downsized to simulate higher-resolution simulations at regional scales. Climate models are used by some research groups worldwide, including the Bureau of Meteorology and CSIRO in Australia.

The accuracy of climate models may be tested via a process known as 'hindcasting'. In common practice, the climate model is run from the present time back into the past (usually back to when reliable instrumental measurements began). The model outputs (simulations) can then be compared with observed climate data, e.g. air temperatures. This hindcasting approach can be used to verify how well the simulated and observational data correspond to assess the validity of the models. If a climate model performs well in the hindcasting process, scientists may use the model (or GCM) to project future climates with high confidence. A climate projection is the simulated response of the climate system to a scenario of future emissions or concentrations of GHGs and aerosols derived using climate models (Arias et al., 2021). Hence, to project climate into the future, the climate forcings (see ERFs: Table 2.5) are adjusted to simulate a plausible future GHG emissions scenario (see SSPs: Table 2.6). Future scenarios are typically based on population change, land use change, urbanisation and social and economic growth and development (see Chapter 4).

The latest generation of climate model simulations are part of the World Climate Research Programme's Coupled Model Inter-comparison Project Phase 6 (CMIP6) and form the basis of the critical line evidence for the IPCC's latest assessment report (Lee et al., 2021). The objective of the CMIP is to understand better past, present and future climate changes arising from natural, unforced variability or in response to changes in RF in a multi-model context (Pascoe et al., 2020). This understanding includes assessments of model performance during the historical period (hindcasting) and quantifications of the causes of the spread in future projections. CMIP6 includes new and higher-resolution representations of physical, chemical and biological processes and improved simulations of the recent mean climate of most large-scale indicators of climate change (Arias et al., 2021). According to Eyring et al. (2021), 'the CMIP6 model ensemble reproduces the observed historical global surface temperature trend and variability with biases small enough to support detection and attribution of human-induced warming' (*very high confidence*).

Shared socio-economic pathways

The latest IPCC climate change assessment report (Intergovernmental Panel on Climate Change, 2021, Arias et al., 2021) refers to a set of five new emissions scenarios or pathways (Table 2.6). These shared socio-economic pathways (SSPs) explore the climate response to a broader range of GHGs, land use and air pollutant futures. The SSPs replace the Representative Concentration Pathways (RCPs) scenarios referenced in the previous IPCC assessments

(Intergovernmental Panel on Climate Change, 2013). This set of emissions drives climate model projections of changes in Earth's climate system in the near term (2021–2040), mid-term (2041–2060) and long term (2081–2100). The SSPs comprise the new scenario framework, developed by Riahi et al. (2017), to facilitate the integrated analysis of future climate impacts, vulnerabilities, adaptation and mitigation (see Figure 3.1, Chapter 4).

Each of the five SSPs has a narrative (Table 2.6) describing alternative socio-economic developments, including sustainable development, regional rivalry, inequality, fossil-fuelled development and middle-of-the-road development (Riahi et al., 2017). The SSPs' long-term population and economic projections (Table 2.6) represent a wide uncertainty range consistent with the scenario literature. For example, future annual emissions of CO_2 will be about 75 Gt/year by 2050 under the fossil-fuelled development scenario (RCP5–8.5), rising to near 130 Gt/year by 2100 (Intergovernmental Panel

Table 2.6 Five illustrative shared socio-economic pathways (SSPs) (emission scenarios) and projected changes in global temperature for three 20-year time periods

Shared socio-economic pathway (SSP) and its effective radiative forcing (ERF) of the climate system in W/m² Emissions trend in parentheses	Changes in global surface temperature (° C) for Selected 20-year time periods. best estimate and very likely range in parentheses
SSP1–1.9 (ERF = 1.9): *Sustainability – Taking the Green Road* (low challenges to mitigation and adaptation) (very strongly declining emissions)	Near term: 1.5 (1.2–1.7) Mid-term: 1.6 (1.2–2.0) Long term: 1.4 (1.0–1.8)
SSP1–2.6 (EFR = 2.6): *Middle-of-the-Road* (medium challenges to mitigation and adaptation) (strongly declining emissions)	Near term: 1.5 (1.2–1.8) Mid-term: 1.7 (1.3–2.2) Long term: 1.8 (1.3–2.4)
SSP2–4.5 (ERF = 4.5): *Regional Rivalry – A Rocky Road* (high challenges to mitigation and adaptation) (slowly declining emissions)	Near term: 1.5 (1.2–1.8) Mid-term: 2.0 (1.6–2.5) Long term: 2.7 (2.1–3.5)
SSP3–7.0 (ERF = 7.0): *Inequality – A Road Divided* (low challenges to mitigation, high challenges to adaptation) (stabilising emissions)	Near term: 1.5 (1.2–1.8) Mid-term: 2.1 (1.7–2.6) Long term: 3.6 (2.8–4.6)
SSP5–8.5 (ERF = 8.5): *Fossil-fuelled Development – Taking the Highway* (high challenges to mitigation, low challenges to adaptation) (rising emissions)	Near term: 1.6 (1.3–1.9) Mid-term: 2.4 (1.9–3.0) Long term: 4.4 (3.3–5.7)

Sources: Riahi et al. (2017), Intergovernmental Panel on Climate Change (2021).
Projected changes in global temperatures (°C) are shown for the near term (2021–2040), mid-term (2041–2060) and long term (2081–2100), relative to the 1850–1900 baseline.
Notes: The SSPs drives climate model projections of changes in the climate system. These projections account for solar activity and background forcing from volcanoes (see discussions above).

on Climate Change, 2021). Under the regional rivalry scenario (SSP2–4.5), CO_2 emissions will be about 45 Gt/year by 2050 (like 2015 rates), declining to near 10 Gt/year by 2100. In the case of the middle-of-the-road scenario (SSP1–2.6), CO_2 emissions will decrease from 40 Gt/year (2015 rate) to about 22 Gt/year by 2050 and decline further to near −10 Gt/year by 2100. Under this GHG emissions scenario, decarbonisation of the atmosphere (or net-zero emissions) will commence in the 2080s (Intergovernmental Panel on Climate Change, 2021). In the case of the very ambitious SSP1–1.9 scenario (Table 2.6), net-zero CO_2 emissions will begin in the 2050s, with net CO_2 emission rates reaching about −15 Gt/year by 2100. Under scenarios with increasing CO_2 emissions (Table 2.6), the ocean and land carbon sinks (Figure 2.2) are projected to be less effective at reducing the accumulation of CO_2 in the atmosphere (Arias et al., 2021).

Projections of future climates

Global warming will cause significant changes in Earth's climate system over the coming decades, and regional variations in many climate indicators are *highly likely* and are summarised below. More detailed discussions may be found in the latest IPCC Sixth Assessment Working Group 1 Reports (Intergovernmental Panel on Climate Change, 2021, Arias et al., 2021). Many of these future changes become amplified in direct relation to increasing global warming (Lee et al., 2021). These include increases in the frequency and intensity of hot extremes, marine heatwaves, heavy precipitation events, agricultural and ecological droughts in some regions, the frequency of intense tropical cyclones and reductions in Arctic sea ice and snow cover and permafrost. Additionally, global warming is projected to intensify the global water cycle and its variability, with implications for global monsoon precipitation and the severity of dry and wet events (Arias et al., 2021). Due to past and future GHG emissions, many changes in the climate system are irreversible for centuries to millennia, especially changes in the ocean, ice sheets and global sea level (Lee et al., 2021, Arias et al., 2021).

GLOBAL MEAN SURFACE AIR TEMPERATURE

Anthropogenic forcing has warmed the climate system since 1880 (Figure 2.6c, Table 2.5). Future warming is *certain*, but the amount and rate of warming are highly dependent on future emissions of well-mixed GHGs (Table 2.6) and – to a lesser extent – emissions of aerosols from natural and anthropogenic sources (Forster et al., 2021). Future changes in land cover and land use will also be significant through their known effects on the surface radiation balance (Figure 2.1). Compared to 1850–1900, projected global mean surface air temperature (GSAT) over 2081–2100 is *very likely* to be higher by 1.0°C–1.8°C for the very low GHG emissions scenario (SSP1–1.9), by 2.1°C–3.5°C for the intermediate scenario (SSP2–4.5) and by 3.3°C–5.7°C for the very high GHG emissions scenario (SSP5–8.5) (Lee et al. 2021). The last time GSAT was sustained at or above 2.5°C higher than 1850–1900 was over 3 Ma ago (see Box 2.1) (*medium confidence*). GSAT will continue to increase until at least 2050 under all emissions scenarios considered (Table 2.6).

The Paris Agreement targets aim to hold the increase in the GSAT to well below 2°C above pre-industrial levels and pursue efforts to limit the temperature increase to 1.5°C above pre-industrial levels (see Chapter 4, the 'Paris Agreement' section). It is, therefore, possible to project when these upper and lower Paris GSAT targets may be reached for the five SSPs (Table 2.6). In the near term (2021–2040), the 1.5°C global warming level is *very likely* to be exceeded under the very high GHG emissions scenario (SSP5–8.5), *likely* to be exceeded under the intermediate and high GHG emissions scenarios (SSP2–4.5 and SSP3–7.0) and *more likely than not* to be exceeded under the low GHG emissions scenario (SSP1–2.6) (Lee et al. 2021). For the very low GHG emissions scenario (SSP1–1.9), it is *more likely than not* that the GSAT would decline back to below 1.5°C by 2100, with a temporary overshoot of no more than 0.1°C above 1.5°C global warming (Arias et al. 2021). However, achieving such a very low emissions pathway will require significant international cooperation and commitment to rapidly transition to a net-zero carbon global economy, which will also need mechanisms to 'decarbonise' the atmosphere and ocean.

There will be regional differences in mean surface air temperature changes (Arias et al., 2021, Lee et al., 2021). Notably, it is *virtually certain* that the land surface will continue to warm more than the ocean surface (likely 1.4–1.7 times more). The land/ocean warming differential will also drive other complex changes in Earth's climate system, such as stronger sea breezes and extreme rainfall events over land. Some of the most significant changes in regional mean surface air temperatures occur in higher latitudes. Hence, it is *virtually certain* that the Arctic will continue to warm more than the GSAT, with *high confidence* its warming rate will be above two times the GSAT rate (Lee et al., 2021). The Arctic is also projected to experience the highest increase in the temperature of the coldest days, at about three times the rate of global warming (*high confidence*). In addition, some mid-latitude and semi-arid regions and the South American Monsoon region are projected to see the highest increase in the temperature of the hottest days, at about 1.5–2 times the GSAT rate (*high confidence*) (Arias et al., 2021).

CARBON CYCLE CHANGES

Global warming is expected to lead to significant changes in Earth's carbon cycle (Intergovernmental Panel on Climate Change, 2021). The natural land and ocean carbon sinks (see Figure 2.2) are projected to take up – in absolute terms – a progressively more significant amount of CO_2 under higher compared to lower emissions scenarios. However, the proportion of emissions taken up by land and ocean sinks decreases with increasing cumulative CO_2 emissions (Lee et al., 2021, Arias et al., 2021). Based on the CMIP6 model projections (Arias et al., 2021) for the five SSP scenarios (Table 2.6), under the intermediate scenario that stabilises atmospheric CO_2 concentrations this century (SSP2–4.5), the rates of CO_2 taken up by the land and ocean (Figure 2.2) are projected to decrease in the second half of the 21st century (*high confidence*). However, under the very low and low GHG emissions scenarios (SSP1–1.9, SSP1–2.6), where CO_2 concentrations peak and decline during the 21st century, land and ocean begin to take up less carbon in response to declining atmospheric CO_2 concentrations (*high confidence*) and convert into a weak net carbon source by 2100 under SSP1–1.9 (*medium confidence*). Under high CO_2 emissions pathways, it is *very likely* that the land carbon sink will grow more slowly due to warming and drying from the mid 21st Century, but it is *very unlikely* that it will 'flip' from being a sink to a source before 2100 (Arias et al., 2021). This process is projected to result in a higher proportion of emitted CO_2 remaining in the atmosphere into

the 22nd century (*high confidence*). Finally, additional ecosystem responses to warming not yet fully included in the CMIP6 climate models (Arias et al., 2021), such as CO_2 and CH_4 emissions from wetlands, permafrost thaw and wildfires, would further increase concentrations of both these gases in the atmosphere (*high confidence*). The combined effects of all these emissions would undoubtedly lead to an even stronger positive forcing of the climate system.

PRECIPITATION AND WATER CYCLE CHANGES

The warming of Earth's climate system is already driving global and regional changes in precipitation (rain, snow, hail, fog), and profound changes are projected to continue even with lower emission scenarios (Table 2.6). For example, heavy precipitation events will *likely* intensify and become more frequent in most regions with additional global warming (Lee et al., 2021). At the global scale, extreme daily precipitation events are projected to intensify by about 7% for each 1°C of global warming (Arias et al., 2021), as warm air contains more water by volume (*high confidence*).

Significant changes may be expected in precipitation over land areas and these are likely to increase faster than over oceans, even for low emission scenarios (Lee et al., 2021). Relative to 1995–2014, the average annual global land precipitation is projected to increase by 0%–5% under the very low GHG emissions scenario (SSP1–1.9), 1.5%–8% for the intermediate GHG emissions scenario (SSP2–4.5) and 1%–13% under the very high GHG emissions scenario (SSP5–8.5) by 2081–2100 (Arias et al., 2021). There is *high confidence* that the global water cycle will intensify as global temperatures rise (Lee et al., 2021). Hence, precipitation and surface water flows are projected to become more variable over most land regions within seasons (*high confidence*) and from year to year (*medium confidence*).

Global warming is projected to exacerbate extremes in the water cycle (Intergovernmental Panel on Climate Change, 2021). Warming will intensify 'very wet' and 'very dry' weather and climate events and seasons, with implications for flooding or drought (*high confidence*). However, the location and frequency of these events depend on projected changes in regional atmospheric circulation, including monsoons and mid-latitude storm tracks (Arias et al., 2021). In particular, it is *very likely* that rainfall variability related to the El Niño-Southern Oscillation (Figure 2.4b) will be amplified by the second half of the 21st century under the SSP2–4.5, SSP3–7.0 and SSP5–8.5 emission scenarios (Table 2.6).

As Earth's climate system warms, the tropics will expand poleward, resulting in changes in global climate zones (Turton, 2017). Hence, the portion of the worldwide land area experiencing detectable increases or decreases in seasonal mean precipitation is projected to increase (*medium confidence*). Regional differences in precipitation are also *likely* to occur as the tropics expand. Precipitation will *likely* increase over high latitudes, the equatorial Pacific, but *likely* decrease over parts of the subtropics and some limited areas in the tropics under the SSP2–4.5, SSP3–7.0 and SSP5–8.5 scenarios (Arias et al., 2021). Significantly, higher latitudes and mountainous regions have *high confidence* in an earlier onset of spring snowmelt, with higher peak flows at the expense of summer flows in the snow-dominated areas of the world (Lee et al., 2021).

Globally, monsoon precipitation is projected to increase mid- to long term (Lee et al., 2021), particularly over South and Southeast Asia, East Asia and West Africa apart from the far west Sahel region (*high confidence*). However, the monsoon season is projected to have a delayed onset over North and South America and West Africa (*high confidence*) and a delayed retreat over West Africa (*medium confidence*).

OCEANIC CHANGES

There is *high confidence* that steady increases in GHG emissions since 1750 (Table 2.1, Figure 2.5) has assigned the global ocean to sustained future warming (Arias et al., 2021). From now on, projected changes in ocean warming range from 2 to 4 (SSP1–2.6) to 4–8 times (SSP5–8.5) the rate of temperature change measured from 1971–2018 (Lee et al., 2021).. Global warming will also affect other oceanic change indicators (Arias et al., 2021). There is *high certainty* that future global warming will affect upper ocean temperature, salinity stratification and ocean acidification. There is also *high confidence* that ocean deoxygenation will continue to increase in the 21st century, at rates dependent on future emissions (Table 2.6). Lee et al. (2021) assert that these oceanic changes are irreversible

on centennial to millennial time scales. There is *very high confidence* that there will be continued changes in global ocean temperature and deep ocean acidification, while there is *medium confidence* for changes in ocean oxygen levels.

SEA-LEVEL CHANGES

It is *virtually certain* that the GMSL will continue to rise over the 21st century and beyond (Intergovernmental Panel on Climate Change, 2021). Rates of sea-level rise will depend on future GHG emissions (Table 2.6) (Arias et al., 2021). Relative to 1995–2014, the *likely* GMSL rise by 2100 is 0.28–0.55 m under the very low GHG emissions scenario (SSP1–1.9), 0.32–0.62 m under the low GHG emissions scenario (SSP1–2.6), 0.44–0.76 m under the intermediate GHG emissions scenario (SSP2–4.5) and 0.63–1.01 m under the very high GHG emissions scenario (SSP5–8.5). These sea-level projections are conservative, as they do not factor in deep uncertainties regarding ice sheet processes, such as the rapid melting of the Greenland and West Antarctica ice sheets (Arias et al., 2021). Hence, there is *low confidence* that GMSL rise will approach 2 m by 2100 and 5 m by 2150 under a very high GHG emissions scenario (SSP5–8.5) (Lee et al., 2021). Finally, there is *high confidence* that the sea level will rise for centuries to millennia. Deep ocean warming and ice sheet melt will remain elevated for thousands of years (Intergovernmental Panel on Climate Change, 2021).

HOT AND COLD EXTREME EVENTS

While there is a big focus on projected increases in GSAT, with every additional increment of global warming, changes in extremes continue to become more significant and more widespread (Intergovernmental Panel on Climate Change, 2021). Hence, it is *very likely* that for each additional 0.5°C of global warming, there will be discernible increases in the intensity and frequency of hot extremes, including heatwaves (Lee et al., 2021). There will be significant regional differences in extreme climate events (Arias et al., 2021). Some mid-latitude and semi-arid regions and the South American Monsoon region are projected to see the highest increase in the temperature of the hottest days, at about 1.5–2 times the global average (*high confidence*). The most extreme changes are expected in the Arctic (Lee et al. 2021), which is projected to experience the highest temperature increase for the coldest days, at about three times the global average (*high confidence*). With continued global warming, the frequency of marine heatwaves will increase, particularly in tropical oceans (*high confidence*) and the Arctic Ocean (*medium confidence*).

CRYOSPHERE CHANGES

As Earth's climate system warms, the global retreat of sea ice, land glaciers and ice sheets are projected to increase this century and beyond, but the retreat rates will depend on future GHG emissions (Table 2.6). The Arctic will continue to be the focus of rapid climate change (Arias et al., 2021). Additional warming is projected to amplify further permafrost thawing and loss of seasonal snow cover, land ice and Arctic sea ice (*high confidence*). Perhaps most alarming is that under the five illustrative scenarios (Table 2.6), the Arctic is *likely* to be practically sea ice-free in September (month of minimum sea ice) at least once before 2050. Conversely, there is *low confidence* in the projected decrease of Antarctic sea ice (Lee et al., 2021). In the longer term, loss of Arctic permafrost carbon following permafrost thaw is considered irreversible at centennial timescales (*high confidence*).

In contrast, continued ice loss over the 21st century is *virtually certain* for the Greenland ice sheet and likely for the Antarctic ice sheet (Lee et al., 2021). There is *high confidence* that total ice loss from the Greenland ice sheet will increase with cumulative GHG emissions (Arias et al., 2021). Meanwhile, mountain and polar glaciers are committed to melting for decades or centuries (*very high confidence*).

CHANGES IN TROPICAL AND EXTRA-TROPICAL CYCLONES

There have already been latitudinal changes in the mean tracks of tropical and extra-tropical cyclones (storms), driven mainly by an expanding tropical zone (Turton, 2017). Changes in both weather systems are expected to continue (Intergovernmental Panel on Climate Change, 2021). There is *high confidence* that the proportion of intense (severe) tropical cyclones (Categories 4–5) and peak wind speeds of the most intense tropical cyclones are projected to increase globally with increasing warming (Lee et al., 2021). However, in some tropical cyclone regions (including Australia), cyclone frequency is expected to decline (Turton, 2019a). In response to the expansion of the tropics (Turton 2017), a southward shift and intensification of Southern Hemisphere summer mid-latitude (extra-tropical) storm tracks, and associated precipitation is *likely* in the long term, especially under high GHG emissions scenarios (SSP3–7.0, SSP5–8.5). However, the near-term effect of stratospheric ozone recovery counteracts these changes (*high confidence*). Chapter 3 discusses the implications of this long-term climate shift for winter and spring rainfall in southern Australia. Meanwhile, there is *medium confidence* in a continued poleward shift of storms and precipitation in the North Pacific. At the same time, there is *low confidence* in projected changes in the North Atlantic storm tracks (Arias et al., 2021).

COMPOUND EXTREME EVENTS

It is well known that changes in climate-driven extremes continue to become more prominent with every additional increment of global warming. There is *high confidence* that there will be an increasing occurrence of extreme events with global warming, which will be unprecedented in the observational record (Arias et al., 2021). Extreme climate events will occur even if collective efforts maintain global warming at the lower end of the Paris targets (1.5°C above pre-industrial). Every additional 0.5°C of global warming causes discernible increases in the intensity and frequency of hot extremes, including heatwaves (*very likely*), and heavy precipitation (*high confidence*), as well as agricultural and ecological droughts in some regions (*high confidence*) (Lee et al., 2021). Lastly, there is *moderate confidence* that with every additional 0.5°C of global warming, there will be more frequent and severe meteorological and hydrological droughts (Arias et al., 2021).

SUMMARY

- Anthropogenic climate change is attributed directly or indirectly to human activity that alters the composition of the global atmosphere and which is in addition to natural climate variability observed over comparable periods.

- Today, the levels of well-mixed GHGs are unprecedented in the past 800,000 years and highly likely to be the highest atmospheric concentrations in the last 2 million years. Carbon dioxide (CO_2) is the most common GHG (accounting for about 75% of all anthropogenic GHG emissions), and it has an atmospheric lifetime of more than 300 years. The combustion of coal, natural gas and oil for electricity is the largest source of global GHGs, but proportions of fossil fuel emissions vary by country.

- Natural climate variability is driven by a range of external drivers (forcings) that include changes caused by the Earth's orbital position relative to the Sun, changes in solar output and volcanic eruptions. These drivers vary from multi-millennial to decadal scales. The climate system's most significant internal driver of climatic variability is the ENSO, which has a periodicity of 3–7 years. However, anthropogenic forcing of the climate system is amplifying the effects of natural climate variability resulting in more extreme weather events globally.

- The study of past climates (paleoclimatology) provides a sound basis for evaluating current rates of change in key climate indicators, primarily driven by human activities such as burning fossils fuels and land use change. Rates of change in CO_2 concentrations per century today are much faster than those experienced over the past 56 million years.

- The scale of recent changes in the climate system and the present state of many aspects of the climate system are unprecedented over many centuries to many thousands of years. Examples include significant changes in global mean surface temperature, changes in the carbon cycle, changes in precipitation and the water cycle, oceanic changes, sea level changes, changes in hot and cold extreme events, cryosphere changes, changes in tropical cyclones and extra-tropical cyclones, and changes in compound extreme events.

- Future changes in climate are unavoidable, but the amount of change will depend on future emissions of GHGs and the sensitivity of the climate system to rising GHGs.

- Global warming will cause significant changes in the climate system over the coming decades, and regional variations in key climate indicators are highly likely (e.g. across Australia). These changes include increases in the frequency and intensity of hot extremes, marine heatwaves, heavy precipitation events, agricultural and ecological droughts in some regions, intense tropical and extra-tropical cyclones, and reductions in Arctic sea ice and snow cover and permafrost. Additionally, global warming is projected to intensify the global water cycle and its variability, with implications for global monsoon precipitation and the severity of dry and wet events. Many of these future changes become amplified in direct relation to increasing global warming.

- Due to past and likely future GHG emissions, many climate system changes are irreversible for centuries to millennia, with significant changes expected in the oceans, ice sheets and global sea levels.

- This chapter has provided a global perspective of natural climate variability and human-caused climate change. The next chapter examines natural climate variability and human-induced climate change in Australia to provide background context to the remaining chapters in the book.

3

The Climate of Australia: Past, Present and Future

In Australia, average temperatures have risen almost one degree since 1910, and each decade since the 1940s has been warmer than the one before. That warming is real. Its consequences are real. And it will change our lives in real and practical ways.

Julia Gillard AC (2011), former Australian Prime Minister

INTRODUCTION

Australia's climate (Figure 3.1) has changed significantly since its geological separation from Gondwanaland, some 45 million years (Ma) before the present (BP). Gondwanaland, or 'Gondwana', is the southern half of the Pangaean supercontinent that existed some 300 Ma BP (Meert, 2011). It was composed of the major continental blocks (plates) of South America, Africa, Madagascar, Arabia, Sri Lanka, India, Antarctica and Australia. This chapter will first provide a brief overview of Australia's climate from the early Eocene (56 Ma BP) to the Holocene epoch that commenced about 12 thousand years (ka) BP. Second, it will examine the key drivers of Australia's present climate at different time scales, followed by discussing its main climate types. Third, it will evaluate changes in key climate indicators since reliable records began in 1910. The final section will describe projected changes in key climate indicators over 21st century for a range of shared socio-economic pathways (SSPs) (see Chapter 2, Table 2.6).

PAST CLIMATES

Understanding Australia's past (or paleo) climates is necessary to provide context for discussions of contemporary climate patterns and processes and projected future changes due to anthropogenic climate drivers. These are called global 'paleo-reference' periods (see Chapter 2, Box 2.1), which represent colder and warmer times than today and periods of rapid climate change – many with informative parallels to projected climate futures (see discussion below). They also provide background for appreciating the contemporary biogeography of Australia – that is essentially a result of significant changes in its climate extending back over tens of Ma BP (see Chapter 5). For example, the dominance of 'pyrophytic' (fire-tolerant and fire-dependent) vegetation across most of Australia is a consequence of a general drying climate over the past 10 Ma. While Australia has a deep geological history – extending back some 4.4 billion years – for contextual reasons, the chapter only summarises significant environmental changes over the past 56 Ma (the beginning of the Eocene epoch).

DOI: 10.1201/9781003189909-3

Figure 3.1 Australia showing state borders, elevation, main deserts, lakes and rivers, transport networks and places. Creative Commons, 2010. 1:20M Australia General Reference Map (A4). Geoscience Australia, Canberra. http://pid.geoscience.gov.au/dataset/ga/65187.

In the early Eocene, Gondwanaland had almost broken up into separate landmasses – apart from Australia, Antarctica and South America that remained connected. Globally, the early Eocene climatic optimum (53–49 Ma BP) was a prolonged 'hothouse' period with atmospheric CO_2 concentrations more than 1,000 ppm (see Chapter 2, Box 2.1), much like the high emissions (SSP5–8.5) long-term (2081–2100) pathway projection (see Chapter 2, Table 2.6). While Australia lay close to the South Pole at that time, Earth's climate was much warmer and wetter than today (Sloan and Rea, 1996), and Gondwanaland remained free of ice and snow. Lush near-tropical rainforests dominated the ancient landscape, even surviving within the Antarctic Circle.

At the beginning of the Miocene epoch (23.3 Ma BP), Gondwanaland had broken up, and Australia was drifting slowly northwards (about 7 cm per year) towards the Equator, together with the islands of New Guinea on its northern flank. The Miocene was a period of considerable climatic and environmental change across the newly separated continent of Australia. The epoch began with a warm and wet climate until the mid-Miocene (Martin, 2006). Globally, the Miocene Climatic Optimum (16.9–14.97 Ma BP) was another prolonged warm period with atmospheric CO_2 levels of 400–600 ppm, much like the medium emissions (SSP2–4.5) long-term (2081–2100) pathway projection (see Chapter 2, Table 2.6). However, global conditions became much colder and drier by the late Miocene (10.4 Ma BP), with the rapid accumulation of ice at both poles and associated sea-level falls.

The northern tectonic drift of the Australian continent into the sub-tropics and tropics essentially negated the effects of general global cooling, allowing many Gondwanan lifeforms to survive in moister, fire-free areas (climate refugia) in the east and west. However, as the continent moved north, larger land areas began to be affected by dominant sub-tropical high-pressure systems, resulting in a general drying trend. Hence, the late Miocene was a time of significant desiccation. The predominant vegetation in central Australia was open woodland and gallery woodland adjacent to watercourses (Mao and Retallack, 2019), while sclerophyll forests were prevalent across the south-eastern inland (Martin, 2006). There was widespread extinction of many plant and animal groups during this time,

while forms better adapted to a drying environment evolved – including many that now comprise Australia's modern flora and fauna (see Chapter 5).

Globally, the Pliocene (5–1.6 Ma BP) was an epoch known for the overall cooling of atmospheric and oceanic temperatures. During this epoch, the global climate was highly variable due to the expansion and contraction of the Northern and Southern Hemisphere ice sheets – in response to the orbital forcing of the climate system (see Chapter 2, Table 2.4). Globally, the mid-Pliocene warm period experienced atmospheric CO_2 concentrations like those over the last decade. However, the Arctic was much warmer than today, and tropical temperatures were slightly warmer (see Chapter 2, Box 2.1). In the early Pliocene, the climate in southeast Australia was warm and wet with a summer rainfall peak, but by the late Pliocene, conditions were cooler and drier with a winter rainfall peak (Gallagher et al., 2003). This pattern suggests a change from a tropical (monsoon) summer dominated rainfall regime to one governed by extra-tropical winter rain-bearing weather systems – much like the contemporary climate of the region.

Pleistocene Australia (1.6 Ma–11 ka BP) was a time of considerable environmental change. Like other major continents, it was an epoch where mega-fauna roamed the landscape and humans began migrating into new regions. There are varying estimates for how long Aboriginal and Torres Strait Islander Peoples have lived on the continent, with most studies confirming at least 60,000 years. While Australia was more-or-less in the same geographical position as today (Figure 3.1), the Pleistocene glacial cycles (see Chapter 2, Figure 2.3) reduced sea level by up to 130 m below present levels (Ludt and Rocha, 2014). This fall in sea level allowed temporary land bridges to form between the mainland, Tasmania and New Guinea. Due to high-quality atmospheric chemistry records found in Antarctic ice cores, it is possible for scientists to reconstruct global climates for large portions of the Pleistocene. These long records show that global average temperatures varied by up to 8°C between the glacial/interglacial cycles (see Chapter 2, Figure 2.6a).

During the Last Ice Age (~20 ka BP), Australia was about 5°C–6°C cooler than today and endured a dry climate over much of the continent (see Chapter 2, Figure 2.3). The Last Deglacial

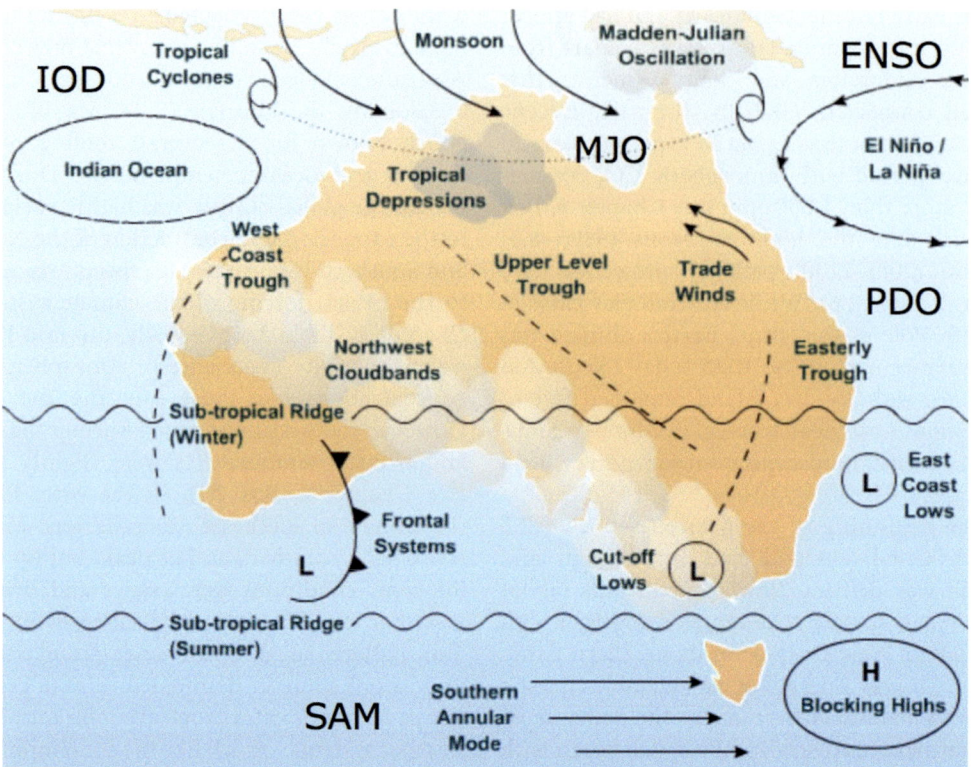

Figure 3.2 Schematic representation of the main drivers of climate variability in the Australian region. The dominant features are IOD, MJO, ENSO) and PDO in the tropics, and SAM in the extra-tropics. Reproduced by permission of the Bureau of Meteorology, © 2022 Commonwealth of Australia.

Transition (18–11 ka BP) was a period of warming that saw decreases in the Northern Hemisphere ice sheets and rapid sea-level rise (see Chapter 2, Box 2.1). Mega-fauna populations on all the continents (including Australia) declined or went extinct during this time of rapid environmental change, driven by climate and human activities – such as hunting and widespread use of fire to manage the landscape. The Holocene epoch commenced about 12 ka ago and ushered in a long period with atmospheric CO_2 concentrations that only ranged from 265 to 285 ppm until the advent of the industrial revolution in 1750 (see Chapter 2, Box 2.1).

PRESENT CLIMATES

Australia's geographical location (Figure 3.1) means that it is influenced – at a range of spatial and temporal scales – by equatorial, tropical and polar air masses of both maritime and continental origin (Sturman and Tapper, 2006). The contemporary climate of Australia is affected by several atmospheric and oceanic drivers at a range of spatial and temporal scales (Figure 3.2). Understanding these key drivers and how they operate and interact is essential for explaining Australia's highly variable climate and discussing future regional climate change.

Climate drivers

PACIFIC DECADAL OSCILLATION

The Pacific Decadal Oscillation (PDO) directly influences Australia's climate. Several studies emphasise a strong tendency for PDO impacts in the Southern Hemisphere, with significant surface climate anomalies over the mid-latitude South Pacific Ocean, Australia and South America (Mantua and Hare, 2002). Some have described the PDO as a long-lived El Niño-like pattern (see Chapter 2, Figure 2.4b) of Pacific climate variability (Zhang et al., 1997). However, while ENSO is primarily an interannual phenomenon (see the 'El Niño Southern Oscillation' section below), the

PDO is decadal in scale (Deser and Trenberth, 2016). Hence, understanding the PDO requires at least 50 years of data, and even then, it may be challenging to distinguish PDO-driven climate anomalies from ENSO and other climate drivers (Mantua and Hare, 2002). The PDO has two phases: cold (negative) and warm (positive). During a cold phase, ocean temperatures in the tropical Pacific are lower, and Australia's decadal rainfall tends to be above average (Li et al., 2019).

Conversely, during a warm phase, rainfall is below average. Evidence suggests that the PDO can also amplify the effects of ENSO droughts worldwide (Li et al., 2019, Nguyen et al., 2021). In eastern Australia – a global El Niño drought hotspot – effects of El Niño droughts are generally amplified and protracted when they occur during a warm phase of the PDO (Nguyen et al., 2021).

EL NIÑO SOUTHERN OSCILLATION

Due to Australia's location in the western Pacific, El Niño and La Niña events (Figure 3.2) play significant roles in driving its weather and climate from year to year (Nicholls et al., 1996). Australia generally experiences below-average rainfall during El Niños and above-average during La Niñas (Wang and Hendon, 2007). The shift in rain away from the western Pacific, associated with El Niño (see Chapter 2, Figure 2.4b), means less rainfall during the Austral winter-spring (June-October), particularly across the eastern and northern parts of the continent (Bureau of Meteorology, 2021). Nine of the ten driest winter-spring periods on record for eastern Australia have occurred during El Niño years (Bureau of Meteorology, 2021).

In the Murray-Darling Basin (Figure 3.1), winter-spring rainfall averaged over all El Niño events since 1900 was 28% lower than the long-term basin-wide average, with the severe droughts of 1982, 1994, 2002, 2006 and 2015 all associated with El Niño events (Bureau of Meteorology, 2021). However, not all El Niños produce widespread drought in northeast and southeast Australia. During the extreme El Niño event of 1997–1998, below-average rainfall was generally confined to coastal south-eastern Australia and Tasmania, while the relatively weak El Niño event of 2002–2003 saw widespread and significant drought in eastern Australia (Bureau of Meteorology, 2021).

Research has shown that not all El Niños produce the same rainfall patterns across the continent

(e.g. Taschetto and England, 2009). Their study characterised two kinds of El Niño phenomena: (1) classical El Niño or non-Modoki[1] El Niño events, and (2) El Niño Modoki events. With classical El Niños, peak sea surface temperature (SST) anomalies[2] appear in the eastern Pacific (see Figure 2.4b). During Modoki El Niño events, distinct warm SST anomalies occur in the central Pacific and weaker cold anomalies in the west and east (Taschetto and England, 2009). The same study showed that Modoki and non-Modoki El Niño events exhibited a marked difference in rainfall impact over Australia. Notably, classical El Niños are associated with a significant reduction in rainfall over north-eastern and south-eastern Australia. At the same time, El Niño Modoki events appear to drive a large-scale decrease in rainfall over north-western and northern Australia.

La Niña events are associated with a stronger than average Walker Circulation in the Pacific and related upwelling of cold water off the coast of South America, resulting in a cold pool (see Chapter 2, Figure 2.4a). The enhanced trade winds associated with the Walker Circulation push warm surface waters even further into the western Pacific and to the north of Australia (Bureau of Meteorology, 2021). This setup creates the ideal atmospheric conditions for low pressure, rising air, widespread rainfall and cloudiness – enhanced by the above-average SSTs. When SST anomalies are strongly positive (warm) in the western Pacific, along with a strongly positive Southern Oscillation Index[3] (SOI), then the more extensive the rainfall response over northern and eastern Australia (Power and Callaghan, 2016). Notably, the wettest years on record for Australia occurred during the strong 1973–1974 and 2010–2012 La

---

[1] Modoki is Japanese for "same but different".

[2] The term temperature anomaly means a departure, plus or minus, from a reference value or long-term average value. A positive anomaly indicates that the observed temperature was warmer than the reference value, while a negative anomaly indicates that the observed temperature was cooler than the reference value.

[3] The Southern Oscillation Index (SOI) gives an indication of the development and intensity of El Niño or La Niña events in the Pacific Ocean (see Figure 2.4). The SOI is calculated using the pressure differences between Tahiti and Darwin (BOM, 2021).

Niña events and were associated with this type of atmosphere-ocean setup. The increased rainfall in the western Pacific associated with La Niña usually means above-average winter-spring rainfall for Australia, particularly in the east and north. According to Bureau of Meteorology (2021), the six wettest winter-spring periods on record for eastern Australia occurred during known La Niña years. In the Murray-Darling Basin (Figure 3.1), winter-spring rainfall averaged over all 18 La Niña events (including multi-year events) since 1900 was 22% higher than the long-term average, with the severe floods of 1955, 1988, 1998 and 2010 all associated with La Niñas (Bureau of Meteorology, 2021). In September 2022, a rare triple La Niña event was occurring in the Pacific, that began back in September 2021.

While El Niños largely depress winter-spring rainfall across eastern Australia, effects of La Niñas often continue from winter-spring into the Austral summer (November–March). In eastern Australia, the average December–March rainfall during La Niña events is 20% higher than the long-term average, with eight of the ten wettest such periods on record occurring during La Niña years (Bureau of Meteorology, 2021). Historically, eastern Australia often experiences severe flooding during La Niña summers. The devastating Brisbane floods of February 1893, January 1974, January 2011 and February 2022 were all associated with strong La Niña events. Finally, some La Niñas coincide with a negative Indian Ocean Dipole (IOD), and rainfall is above average over large parts of Australia (see IOD below). For example, during the strong La Niña in 2010, a negative IOD also prevailed (Figure 3.2). These combined phenomena produced heavy winter-spring rainfall and widespread flooding across eastern Australia (Bureau of Meteorology, 2021).

El Niño events tend to produce above-average maximum temperatures across most of southern Australia, particularly during the second half of the year (Bureau of Meteorology, 2021). Before 2013 (a neutral ENSO year), Australia's two warmest years for seasonal daytime temperatures for winter (2009 and 2002), spring (2006 and 2002) and summer (1982–1983 and 1997–1998) had all occurred during an El Niño (Bureau of Meteorology, 2021). In general, decreased cloud cover during El Niños results in above-average daytime temperatures, particularly in the spring and summer months.

Higher daytime temperatures amplify the effect of lower rainfall by increasing evaporative demand that adversely affects natural ecosystems (see Chapter 5) and agriculture (see Chapter 6). By comparison, La Niña events mean below-average maximum temperatures in Australia. This pattern is mainly due to increased cloud cover that keeps daytime temperatures below average and traps heat, resulting in warmer night-time temperatures (Bureau of Meteorology, 2021). The lack of cloud cover and below-average minimum temperatures during El Niño events increases the risk of winter and spring frosts in southern and inland Australia.. Parts of southern New South Wales and northern Victoria (Figure 3.1) can experience 15%–30% more frost days than the historical average (Bureau of Meteorology, 2021). In contrast, the frost risk is lower for these regions during La Niñas due to warmer minimum temperatures that prevent freezing

In Australia, extreme heat events (e.g. heatwaves, hot spells) can occur during El Niño and La Niña events, but with essential differences in their intensity, duration and geographical extent. El Niños are associated with wide-area heatwaves (as indicated by a higher national area-average temperature), single-day extremes at specific point locations and long-duration warm spells (Bureau of Meteorology, 2021). By comparison, southern coastal areas incur fewer individual daily heatwaves during La Niñas but an increased frequency of prolonged warm spells. For example, of the 32 Victorian heatwaves between 1989 and 2009, 17 occurred during La Niñas, while only six occurred during El Niños (Parker et al., 2014).

During classic El Niño events, there are fewer tropical cyclones in the Australian region, as areas of convection tend to shift near and east of the International Dateline (see Chapter 2, Figure 2.4b). This pattern is particularly the case for northeast Queensland, where cyclones are half as likely to cross the coast during El Niños compared to neutral years (Bureau of Meteorology, 2021). Hence, during El Niños, there is a decreased likelihood of significant damage and flooding related to strong winds, high seas and heavy rains associated with tropical cyclones. Conversely, during La Niñas, there are typically more tropical cyclones in the Australia region, with twice as many making landfall than during El Niño events on average (Bureau of Meteorology, 2021). During La Niña events, the first cyclone to cross the Australian coast tends to occur earlier in the season due to the onset of the northern Australia monsoon. One study (Kuleshov et al., 2008) showed that the only years with multiple severe tropical cyclone landfalls in Queensland had been La Niña years.

The northwest monsoon plays an important role in northern Australia's weather and climate between November and April when most of the region's annual rainfall occurs (Figure 3.2). However, the monsoon onset in the continent's north is generally 2–6 weeks later during El Niños than in La Niñas. Long-term data showing rainfall across the tropical north of Australia is typically well-below-average during the early part of the wet season for El Niño events but close to average during the latter part of the wet season (Bureau of Meteorology, 2021). However, during La Niñas, it is typically above average during the early part of the wet season, while only slightly above average throughout the latter.

Bushfire risk is prevalent across most regions of Australia at different times of the year (see Chapter 5, Box 5.3). The risk is generally high in the continent's southeast following an El Niño event, especially when combined with a positive IOD event (see IOD below). Extreme summer bushfires, including Ash Wednesday (16 February 1983) and the 2002–2003 and 2006–2007 fire seasons, followed an El Niño event (Bureau of Meteorology, 2021). However, not all significant bushfires follow El Niños, such as the spring bushfires in the Blue Mountains during October 2013 that occurred during a neutral ENSO year. It is rare for extreme bushfires to follow La Niñas, such as the Black Saturday bushfire event (7 February 2009) that occurred after a weak La Niña. On that occasion, a strongly positive IOD was the dominant climate driver for that extreme fire weather event (see IOD below).

Snowpacks in the Australian Alps are essential for maintaining stream flows into the southern part of the Murray-Darling River system (Figure 3.1) during the spring and early summer months. El Niño events tend to produce lower than average snow depths in Australia's alpine regions because of the generally lower winter rainfall in the southeast of the continent (Bureau of Meteorology, 2021). For example, the average peak snow depth at Spencer's Creek is 35 cm lower during El Niños and the season length (i.e. the period

with snow depths greater than 100 cm) is 2.5 weeks shorter than average. By comparison, La Niñas tend to have increased snow depths in Australia's alpine regions than El Niños but are less reliable than neutral years (Pepler et al., 2015). While La Niña usually brings above-average precipitation, temperatures can sometimes be too high, and hence it may fall as rain rather than snow at high elevations, reducing snow depths (Bureau of Meteorology, 2021). This pattern has become more common in recent decades, with seven of the past eight La Niña events producing lower than average maximum snow depths (Bureau of Meteorology, 2021).

INDIAN OCEAN DIPOLE

Much like ENSO in the Pacific Ocean, Indian Ocean SSTs also affect rainfall and temperature patterns across Australia. The IOD refers to an ENSO-like ocean-atmospheric phenomenon that occurs across the tropical Indian Ocean (Figure 3.2) that generally lasts from 2 to 7 months (Bureau of Meteorology, 2021). The IOD has three phases that drive Australia's weather patterns. These events usually start in May or June, peaking between August and October and then rapidly declining with the arrival of the northwest monsoon later in the year (Bureau of Meteorology, 2021). When the IOD is in a positive phase, westerly winds weaken (and easterly winds strengthen) along the Equator, allowing warm water to shift towards East Africa. As the westerly winds decline, cool water upwells from the deep ocean in the East Indian Ocean to the northwest of Australia. With warm SSTs in the west and cool SSTs in the east, there is less moisture than usual in the atmosphere to the northwest of Australia. Hence, the path of extra-tropical weather systems from the southwest is affected, with less rainfall and higher than average temperatures over parts of Australia during Austral winter and spring.

Conversely, when the IOD is negative, westerly winds strengthen along the Equator, allowing warmer than average SSTs to concentrate off northwest Australia and cooler than average SSTs in the western Indian Ocean. A negative IOD typically produces above-average winter-spring rainfall over parts of southern Australia. The warmer waters near northwest Australia provide more moisture to weather systems crossing the country. Finally, there is little influence on Australian weather patterns during a neutral IOD phase, and other large-scale drivers are more important.

The IOD interacts with ENSO in some years, but its effects depend on the phases of both climatic systems (Bureau of Meteorology, 2021). When a strong El Niño coincides with a positive IOD, the two phenomena can reinforce their drying influences over much of the south and southeast of Australia, including Tasmania (such as 1982, the driest year on record for southeast Australia). Similarly, when La Niña coincides with negative IOD, the chance of above-average winter-spring rainfall typically increases over the same greater region (such as 1974, the wettest year on record for Australia).

SOUTHERN ANNULAR MODE

The Southern Annular Mode (SAM) describes the (non-seasonal) north-south drift of rain-bearing westerlies and weather systems in the Southern Ocean from their usual (neutral) location (Figure 3.2). Strong circumpolar westerly winds blow almost continuously in the Southern Hemisphere's mid- to high latitudes. This zone of westerly winds is also associated with extra-tropical cyclones (storms) and cold fronts that move from west to east, bringing rainfall to southern Australia. SAM is a relatively short-lived and somewhat random phenomenon that lasts from a week to several months (Bureau of Meteorology, 2021). Like ENSO and IOD, SAM has three phases: positive, neutral and negative, and its effects vary between the Austral winter and summer months. During a positive phase, the westerly winds are further south than average (Figure 3.2), while they are further north during a negative phase of SAM (Bureau of Meteorology, 2021, Holgate et al., 2020).

A positive SAM phase typically brings more rainfall into southeast Australia in summer, including inland areas (Holgate et al., 2020). This pattern is due to the increased onshore flow from the moist southeast trade winds (Figure 3.2). Notably, there is a reduced chance of extreme heat in spring and early summer with enhanced trade winds. La Niña conditions often amplify a positive SAM (Holgate et al., 2020). By comparison, a positive SAM during winter tends to favour more rain in the east and more East Coast Lows.[4] However, less rainfall occurs in parts of the far

south of Australia (including Tasmania) due to the southern displacement of the westerly winds. This phase also results in a below-average snow season in alpine areas.

In summer, a negative SAM phase is often associated with lower rainfall over most of southeast Australia. This pattern is due to weaker southeast trade flow in the southwest Pacific Ocean (Figure 3.2). However, western Tasmania tends to have higher than average rainfall due to the more northern drift of the moist westerly winds (Bureau of Meteorology, 2021). This setup also brings a greater chance of spring heatwaves in southern Australia. Finally, a negative SAM is typically enhanced if it coincides with an El Niño event (see ENSO above). Given the more northern location of westerly winds, a negative SAM in winter enhances rainfall in the southwest of the continent, the far southeast of South Australia, for most of Victoria and all of Tasmania (Figure 3.1). Areas sheltered from stronger westerly winds typically receive less rainfall (Bureau of Meteorology, 2021), including western fringes of the Great Australia Bight, parts of northern and central New South Wales and the coastal edge of eastern Queensland (Figure 3.1). This negative phase also favours above-average snowfalls in alpine areas of Australia.

MADDEN-JULIAN OSCILLATION

The Madden-Julian Oscillation (MJO) was first discovered in the early 1970s by American meteorologists' Drs Roland Madden and Paul Julian when studying wind and pressure patterns across the global equatorial belt (Madden and Julian, 1971). Their initial analysis detected regular oscillations in winds (as defined from departures from average) between Singapore and Canton Island in the west-central equatorial Pacific Ocean. While different phases of SAM primarily affect southern Australia (see SAM above), the MJO is an essential short-term driver of climate variability over the northern tropical region from about Broome in the west to Townsville in the east (Figure 3.1). While the MJO can occur at any time of the year, it mainly influences the northern wet season from October to April (Bureau of Meteorology, 2021). It manifests as a pulse of wind, enhanced cloud and rainfall that cycles eastwards around the globe near the Equator, with a periodicity of 30–60 days (Figure 3.2).

---

[4] These are very intense low-pressure systems characteristic of the eastern coastline of Australia (BOM, 2021).

The MJO has a typical three-phase (4–5 week) cycle, affecting northern Australia's summer weather conditions (Bureau of Meteorology, 2021). Phase 1 (week 1) emerges over tropical Africa in the far western Indian Ocean. During this phase, there is less rainfall over northern Australia, and easterly trade winds are dominant. For northern Australia, this is an 'inactive' phase of the MJO. Through phase 2 (week 2–3), the MJO enters Australian longitudes and enhances rainfall over northern Australia, Indonesia and the southwest Pacific. This 'active' phase of the MJO generally enhances the northwest monsoon flow (Bureau of Meteorology, 2021). Conditions become very favourable for developing tropical cyclones in the eastern Indian Ocean, Timor Sea, Arafura Sea, Gulf of Carpentaria and western Coral Sea (Figures 3.1 and 3.2). Phase 3 (weeks 4 and 5) witnesses the shifting of the MJO further east, away from Australia, enhancing rainfall over the central equatorial Pacific Ocean (Bureau of Meteorology, 2021). Once again, the MJO enters an 'inactive' phase over northern Australia, and the monsoon westerlies are replaced by easterly winds and widespread rainfall declines. Maximum (afternoon) temperatures across Australia north tend to be below average during the active MJO phase due to greater cloud cover, while minimum (overnight) temperatures tend to be above average. Due to lower cloud cover and humidity, the opposite diurnal temperature patterns occur in inactive phases.

Climate types

Australia's size and geographical location (Figure 3.1) mean it experiences a wide variety of climates, including its hot tropical north and cool temperate south. This section discusses its main climate types, according to the Köppen–Geiger system (Beck et al., 2018). This system classifies climate into five main classes and 30 sub-types based on threshold values and seasonality of monthly air temperature and precipitation (Peel et al., 2007). This classification treats vegetation as 'crystallised, visible climate' (Köppen, 1936) and aims to empirically map biome distributions worldwide based on common vegetation characteristics, such as savannah. Hence, its use as a mapping tool is because the climate of a region is a significant driver of global vegetation distribution (Beck et al., 2018).

The most recent (1980–2016) global map of the Köppen–Geiger climate classification is an ensemble of four high-resolution, topographically corrected climatic maps (Beck et al., 2018). At a regional level, 17 climate sub-types occur for Australia (Table 3.1) at a 1 km resolution in Figure 3.3. Australia represents all climate classes[5] (except Polar E) but not all sub-types (Figure 3.3).

Australia has the complete representation of the Tropical (A) climate class (Table 3.1), located across the northern flank of the continent (Figure 3.3). This northern region is dominated by the tropical savannah (Aw) sub-type, stretching from just north of Broome in the Kimberly region of Western Australia to far eastern Queensland to the north of Rockhampton (see Figure 3.1). Tropical rainforest (Af) and tropical monsoon (Am) sub-types are restricted to a small near-coastal area of the wet tropics bioregion – from about 75 km north of Townsville to about 100 km north of Cairns (Figure 3.1). This remarkably small bioregion is less than 0.3% of the total land area of Australia (Stork and Turton, 2008).

Semi-permanent sub-tropical high-pressure systems dominate Australia's climate (Turton, 2017). Together with a lack of significant topographic relief (Figure 3.1), it is hardly surprising that two-thirds of the continent comprises the arid (B) climate class (Figure 3.3), making it the driest permanently inhabited continent in the world. The hot desert climate sub-type (BWh) occurs in the centre of Australia and extends to the coast in the northwest. By comparison, the cold desert climate sub-type (BWk) is much less extensive, extending to the south of the hot desert area and reaching the coast in the Great Australia Bight (Figure 3.1). Hot steppe (BSh) is the second most dominant sub-type in Australia (Figure 3.3), extending west to east from the coast south of Broome into the northeast inland and almost reaching the coast near Townsville (Figure 3.1). It then extends throughout inland Queensland (west of the Great Divide) into northern areas of New South Wales. A small area of this sub-type also occurs in Western Australia (Figure 3.3). The cold steppe (BSk) sub-type occurs in southwest Australia to the east of Albany and in parts of South Australia, western Victoria and southern inland New South Wales (Figure 3.1).

The temperate (C) climate class and its sub-types dominate southeast, southern and southwest

[5] Australia's Antarctic territories are not included in this assessment.

Table 3.1 The Köppen–Geiger climate classification system, including definitions of criteria

First	Second	Third	Description	Criterion
A			Tropical	Not (B) and $T_{COLD}>$or$=18°C$
	f		• Rainforest	$P_{DRY}>$or$=60$ mm
	m		• Monsoon	Not (Af) and $P_{DRY}>$or$=(100-MAP/25)$
	w		• Savannah	Not (Af) and $P_{DRY}<(100-MAP/25)$
B			Arid	$MAP<10×P_{THRESHOLD}$
	W	h	• Desert	$MAP<5×P_{THRESHOLD}$
	S	k	• Steppe	$MAP>$or$=5×P_{THRESHOLD}$
			• Hot	$MAT>$or$=18°C$
			• Cold	$MAT<18°C$
C			Temperate	Not (B) and $T_{HOT}>10°C$ and $0°C<T_{COLD}<18°C$
	s		• Dry summer	$P_{SDRY}<40$ mm and $P_{SWET}<P_{WWET}/3$
	w		• Dry winter	$P_{WDRY}<P_{SWET}/10$
	f		• Without dry	Not (Cs) or (Cw)
		a	season	$T_{HOT}>$or$=22°C$
		b	• Warm summer	Not (a) and $T_{MON10}<4$
		c	• Cold summer	Not (a or b) and $1<$or$=T_{MON10}<4$
			• Cold summer	
D			Cold	Not (B) and $T_{HOT}>10°C$ and $T_{COLD}<$or$=0°C$
	s		• Dry summer	$P_{SDRY}<40$ and $P_{SDRY}<P_{WWET}/3$
	w		• Dry winter	$P_{WDRY}<P_{SWET}/10$
	f		• Without dry	Not (Ds) or (Dw)
		a	season	$T_{HOT}>$or$=22°C$
		b	• Hot summer	Not (a) and $T_{MON10}>$or$=4$
		c	• Warm summer	Not (a, b or d)
		d	• Cold summer	Not (a or b) and $T_{COLD}<-38°C$
			• Very cold summer	
E	T		Polar	Not (B) and $T_{HOT}<$or$=10°C$
	F		• Tundra	$T_{HOT}>0°C$
			• Frost	$T_{HOT}<$or$=to 0°C$

Source: Beck et al. (2018).

Notes: Summer (winter) is the 6-month period that is warmer (colder) between April–September and October–March.

MAP, mean annual precipitation (mm/year); MAT, mean annual air temperature (°C); P_{DRY}, precipitation in the driest month (mm/month); P_{SDRY}, precipitation in the driest month in summer (mm/month); P_{SWET}, precipitation in the wettest month in summer (mm/month); P_{WDRY}, precipitation in the driest month in winter (mm/month); P_{WWET}, precipitation in the wettest month in winter (mm/month); $P_{THRESHOLD}$, $2×MAT$ if >70% of precipitation falls in winter, $P_{THRESHOLD}$ ($2×MAT+28$) if >70% of precipitation falls in summer, otherwise $P_{THRESHOLD}$ ($2×MAT+14$). T_{COLD}, the air temperature of the coldest month (°C); T_{HOT}, the air temperature of the warmest month (°C); T_{MON10}, the number of months with air temperature >10°C (unitless).

Australia and Tasmania (Figure 3.3). All of Australia's major capital cities (except Darwin) and the most significant number of people (>90%) reside in this greater temperate region (see Chapter 7). Australia has essential agricultural and biologically significant areas with *Csa* and *Csb* (Mediterranean) sub-types (Figure 3.3). These have been identified globally as regions with historical and future significant winter and spring rainfall declines (see Chapter 2).

Mediterranean climates have either hot (or warm) dry summers and mild, wet winters (Table 3.2). They occur in the southwest of Western Australia, in southern parts of South Australia and in the far southwest of Victoria (Figure 3.1).

Australia contains a small region with the *Cwa* sub-type (Figure 3.3) in northeast Queensland, from inland from about Cairns to near Rockhampton (Figure 3.1.) This sub-type is

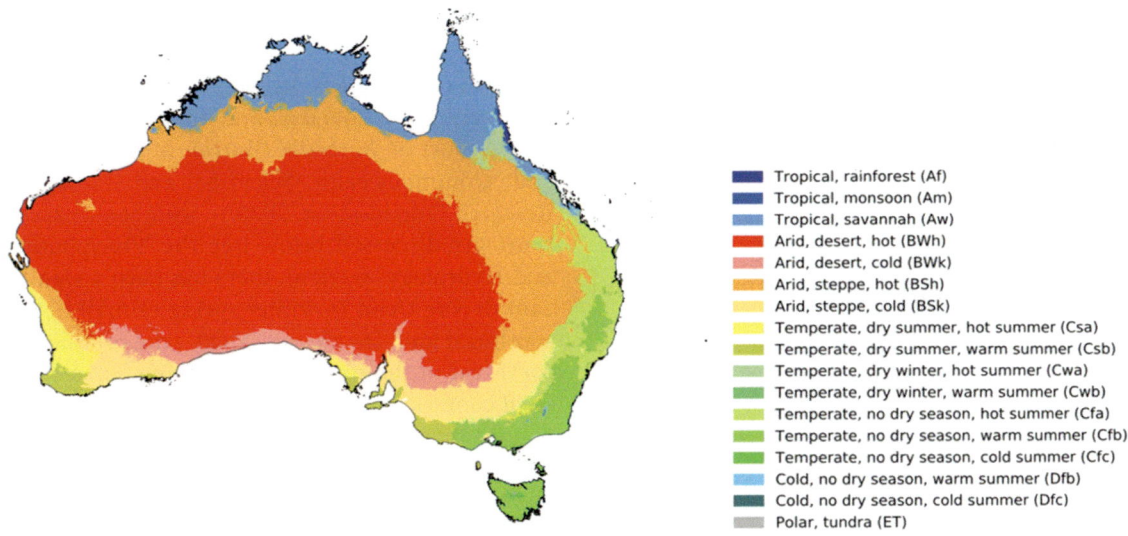

Figure 3.3 The Köppen–Geiger climate classification of Australian climates based on the 1980–2016 baseline data (see Table 3.2). The map has been topographically corrected to a 1 km resolution. (Beck, H.E., N.E. Zimmermann, T.R. McVicar, N. Vergopolan, A. Berg, E.F. Wood: Present and future Köppen–Geiger climate classification maps at 1-km resolution, *Scientific Data* 5:180214, doi:10.1038/sdata.2018.214 (2018). Creative Commons.)

characterised by its dry winters and warm to hot wet summers (Table 3.1). A large region of southeast Australia and Tasmania experiences a 'classic' temperate climate, where there is no distinct seasonality in rainfall, but there are differences in average summer temperatures (Table 3.1). Three sub-types occur in this broad region (Figure 3.3). The *Cfa* (warm summer) sub-type occurs east of the Great Divide in Queensland, south of Rockhampton to the near north of Sydney in New South Wales (Figure 3.1). The *Cfb* (warm summer) sub-type occurs in elevated parts of New South Wales, to the east of the Great Divide. It then extends to a large area of southern New South Wales, most southern Victoria and Tasmania. However, the *Cfc* (cold summer) sub-type is restricted to Tasmania, New South Wales and Victoria (Figure 3.3). Finally, for the Cold (*D*) climate class, only one sub-type (*Dfb*) occurs in a tiny elevated area of the Australian Alps, to the southeast of Canberra (Figure 3.1). This sub-type is characterised by having cold summers and evenly distributed monthly precipitation in the form of rain in summer and mixed rain and snow in winter (Table 3.1).

HISTORICAL CHANGES IN CLIMATE

Reliable instrumental weather records began in Australia in 1910, and this is the baseline year for studies of historical climate variability and change (Bureau of Meteorology and Commonwealth Scientific and Industrial Research Organisation, 2020). Globally, the country is renowned for its highly variable climate, mainly due to its high sensitivities to a range of short- to medium-term climate drivers described above (Figure 3.2). Over the past 110 years, changes in Australia's climate have more-or-less paralleled global patterns (see Chapter 2, the 'Changes in the Modern Era' section), but with significant regional differences. This section evaluates key regional climate indicator changes over the past 110 years (Bureau of Meteorology and Commonwealth Scientific and Industrial Research Organisation, 2020, Lawrence et al., 2022). There are different confidence levels in science (given in italics), depending on the climate indicator (see Chapter 2, the 'Dealing with certainty and uncertainty' section, and Box 2.2).

Changes in key regional climate indicators

TEMPERATURE AVERAGES AND EXTREMES

Australian land areas have warmed by around 1.4°C between 1910 and 2020 (*very high confidence*). There is *high confidence* that annual temperature changes have emerged above natural variability in all land regions (Lawrence et al. 2022). There has also been more significant observed warming over central Australia than in coastal regions, which are tempered by their proximity to the ocean. In line with global trends (see Chapter 2, Figure 2.6c), most warming has occurred since 1950 – with every decade since then being warmer than the ones before (Bureau of Meteorology and Commonwealth Scientific and Industrial Research Organisation, 2020). As of January 2021, Australia's warmest year on record was 2019 (2020 globally), and the 7 years from 2013 to 2019 all rank in the nine warmest years on record. While rising GHGs are the primary driver of background global warming (see Chapter 2, Table 2.5), temperatures vary from year to year across the country due to natural climatic drivers. These drivers include El Niño and La Niña events, solar variations and episodic volcanic eruptions (see ENSO above and Chapter 2).

In recent decades, warming has been observed across Australia in all months, with both day (maximum) and night-time (minimum) temperatures increasing (Bureau of Meteorology and Commonwealth Scientific and Industrial Research Organisation, 2020). For example, very high monthly maximum temperatures that occurred nearly 2% of the time in 1960–1989 and over 4% of the time in 1990–2004 now occur over 12% of the time (2005–2019). This change is more than a six-fold increase over the 60 years (Bureau of Meteorology and Commonwealth Scientific and Industrial Research Organisation, 2020) and is beyond natural variability alone. There is also *high confidence* that heat extremes have increased and cold extremes have decreased across the country (Lawrence et al., 2022). In terms of national daily average maximum temperatures, 33 days exceeded 39°C in 2019 alone, more than the total observed (24 days) over the multi-decadal period from 1960 to 2018 (Bureau of Meteorology and

Commonwealth Scientific and Industrial Research Organisation, 2020).

As the climate has warmed in recent decades, the frequency of cold days and nights has declined across most of Australia (Bureau of Meteorology and Commonwealth Scientific and Industrial Research Organisation, 2020) – with some exceptions. As discussed below (see the 'Precipitation and Hydrology' section), there has been a significant drying trend in winter-spring rainfall in the southwest and – to a lesser extent – the continental southeast of the country. This drying has created ideal conditions for extremely cold nights in those regions due to less cloudy and dry winter nights. However, the number of frost days in these parts has remained unchanged since the 1980s (Bureau of Meteorology and Commonwealth Scientific and Industrial Research Organisation, 2020).

PRECIPITATION AND HYDROLOGY

Australia is the driest permanently inhabited continent globally (Figure 3.3). Any changes in precipitation (rainfall and alpine winter snowfall) are *likely* to have significant environmental, economic and social implications (see Chapters 5–7). As discussed above, rainfall across Australia is highly variable and affected by natural climate drivers such as ENSO, IOD and SAM. However, despite very high variability from year to year, long-term trends in rainfall for some regions of Australia are evident (Bureau of Meteorology and Commonwealth Scientific and Industrial Research Organisation, 2020).

April to October rainfall in the southwest (Figure 3.1) has declined by 16% since 1970, while winter (May to July) rainfall has declined by nearly 20% over the same period. This decline is a significant change for a region that experiences a classic Mediterranean (dry summer/wet winter) climate (see *Csa* and *Csb* climate sub-types in Figure 3.3). It is now *highly likely* that this significant decline in rainfall across the southwest from 1910 to 2019 is attributable to human influence (Lawrence et al. 2022). Similarly, in the continental southeast (Figure 3.1), there has been a decline of around 12% in April to October rainfall since the late 1990s. Meanwhile, summer (October–April) rain across northern Australia has increased since the 1970s (Bureau of Meteorology and Commonwealth Scientific and Industrial Research Organisation, 2020). There has also been an observed increase in heavy rainfall events across the north and a decline in snow cover and depth in the alpine areas of the southeast mainland and Tasmania since the late 1950s (Lawrence et al., 2022).

Hydrological gauges are a valuable indicator of climate change's long-term effects on streamflow, as they provide continuous monitoring over at least 30 years. Three-quarters of Australia's hydrological reference stations show a declining trend in streamflow, with a quarter of these showing a statistically significant declining trend (Bureau of Meteorology and Commonwealth Scientific and Industrial Research Organisation, 2020). Notably, the observed long-term reduction in rainfall across many parts of southern Australia has led to reduced streamflow (see Chapter 5, Box 5.4). Declines in annual median streamflow have occurred in the Murray-Darling Basin, the southwest coast, the South Australian Gulf and the southeast (Victoria and New South Wales) drainage divisions (Figure 3.1). By comparison, in the Tanami-Timor Sea Coast drainage division in northern Australia, which includes Darwin and much of the Northern Territory (Figure 3.1), there has been an increasing trend in annual median flows at about 67% of the gauges since 1975 (Bureau of Meteorology and Commonwealth Scientific and Industrial Research Organisation, 2020). This trend is due to increased wet season rainfall across northwest Australia.

Australia's largest food production region – the Murray-Darling Basin – has experienced significant declines in streamflow since records began in 1970 (Bureau of Meteorology and Commonwealth Scientific and Industrial Research Organisation, 2020), and patterns vary across the basin (Figure 3.1). In the northern basin, 94% of the gauges show a declining trend in streamflow, compared with about 75% for those in the southern basin.

FIRE WEATHER

Australia has always been a global bushfire 'hotspot', mainly due to its highly variable rainfall, hot summers and dominance of 'pyrophytic' vegetation (e.g., eucalypts and acacias). The Forest Fire Danger Index (FFDI) is employed to forecast bushfire weather risk (see Chapter 5, Box 5.3). The FFDI indicates the fire danger based on temperature, rainfall, humidity and wind speed (McArthur, 1967). Based on this index, there is *medium confidence* that the frequency of extreme

fire weather days has increased, and the fire season has become longer since 1950 at many locations – especially in southern Australia (Lawrence et al., 2022). These FFDI increases are particularly evident during spring and summer and are associated with an earlier start to the southern Australian fire weather season (Bureau of Meteorology and Commonwealth Scientific and Industrial Research Organisation, 2020). Climate change contributes to these changes in fire weather by affecting temperature, relative humidity and associated changes to the fuel moisture content (Lawrence et al., 2022). The increased frequency, intensity and longevity of extreme heat due to global warming have exacerbated extreme fire weather risk and bushfires in southern and eastern Australia (see Chapter 5, Box 5.3).

OCEANIC CONDITIONS

Average SST around the Australian region has warmed by more than 1°C since 1900, with eight of the ten warmest years on record occurring since 2010 (Bureau of Meteorology and Commonwealth Scientific and Industrial Research Organisation, 2020). The warmest year on record was 2016 – associated with one of the strongest negative IOD events on record (see IOD above). A strong El Niño event that year also amplified the ocean warming (see ENSO above). The most significant ocean warming in the Australian region since 1970 has occurred around south-eastern Australia and Tasmania (Bureau of Meteorology and Commonwealth Scientific and Industrial Research Organisation, 2020). There is now *very high confidence* that the East Australian Current is extending further south (Ridgway and Hill 2009), creating an area of more rapid warming in the Tasman Sea (Figure 3.1). This warming rate is now twice the global average (Lawrence et al., 2022). There has also been warming across large areas of the Indian Ocean region to the southwest of Australia.

Warming of the ocean has contributed to longer and more frequent marine heatwaves (*high confidence*), including those resulting in mass coral bleaching of parts of the Great Barrier Reef (GBR) in 2016, 2017 and 2020 (Hughes et al., 2021). Another significant marine heatwave and resultant coral bleaching event affected the GBR in February–March 2022. Other adverse oceanic warming impacts include the depletion of kelp forests and seagrasses, a poleward shift in some marine species and increased disease occurrence (see Chapter 5). Marine heatwaves are defined as periods when temperatures are in the upper range of historical baseline conditions for five days or more (Bureau of Meteorology and Commonwealth Scientific and Industrial Research Organisation, 2020). Unlike terrestrial heatwaves, those in the ocean often persist much longer, occasionally lasting multiple months or even years (see Chapter 2).

SEA LEVEL

The global mean sea level has risen by about 25 cm since 1880 (see Chapter 2), with nearly half of this increase happening since 1970. Sea-level change varies globally and regionally, partly due to the natural variability of the climate system from the effect of climate drivers, such as El Niño and La Niña (see Chapter 2, Figure 2.4 and ENSO above). The rates of sea-level rise to the north and southeast of Australia have been significantly higher than the global average, while rates of sea-level rise along the other coasts of the country have been closer to the worldwide average (Bureau of Meteorology and Commonwealth Scientific and Industrial Research Organisation, 2020). Consequently, sandy shorelines have retreated in many locations, contributing to increased coastal flooding in some densely populated areas (Lawrence et al. 2022, see Chapter 7).

TROPICAL CYCLONES

Tropical cyclones may affect the Australian region anytime between November–April each Austral summer. On average, about 13 cyclones form in the greater Southeast Indian/Southwest Pacific region, including northern Australia, southern Papua New Guinea, Solomon Islands, Vanuatu, New Caledonia, Fiji and Samoa (Turton, 2019a). Cyclone activity varies substantially from year to year, partly due to the influence of large-scale climate drivers like ENSO and IOD (see above). For example, cyclones generally decline with El Niño and increase with La Niña. Historically, tropical cyclones have impacted coastal areas of Australia north of about 30° South (Figure 3.1) but are more common in tropical latitudes where SSTs are favourable for their formation (see the 'Madden-Julian Oscillation' section above). There has been a decrease in the frequency (number) of tropical cyclones observed in the northern Australian region since 1982 (Bureau of Meteorology and Commonwealth Scientific

and Industrial Research Organisation, 2020). The trend in cyclone intensity is more complex to quantify across the greater north region (Turton, 2019a). However, cyclone activity is increasing in southeast Queensland, northern New South Wales and some parts of Western Australia (Australian Academy of Science, 2021). This pattern is due to the observed southward shift in cyclone activity under the expanding tropics (Turton, 2017, Sharmila and Walsh, 2018). This trend brings increasing cyclone risk to more highly populated sub-tropical locations, e.g. highly populated areas of southeast Queensland and northern New South Wales.

The tropical cyclone climatology of the southwest Pacific basin, part of Australia's eastern cyclone region (Figure 3.1), is well understood (Diamond et al., 2013). Over the period 1970–2010, the basin experienced an average of 13 cyclones per season (November–April), with an average of 6.4 cyclones attaining severe status each season (i.e. Categories 3–5 on the Australian scale[6]). The latter half of the study period (1990–2010) experienced fewer cyclones, but the proportion of severe ones was significantly higher (Diamond et al., 2013). While there appears to be a transition to fewer but more intense tropical cyclones in the southwest Pacific, the forward speed of cyclones in the same region has also slowed by 15%–20% over 1949–2016 (Kossin 2018). These changes are *likely* to affect human communities, infrastructure, agriculture and ecosystems in northern Australia (see Chapters 5–7).

COMPOUND EXTREME EVENTS

Climate change influences the frequency, magnitude and effects of many types of extreme weather and climate events (see Chapter 2). When such events occur consecutively within a short period of each other or when multiple types of extreme events coincide, the combined effects often amplify the overall severity (Lawrence et al., 2022). Heatwaves can significantly affect ecological and agricultural systems when combined with prolonged drought stress. Extreme events are more likely when natural

---

[6] The Saffir–Simpson scale (USA) uses 1-minute average winds for hurricanes and typhoons, while 10-minute averages are used in the Australian tropical cyclone categories. Hence, the intensity definitions of the two cyclone intensity scales will differ by about 10% (Turton, 2019a).

climate variability acts to amplify the background influence of climate change (Bureau of Meteorology and Commonwealth Scientific and Industrial Research Organisation, 2020). Record-breaking extreme heat and record-breaking fire weather are more likely when the ENSO or IOD favour warmer and drier conditions in Australia (see El Niño and positive IOD above).

Agricultural and ecological droughts are associated with compound extreme events driven by global warming. There is *high confidence* that the significant decrease in rainfall in the southwest of Australia (Figure 3.1) since the 1970s is due to human influence and increased agricultural and ecological droughts in the region (Lawrence et al., 2022). In comparison, there is *medium confidence* that the observed decrease in rainfall in continental southern Australia has led to increased agricultural and ecological droughts over many parts of the region (Lawrence et al., 2022).

FUTURE CLIMATES

Chapter 2 provided an overview of future changes in Earth's climate system at the global and regional scales and concluded that rates and amounts of change would be largely dependent on future GHG emissions (see Chapter 2, Table 2.6). While future climate change in Australia will more-or-less parallel global trends in the climate system, some essential regional differences will largely explain why Australia is considered a global climate change 'hotspot' (Lawrence et al., 2022).

Projected changes in key regional climate indicators

LARGE-SCALE CLIMATE DRIVERS

Australia's climate is affected by several large-scale phenomena at various spatial and temporal scales (see the 'Climate drivers' section above). The El Niño Southern Oscillation (ENSO) is the most dominant driver of climate variability on interannual timescales across the country and the dominant source of seasonal climate predictability (Lee et al., 2021). Earlier assessments (Intergovernmental Panel on Climate Change, 2013) found that ENSO variability will *likely* remain the dominant mode of interannual climate variability in the future. Meanwhile, ENSO precipitation variability on regional scales

(including Australia) is *likely* to intensify (see Intergovernmental Panel on Climate Change, 2013). However, their assessment noted *low confidence* in projected changes in ENSO variability in the 21st century, mainly due to a strong internal variability component (see Chapter 2, the 'Internal forcing' section). Latest assessments (Intergovernmental Panel on Climate Change, 2021) concur with the previous (Intergovernmental Panel on Climate Change, 2013), showing it is *very likely* that the amplitude of ENSO rainfall variability will intensify in response to global warming over the 21st century. However, there is no robust consensus from CMIP6 climate models (see Chapter 2, the 'Climate models' section) for a systematic change in amplitude of ENSO-SST variability – even under the high-emission scenarios (see Chapter 2, Table 2.6) of SSP3–7.0 and SSP5–8.5 (Lee et al., 2021). Cai et al. (2022) disagree with this finding. Their study compared ENSO-SST variability between the 20th and 21st centuries and showed a robust increase in century-long ENSO-SST variability under four IPCC plausible emission scenarios. This ENSO-SST variability trend means an increased risk of below-average rainfall for northwest, central, south and southeast Australia, including Tasmania. Under global warming, there will *likely* be more extreme positive phases of the IOD (see IOD above) (Cai et al., 2018). This positive IOD trend will enhance the drying effects of ENSO-SST events over most of southern Australia when the two systems are in phase.

The Southern Annular Mode (SAM) is the leading driver of large-scale extra-tropical atmospheric variability in the Southern Hemisphere (including Australia) and influences most of the southern regions of the country (Lee et al., 2021). Although the SAM is a proxy for the location of the mid-latitude westerly wind belt (see the 'Southern Annular Mode' section above), trends in the SAM can also reflect a combination of changes in the Southern Hemisphere jet stream position, width and strength[7] (Lee et al., 2021). To the south of Australia, changes in the atmospheric circulation

---

[7] Jet streams are fast flowing, relatively narrow westerly air currents found in the atmosphere at around 12 km above the surface of the Earth, just under the tropopause. Shifts in the jet stream in Australian longitudes are associated with shifts in the extra-tropical storm tracks, with associated changes in regional weather and rainfall (BOM, 2021).

associated with phases of the SAM influence surface winds and ocean currents in the Southern Ocean (Wang et al., 2022).

Over the instrumental period (see Chapter 2), there is *high confidence* there has been a robust positive trend in the SAM index, particularly since 1970 (Lim et al., 2016, Lee et al., 2021). Similarly, the suite of CMIP6 models shows a near-term tendency towards a more positive SAM index, especially in the Austral winter (May–August). During a positive phase, the westerly winds are further south than usual. This position means lower April–October rainfall in parts of southern Australia due to the southern displacement of the westerly rain-bearing winds. There is *medium confidence* that this recent trend in the SAM has been unprecedented in the past several centuries (Arias et al., 2021). However, there is *high confidence* that stratospheric ozone depletion and GHG increases have contributed to the positive SAM trend during the late 20th century (Intergovernmental Panel on Climate Change, 2021). This pattern is associated with stratospheric ozone depletion prevailing in the Austral summer (November–March), following the seasonal peak in the Antarctic ozone hole[8] in September–October and steady increases in GHG emissions dominating in other seasons (Lee et al., 2021).

CHANGES IN CLIMATE TYPES AND ZONES

Australia's main climate zones (Figure 3.3, Table 3.1) are already shifting in response to climate change (Turton, 2017). The extent of any future changes to the geographical location of climate zones will depend on global GHG emissions (see Chapter 2, Table 2.6). Notably, the temperate dry summer sub-types (*Csa* and *Csb*) in southern and southwest regions are projected to move further south. This southward contraction of these sub-types will have significant adverse impacts on the biologically diverse ecosystems (see Chapter 5) and agricultural systems (see

---

[8] The Antarctic ozone hole is a thinning or depletion of ozone in the stratosphere over the Antarctic each Austral spring. This damage occurs due to the presence of chlorine and bromine from ozone-depleting substances in the stratosphere (see Montreal Protocol) and the specific meteorological conditions over the Antarctic (Department of Agriculture, Water and Environment, 2021).

Chapter 6, the 'Agriculture and Land' section) that depend on this classic Mediterranean climate. In the east and southeast, shifts in climate zones are also *likely*. The Tropical savannah sub-type (*Aw*) is projected to shift further south into the sub-tropics of Queensland, replacing the *Cwa* sub-type (Figure 3.3). Temperate dry winter sub-types (*Cwa* and *Cwb*) are projected to move south into parts of northern and central New South Wales, replacing the temperate (no dry season) sub-type (*Cfa*). The *Cfa* (hot summer) sub-type will shift into southern New South Wales, replacing the *Cfb* (warm summer) sub-type. Lastly, the tiny area of the cold summer sub-type (*Dfb*) is *likely* to disappear entirely under medium-to-high GHG emissions pathways (see Chapter 2, Table 2.6). This reduction will have severe consequences for Australia's winter alpine tourism industry (see Chapter 6, the 'Tourism' section).

CHANGES IN OTHER CLIMATE INDICATORS

Many of the historical changes in Australia's key climate indicators – described above – are expected to follow the same trends with global warming (Table 3.2). However, future GHG emissions pathways will determine the rate and magnitude of changes (see Chapter 2, Table 2.6). The country may expect continued increases in average air temperatures, particularly away from the coast – along with more frequent, prolonged and severe heat events. Most areas of southern Australia, particularly the southwest and continental southeast, may continue the drying trend for winter and spring rainfall. Changes in annual rainfall for northern Australia are *less certain*.

Given the potential strengthening of ENSO cycles (Lawrence et al., 2022), it will mean more pronounced drought and flood events in the future. Drought years will exacerbate fire weather conditions and worsen as baseline temperatures and evapotranspiration rates shift upwards. All parts of the country may expect to witness more extreme rainfall events as the oceans and atmosphere warm. Winter snow cover and depth are projected to decline in alpine areas, with significant implications for spring and early summer runoff in the Murray River system.

Oceans around the country will continue to warm, including the southward movement of the East Australian Current. Warming oceans will mean a much greater risk of marine heatwaves

Table 3.2 Projected changes in Australia's climate during the 21st century

Climate indicator	Projected change and level of confidence
Temperature averages and extremes	Continued increases in air temperatures (*high confidence*). Greater warming in central Australia than coastal regions under all scenarios (*high confidence*). More heat extremes and fewer cold extremes (*high confidence*).
Precipitation and hydrology (Chapter 5, Box 5.4: Freshwater resources)	Heavy rainfall and river floods are projected to increase across the entire country (*medium confidence*). Northern Australia: mean rainfall changes are *uncertain*. An increase in heavy rainfall and river flooding projected by mid-century (*medium confidence*). Central Australia: increase in heavy rainfall and river flooding (*medium confidence*). Southern Australia: reduction in mean rainfall, particularly in the cool season (*medium confidence*). Southwest Australia: rainfall *very likely* to continue decreasing under all future scenarios (*high confidence*) (*high model agreement*). Eastern Australia: decrease in mean cool season rainfall, but more extreme rainfall events (*medium confidence*). Snow cover and depth are projected to decrease further (*high confidence*).
Fire weather (see Chapter 5, Box 5.3: Bushfire risk)	The intensity, frequency and duration of fire weather events are projected to increase throughout Australia, particularly southern, southeast and southwest regions (*high confidence*).
Ocean conditions	Continued warming and acidification of the oceans around Australia (*high confidence*). Continued warming in the East Australian Current region of the Tasman Sea is (*very high confidence*). Increased and longer-lasting marine heatwaves that will affect marine environments, such as kelp forests, and raise the likelihood of more frequent and severe bleaching events in coral reefs around Australia, including the Great Barrier and Ningaloo reefs (*high confidence*).
Sea level[a]	Relative sea-level rise to continue in the 21st century and beyond, contributing to increased coastal flooding and shoreline retreat along sandy coasts (*high confidence*).
Tropical cyclones	Fewer tropical cyclones, but a greater proportion projected to be of high intensity, with large variations from year to year (*medium confidence*). Continued southward movement and slowing of forward motion over ocean and land.
Compound extreme events	Compound changes in several climatic impact drivers (e.g. heatwaves, droughts, floods) would be more widespread at 2°C compared to 1.5°C global warming and even more widespread and/or pronounced for higher warming levels (*high confidence*). Sand storms and dust storms are projected to increase throughout Australia (*medium confidence*). Increase in aridity and agricultural and ecological droughts in southern and eastern Australia (*medium confidence*). Increase in meteorological droughts in southern Australia (*medium confidence*).

Source: Bureau of Meteorology and Commonwealth Scientific and Industrial Research Organisation (2020) and Lawrence et al. (2022).

Note: See Chapter 2 (the 'Dealing with certainty and uncertainty' section, and Box 2.2) for detailed explanations of expressions that are given in italics.

[a] These projections have not been updated to include an Antarctic dynamic ice sheet factor which increased global sea-level projections for RCP8.5 by ~10cm (Lawrence et al., 2022).

and intense tropical cyclones and East Coast Lows. Depending on the extent of oceanic warming, the tropical cyclone belt may have continued its southward movement into regions where cyclones are either absent or rare. Increasing ocean acidification and deoxygenation are expected in the waters around the country. Sea level is expected to rise around the country, with some regional differences.

These projections show that Australia is facing many decades of climate change, even if the Paris Agreement lower (1.5°C) and upper (2.0°C) targets are achieved by 2100 (see the 'Paris Agreement' section, Chapter 4). If warming exceeds 2.0°C above pre-industrial levels, Australia will experience greater warming, more significant drying of southern regions and more pronounced climate extremes.

AUSTRALIA AT 3°C PLUS

Australia's leading climate change scholars have published a compelling report detailing the risks of 3°C or more global warming for the country's natural, economic and social systems (Australian Academy of Science, 2021). This report builds on the much more conservative IPCC Special Report on Global Warming of 1.5°C (Intergovernmental Panel on Climate Change, 2018), which aligned with the lower limit of the Paris Agreement warming targets for 2100 (see Chapter 4, the 'Paris Agreement' section). Warming of 3°C (above pre-industrial) is consistent with the SSP3–7.0 emissions pathway (see Chapter 2, Table 2.6). The AAS report provides a daunting glimpse of climate conditions that future generations may expect in Australia during the second half of the 21st century (Box 3.1).

BOX 3.1

The risks to Australia of a 3°C warmer world

Risks for 3°C or more of warming:
- More frequent, longer and more intense heatwaves, relative to 1.5°C and 2°C warming.
- Days above 50°C in Sydney and Melbourne are *very likely* to be regular occurrences.
- The number of days above 35°C is projected to be three times greater by 2070 as compared to today in 15 towns and cities.
- For Queensland's local government areas, one study found heatwaves would happen as often as seven times a year, with events lasting 16 days on average.
  - Weather systems are *likely* to be more energetic with further global warming.
  - Tropical cyclones are *likely* to be stronger and move at slower speeds, delivering greater volumes of water than they do today.
  - Higher sea level will exacerbate levels of coastal inundation associated with storms, which is *highly likely* to increase relative to today.
  - Short duration extreme rainfall would increase relative to the present day for much of Australia.
  - Large hailstorms may increase in frequency and size in the southern regions of Australia.
- Risks from bushfires will increase substantially with *moderate confidence* that the number of extreme fire days will double.
  - Fire risk (driven by record heat, dryness, and fuel, see Chapter 5, Box 5.3) will increase by 30% or more in south-eastern Australia.
- Large decreases in seasonal rainfall will occur in southern Australia.

Source: Australian Academy of Science (2021).
This amount of global warming is consistent with the lower range of the SSP3–7.0 Emissions Pathway (see Chapter 2, Table 2.6).

SUMMARY

- Past changes in Australia's climate provide helpful background to understanding its contemporary biogeography, notably the dominance of fire-tolerant and fire-dependent vegetation. This attribute is the consequence of a general drying trend over much of the country over the last 10 million years.
- Climate variability in Australia is strongly influenced by coupled ocean-atmospheric drivers that operate at different spatial and temporal scales. The main drivers are ENSO, IOD, SAM and MJO. These, often interacting drivers, explain most of the variability in rainfall and land and ocean temperatures from year to year. Australia's size and geographical location in the Southern Hemisphere mean it experiences a diversity of climates, including tropical and temperate types. Roughly two-thirds of the country experiences arid climates, making it the driest permanently inhabited continent globally.
- Australian land areas have warmed by about 1.4°C between 1910 and 2020. There has also been more significant observed warming over central Australia than coastal regions tempered by their proximity to the ocean. In recent decades, warming has been observed across the country, increasing day (maximum) and night-time (minimum) temperatures. At the same time, heat extremes have increased, and cold extremes have declined. Average sea surface temperatures around the country have also warmed by more than 1°C since 1900, with the most significant warming around south-eastern Australia and off northeast Tasmania due to strengthening of the East Australian Current. Background warming has also increased the frequency and duration of marine heatwaves around the country.
- Most parts of southern Australia have witnessed declining rainfall in recent decades. April to October (autumn, winter and spring) rainfall in the southwest has declined by 16% over the past 50 years. Similarly, April to October rainfall in the continental southeast has reduced by 12% over the past 25 years. Notably, the country's largest food production region (Murray-Darling Basin) has experienced declines in streamflow since 1970, particularly in the northern (Darling) basin. There is also evidence that the observed decrease in rainfall and increase in temperatures across continental southern Australia has resulted in an increase in agricultural and ecological droughts in recent decades.
- Over most of the country, the frequency of extreme fire weather has increased, and the bush fire season has become longer since 1950. Climate change contributes to these changes in fire weather by affecting temperature, relative humidity and associated changes in the forest fuel moisture content.
- Sea-level changes around Australia are parallel to the global average. The global mean sea level has risen by about 25 cm since 1880. Nearly half of this increase has occurred over the past 50 years.
- There has been a decrease in the frequency of tropical cyclones in the Australian region since 1982. However, the trend in cyclone intensity is difficult to quantify across the country's northern region.
- Future climate change in Australia will parallel global trends in the climate system, but with regional nuances. The ENSO is the most dominant mode of climate variability on interannual timescales across Australia and the dominant source of seasonal climate predictability. It is *very likely* in the future that ENSO rainfall variability will intensify in response to global warming this century. This means an increased risk of below-average rainfall for northwest,

central, south and southeast Australia, including Tasmania. There will also be more extreme positive phases of the IOD in the future under global warming. Lastly, there is *high confidence* there has been a robust positive trend in the SAM index, particularly since 1970. Model projections show a near-term tendency towards a more positive SAM index, especially in the Austral winter (May–August). During a positive phase, the westerly winds are further south than usual. This position means lower April–October rainfall in parts of southern Australia due to the southern displacement of the westerly rain-bearing winds.

- The country may expect continued increases in average air temperatures, particularly away from the coast – along with more frequent, prolonged and severe heat events. Most areas of southern Australia, particularly the southwest and continental southeast, may continue the drying trend for winter and spring rainfall. Changes in annual rainfall for northern Australia are *less certain*. Oceans around the country will also continue to warm, including the southward movement of the East Australian Current. Warming oceans will mean a much greater risk of marine heatwaves and intense tropical cyclones.

- Sea level is expected to rise around the country in concert with global trends, with some regional differences. Increasing rates of ocean acidification and deoxygenation are *likely* in the waters around the country.

- Australia's main climate types and their zones are *likely* to continue to shift southwards with global warming towards the Southern Ocean. Of most significant concern is the temperate dry summer sub-types (*Csa* and *Csb*) in southern and southwest regions, projected to move further south. This southward contraction of these sub-types will have significant adverse impacts on the biologically diverse ecosystems and agricultural systems that depend on this classic Mediterranean climate.

- Navigating the complexity of information about climate change science is challenging (see also Chapter 2). Nonetheless, there are many robust findings and conclusions about the science. Understanding the science provides the fundamentals for action on climate change through mitigation and adaptation. Notably, awareness of the range of likely changes in climate this century gives the decision-making framework to facilitate practical climate adaptation actions. The next chapter examines how Australia may respond to a changing climate now and in the future.

4

Responding to Climate Change

We are the first generation to feel the effect of climate change and the last generation who can do something about it.

Barack Obama (2014), Nobel laureate, former American president

INTRODUCTION

Chapter 2 examined the science of climate change and concluded that Earth's climate system is influenced by natural and anthropogenic radiation forcing agents at both spatial and temporal scales. With the advent of the Anthropocene (see Chapter 1, Figure 1.1), there is unequivocal evidence that human activities are the dominant driver of global and regional climate change. Rising greenhouse gases (GHGs), driven by the burning of fossil fuels, agricultural production, deforestation and other land-use changes, are causing global climate change (see Chapter 2, Table 2.5). The atmosphere and oceans are warming, and global and regional rainfall patterns are changing in response to these changes, increasing more extreme weather events. Australia's climate (see Chapter 3) has changed significantly since reliable records began in 1910. Notably, there has been an average warming trend of 1.4°C across most of the continent and a general drying trend in the continental southeast and particularly southwest – while there has been a slight wetting trend for parts of northern Australia (BOM and CSIRO, 2020).

This chapter considers how Australia may respond to a changing climate now and in the future. Responding to climate change requires strategies to reduce emissions of GHGs and enhancement of carbon sinks (climate mitigation),

as well as strategies to cope with climate change impacts and risks that are inevitable even with mitigation policies (climate adaptation). Adaptation may also include taking advantage of opportunities that may arise due to climate change. Both climate mitigation and adaptation are relevant in a policy context, and understanding them is central to dealing with climate change.

This chapter begins with a conceptual framework that includes key terms and concepts referred to in the three chapters to follow. The second section discusses the concept of climate-resilient development pathways for natural, economic and social systems, thereby providing context for the more detailed focus on climate change impacts, mitigation and adaptation in Australia (see Chapters 5–7). The third section provides an overview of international climate change conventions, protocols and agreements for which the Australian Government has international obligations. These address climate mitigation and adaptation and include the United Nations Framework Convention on Climate Change (UNFCCC, 1992), the Intergovernmental Science-Policy Platform on Biodiversity and Ecosystem Services (2019), the United Nations 2030 Agenda for Sustainable Development (United Nations, 2021) and its associated Sustainable Development Goals (SDGs). These provide the primary global policy context for addressing climate change at international,

DOI: 10.1201/9781003189909-4

national, regional and local scales now and into the future. The final section provides a high-level overview of the roles and responsibilities of climate change action in Australia.

CONCEPTUAL FRAMEWORK AND KEY DEFINITIONS

Climate variability and climate change are closely related but have distinct definitions and essential differences (see Chapter 2). Both are relevant to this section, as climate change – driven mainly by background anthropogenic global heating of the atmosphere and oceans – is causing more climate extremes across the world (enhanced climate variability). In responding to climate change, it is vital to be aware of discernible changes in climate (trends) that drive more climate extremes as baseline (average) climate conditions shift (Intergovernmental Panel on Climate Change, 2021).

Natural, economic and social systems

Anthropogenic climate change is causing numerous impacts on the integrity, status and trend of natural, economic and social systems (Figure 4.1). However, the patterns and processes of any changes vary widely across the world. Natural systems include terrestrial, freshwater, estuarine, coastal and marine (ocean) ecosystems and their biodiversity and the ecosystem services for human health and well-being (see Chapter 5). Economic systems include agriculture, forestry, fisheries, mining, tourism, manufacturing, construction, energy and transportation (see Chapter 6). Social systems include human health and well-being, built environment and infrastructure, cities and settlements, Indigenous communities, finance and insurance (see Chapter 7). Many of these systems interact and overlap in time and space. For example, natural world heritage sites are essential for biodiversity conservation and ongoing ecological and evolutionary processes. However, they are also crucial for ecotourism, Indigenous cultural values and human spiritual health and well-being (Valentine, 2019).

Natural, economic and social systems vary widely in their climatic coping ranges (Figure 4.1). The coping range is the capacity of systems to accommodate variations in climate conditions due to natural and anthropogenic climate drivers (Intergovernmental Panel

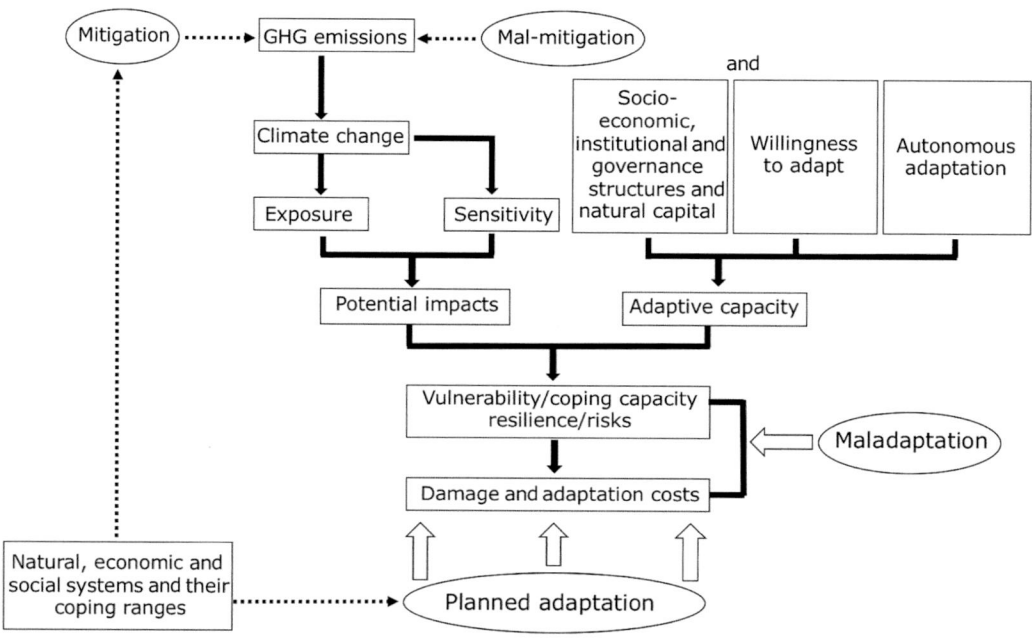

Figure 4.1 Conceptual framework for climate change impacts, vulnerability, climate risks and adaptation options (adapted from Ramieri et al., 2011.)

on Climate Change, 2014b). Some systems have low coping ranges due to their high sensitivity to small temperature rises or changes in rainfall, such as water security, natural ecosystems, coastal communities, agriculture, forestry and sustainable development (Intergovernmental Panel on Climate Change, 2014b). Other systems have moderate to high coping ranges, such as energy security, major infrastructure, tourism and food security. Understanding the coping ranges of essential natural, economic and social systems is crucial for responding to the effects of current and future changes in climate on the integrity, status and trend of such systems (Intergovernmental Panel on Climate Change, 2018, Turton, 2020).

Climate change impacts, exposure and sensitivity

The potential impacts of climate change on natural, economic and social systems (Figure 4.1) will depend on the exposure and sensitivity of these systems to climate change (Intergovernmental Panel on Climate Change, 2018). Potential impacts on various systems may be positive or negative and environmental, economic, social and cultural. Exposure is 'the presence of people; livelihoods; species or ecosystems; environmental functions, services, and resources; infrastructure; or economic, social, or cultural assets in places and settings that could be adversely affected by climate change' (Intergovernmental Panel on Climate Change, 2018). Often defined as the 'external' side of vulnerability, exposure is affected by a combination of the probability and magnitude of climate change (Intergovernmental Panel on Climate Change, 2014b). Sensitivity is the extent to which natural, managed and human systems can absorb impacts without suffering long-term harm or other significant state change and is the 'internal' side of vulnerability (Intergovernmental Panel on Climate Change, 2014b).

Adaptive capacity

When considering the vulnerability of natural, economic and social systems to the actual and potential impacts of climate change (and hence, their risk), it is also essential to evaluate the adaptive capacity of these systems to such changes

(Figure 4.1). Adaptive capacity is the ability of systems, institutions, humans and other organisms to adjust to potential damage, take advantage of opportunities or respond to consequences (Intergovernmental Panel on Climate Change, 2018). In other words, systems can change to make them better equipped to deal with the potential impacts of climate change.

There are three main components to adaptive capacity that need to be considered individually and collectively (Figure 4.1):

1. There is a need to account for autonomous adaptation within natural, economic and social systems. This adaptation does not constitute a conscious response to climate stimuli; instead, it is triggered by ecological and evolutionary changes in natural systems and market or welfare changes in economic and social systems (Intergovernmental Panel on Climate Change, 2014b);
2. Willingness to adapt is an integral part of adaptive capacity and is affected by behavioural, institutional and economic factors within economic and social systems; and
3. The socio-economic, institutional and governance structures, together with natural capital, may influence the adaptive capacity of economic and social systems to climate change.

Countries with well-developed economies and strong institutional and governance structures will generally have a much greater adaptive capacity in the face of climate change than those whose economies and governance structures are far less developed. Natural capital is the world's stock of natural assets, including geology, soil, air, water and biodiversity. Society derives a range of goods and services from natural capital, often called ecosystem services (see Chapter 5, Box 5.2), which maintain human health and well-being (Turton, 2020). Countries and regions that fail to manage their natural capital properly will inevitably have lower adaptive capacities to any adverse climate change impacts now and in the future. Finally, assessments of adaptive capacity are central to understanding the vulnerability of various systems to different drivers of climate and other change agents and for strengthening the adaptive capacity to anticipate future change (Jacobs et al., 2015).

Vulnerability, coping capacity, resilience and risks

The vulnerability of natural, economic and social systems to climate change is central to understanding risks for these systems to likely actual and potential changes (Figure 4.1). Vulnerability is defined as the propensity or predisposition to be adversely affected by climate change. It encompasses a variety of concepts and elements, including sensitivity or susceptibility to harm and lack of capacity to cope and adapt (Intergovernmental Panel on Climate Change, 2018). Vulnerability is widely understood to vary among countries/regions and within communities and change over time (Kienberger et al., 2013). In this context, coping capacity is the ability of individuals, institutions, organisations and systems using available skills, values, beliefs, resources and opportunities to address, manage and overcome adverse conditions (Intergovernmental Panel on Climate Change, 2018).

Together with understanding the coping capacities of various systems to climate change, it is also valuable to know how resilient such systems are likely to be to any changes in the short to medium term. Resilience is

> the capacity of social, economic and environmental [natural] systems to cope with a hazardous event or trend or disturbance, responding or reorganising in ways that maintain their essential function, identity and structure while also maintaining the capacity for adaptation, learning and transformation.
>
> *(Intergovernmental Panel on Climate Change, 2018)*

Systems' coping capacity and resilience to adapt to climate change will differ significantly among countries (e.g. developing versus developed) and regions (e.g. metropolitan versus remote communities). Less-developed countries and remote communities within developed countries will generally have a lower coping capacity and therefore lower resilience to any adverse impacts of climate change.

Risk refers to the potential for adverse consequences of a climate-related hazard or adaptation or mitigation responses to such a hazard on natural, economic and social systems. Risk is 'the potential for adverse consequences where something of value is at stake and where the occurrence and degree of an outcome is uncertain' (Intergovernmental Panel on Climate Change, 2018). Risk develops from the interaction of vulnerability of the affected system, its exposure over time to the hazard, the type of hazard and the likelihood of its occurrence (Figure 4.1). Risk management is defined as 'plans, actions, strategies or policies to reduce the likelihood and magnitude of potential adverse consequences, based on assessed or perceived risks' (Intergovernmental Panel on Climate Change, 2018). Hence, understanding the vulnerability, resilience and risk of systems to current and potential climate change impacts will determine the damage and adaptation costs that will have to be borne by natural, economic and social systems now and into the future (Figure 4.1).

Planned adaptation and maladaptation

Adaptation to climate change occurs as autonomous and planned responses to actual and potential climate change impacts across natural, economic and social systems (Figure 4.1). The goal is to reduce vulnerability to the harmful effects of climate change, such as sea-level encroachment, more intense extreme weather events or food insecurity (Intergovernmental Panel on Climate Change, 2014b). Adaptation works to reduce the risk and vulnerability of systems, thereby enhancing resilience and the coping capacity of systems to climate-driven changes. In economic and social systems, planned adaptation is 'the process of adjusting to actual or expected climate and its effects to moderate harm or exploit beneficial opportunities, such as longer growing seasons or increased crop yields in some regions' (Intergovernmental Panel on Climate Change, 2014b). In natural systems (see Chapter 5), the process of adjustment to actual climate and its effects is mainly autonomous (Figure 4.2). However, human intervention may minimise any expected climate impacts. An example is a policy decision to fund vegetation restoration projects to enhance ecological connectivity between lowland and upland tropical rainforests to allow heat-sensitive plants and animals to move more easily upslope (or downslope) in response to changing conditions.

Other examples include restoring degraded wetlands or mangroves through hybrid combinations of green and grey infrastructure (or soft engineering), such as the construction of artificial wetlands, rainwater harvesting in cities and green roofs or rooftop gardens.

Planned adaptation (Figure 4.1) is, therefore, an adaptation that is 'the result of deliberate policy decisions, based on an awareness that conditions have changed or are about to change and that action is required to return to, maintain, or achieve a desired state' (Intergovernmental Panel on Climate Change, 2014b). Such adaptation measures are conscious policy options or response strategies, often multi-sectoral and aimed at altering the adaptive capacity of systems by facilitating specific adaptations. Hence, climate adaptation typically has five main stages (Jones et al., 2018, Intergovernmental Panel on Climate Change, 2018):

1. Awareness;
2. Assessment;
3. Planning;
4. Implementation; and
5. Monitoring.

Governments, non-government organisations and the private sector have mostly adopted these five stages in their climate adaptation strategies, programmes and policies. Modern approaches to planned climate adaptation overlap strongly with risk management and fostering resilience and sustainable development (Allen et al., 2018).

Planned adaptation of systems to observed and potential climate change has two components, both of which change the fundamental attributes of systems to differing degrees (Allen et al., 2018, see also Figure 4.3). First, where the adaptation response maintains the essence and integrity or process at a given scale, it is considered incremental. Incremental adaptation is a gradual adjustment process, such as a farmer deliberately planting a more drought-tolerant crop to align with declining rainfall in their area (Stafford Smith and Ash, 2011). Second, it may be transformational where the adaptation response changes the fundamental attributes of the socio-ecological system in anticipation of climate change and its impacts, such as the deliberate relocation of an entire community or industry to a new location to avoid rising sea levels (Stafford Smith and Ash,

2011). Under some circumstances, incremental adaptation may amass to result in transformational adaptation (Termeer et al., 2017). Linked social-ecological systems are 'complex adaptive systems composed of many diverse human and non-human entities that interact and adapt to changes in their environment and their environment changes as a result' (Glaser et al., 2008). The functions of such a system arise from the interactions and interdependence of the social and ecological subsystems. The system's structure is characterised by reciprocal feedbacks, emphasising that humans are a part of, not apart from, nature (Berkes and Folke, 1998).

There may be limits to ecosystem-based (autonomous) adaptation or the ability of society to adapt to changing climate (Allen et al., 2018). These are called adaptation limits (Intergovernmental Panel on Climate Change, 2018) and comprise two types:

1. Hard adaptation limits, where no adaptive actions are possible to avoid intolerable risks; and
2. Soft adaptation limits, where options are currently unavailable to avoid intolerable risks through adaptive action.

Adaptive options are the array of strategies and measures available for addressing climate adaptation. These include a wide range of actions that may be structural, institutional, ecological or behavioural (Intergovernmental Panel on Climate Change, 2018).

On occasion, maladaptive actions (maladaptation) may occur (Figure 4.1). These actions may lead to an increased risk of adverse climate-related outcomes, including via increased GHG emissions, increased vulnerability to climate change or diminished welfare, now or in the future (Intergovernmental Panel on Climate Change, 2018). Maladaptation is usually an unintended consequence of human actions.

Mitigation

Mitigation of climate change (Figure 4.1) involves reducing the flow of heat-trapping GHGs into the atmosphere either by reducing sources of these gases or by enhancing the sinks that accumulate and store these gases, such as the oceans, forests and soil (see Chapter 2, Figure 2.2). The mitigation goal is to avoid significant human interference with the climate system and stabilise GHG levels in a timeframe enough to allow ecosystems to adapt naturally to climate change, ensure food security and enable economic development to proceed sustainably (Allen et al., 2018). In climate policy, mitigation measures are technologies, processes or practices that contribute to climate mitigation. Examples include renewable energy, waste reduction, improved public transport, reduced deforestation, reforestation and improved agricultural methods – among many other measures (Allen et al., 2018). Mitigation actions by all countries are urgently required if we are to achieve global policy goals, such as Paris Agreement targets and the SDGs (see below).

The opposite of mitigation is mal-mitigation which includes changes that could reduce GHG emissions in the short term but could inadvertently lock in technology choices or practices that have significant trade-offs for the effectiveness of future adaptation and other forms of mitigation (Allen et al., 2018). In this sense, mal-mitigation and maladaptation are often perverse outcomes of poorly conceived societal responses to managing climate change effectively.

CLIMATE-RESILIENT PATHWAYS AND OPPORTUNITIES

There is now high confidence that average global warming will reach 1.5°C above pre-industrial levels in the early 2030s (Intergovernmental Panel on Climate Change, 2021). Based on existing global emissions reduction policies, the average global warming will be between 2.1°C and 3.9°C by 2100 (Liu and Raftery, 2021). This significant amount of warming means natural, economic and social systems are at very high risk and likely beyond known adaptation limits (Figure 4.1). Responding to rapid anthropogenic climate change requires a two-pronged approach to facilitate climate-resilient pathways and any associated opportunities for society, the economy and the environment (Intergovernmental Panel on Climate Change, 2018). The first critical approach is 'climate mitigation' and the implementation of climate mitigation pathways that address climate risk and aim to enhance any opportunities that may result from mitigation policies, programmes

and projects at local to global scales. The second component is 'planned adaptation' that aims to implement climate-resilient adaptation pathways for natural, economic and social systems that address climate risk and aim to enhance any opportunities that may result with and without climate policy.

Scenarios and pathways

Climate change scenarios and pathways are closely related terms and sometimes applied interchangeably (Rosenbloom, 2017). Both are useful as a means for contextualising future climates and associated climate risks and potential opportunities for natural, economic and social systems (Figure 4.1). Scenarios are plausible descriptions of how the future may develop based on a coherent and internally consistent set of assumptions about fundamental driving forces. Examples include rate of technological change, population growth, governance structures, commodity prices, global markets and their interrelationships (Intergovernmental Panel on Climate Change, 2018). They provide a framework for developing and integrating projections of future GHG emissions, climate change and climate impacts, including assessing their inherent uncertainties (Allen et al., 2018, see Chapter 2 for discussion of uncertainties in a climate change context). Therefore, scenarios help compare climate policy options and associated mitigation and adaptation pathways (Intergovernmental Panel on Climate Change, 2018).

Climate change pathways have a range of types and meanings in the scientific literature and are a progression from an earlier focus on climate change scenarios (Allen et al., 2018, Intergovernmental Panel on Climate Change, 2014b, 2021). They are often used to describe the temporal evolution of a set of scenario features, such as GHG emissions (see Shared Socio-economic Pathways, Chapter 2, Table 2.6) and socio-economic development towards a future state (Allen et al., 2018, Friedlingstein et al., 2021). Pathway concepts range from quantitative and qualitative scenarios or narratives of potential futures to solution-oriented decision-making processes to achieve desirable societal goals (Intergovernmental Panel on Climate Change, 2018). Pathway approaches often consider biophysical, technological, economic and socio-behavioural trajectories and incorporate various dynamics, objectives and actors across different spatial scales (Intergovernmental Panel on Climate Change, 2013, 2018, 2021).

Mitigation and adaptation pathways

A climate mitigation pathway is a temporal evolution of a set of mitigation scenario features, such as GHG emissions and socio-economic development (Intergovernmental Panel on Climate Change, 2018, 2021). For most purposes, mitigation pathways are '1.5°C pathways' to align with the lower end of the Paris Agreement targets, as discussed below (see also SSP1–1.9; Chapter 2, Table 2.6). However, if the climate system overshoots the 1.5°C lower limit in the 2030s, this value may have to be upgraded to the 2°C upper limit or even higher.

Climate adaptation pathways are a series of adaptation choices involving trade-offs between short- and long-term goals and values (Intergovernmental Panel on Climate Change, 2014b, 2018). Such pathways involve sequenced decision-making processes to identify socially relevant solutions (Wise et al., 2014, Allen et al., 2018). They are essentially a temporal series of adaptive learning decision cycles (Figure 4.2), where some chains of decisions lead to maladaptive outcomes (Figure 4.3). However, other alternative pathways remain within adaptive space (Haasnoot et al., 2013, Wise et al., 2014).

Decision cycles (Figure 4.2) have clearly defined decision lifetimes which provide context to the decision-cycle process (Haasnoot et al., 2013, Wise et al., 2014). These include a lead time that requires framing, scoping and resourcing of the decision cycle – through research, consultation and revisions. Community perceptions and reactions are critical components in the decision-cycle process and account for any special interests. The roles of government, regulations and funding are also relevant. These four components drive decision cycles, decision points and consequences over time (Figures 4.2 and 4.3).

Over time, natural, economic and social systems may take various climate adaptation pathways within an adaptive landscape (Figure 4.3). Typically, the system inherited a series of antecedent pathways (A), and these set the baseline for future adaptation pathways, driven by decision cycles over time (Figure 4.2). Such responses involve decision points along the adaptation pathway triggered by thresholds in the managed or planned systems (CoastAdapt, 2017). Some pathways will shift from adaptive space into maladaptive space (B), where following decision cycles (Figure 4.2) and decision points (Figure 4.3), they may either reach a dead end (x) or return to adaptive space (C) through incremental changes within the system. Over time, and assuming changes in climate and other change agents and actors, pathways within adaptive space may be forced into maladaptive space (D), as incremental adaptation decision cycles and decision points cannot avoid maladaptive dead ends. At this stage, transformational pathways (E) return to adaptive space, which requires fundamental changes in the system's attributes (Allen et al., 2018).

Both incremental and transformative adaptation pathways aim to reduce the vulnerability and exposure of systems to the adverse effects of climate change (Figure 4.1). Adaptation pathways are also helpful for considering potential benefits

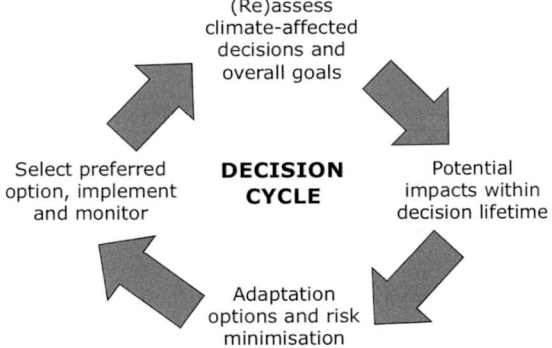

Figure 4.2 The decision cycle for adaptive learning processes over time. Such learning cycles are defined by the decision lifetimes that also change over time (adapted from Wise et al., 2014.)

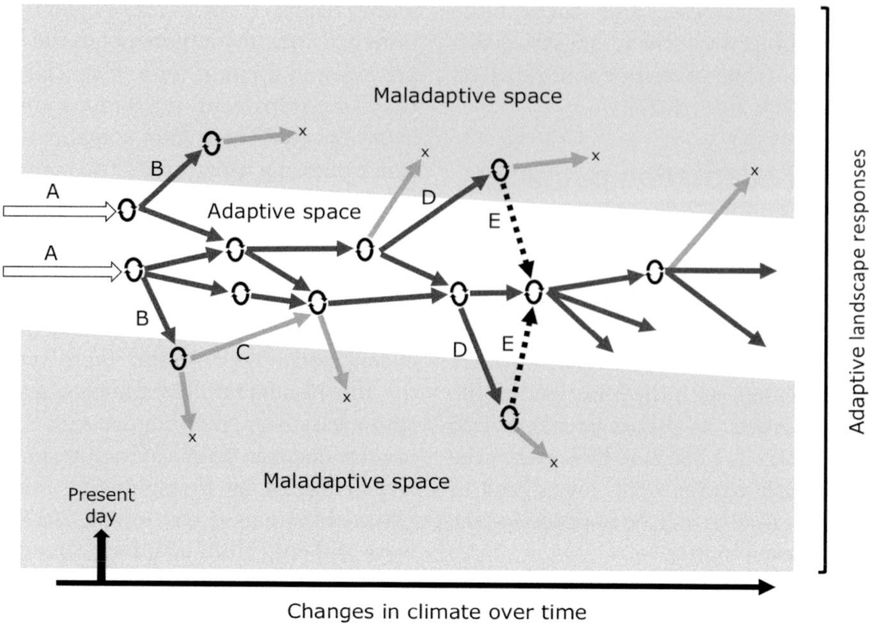

Figure 4.3 Incremental and transformative adaptation pathways within an adaptive landscape affected over time by climate change and other change agents and actors. Circles represent decision points (see Figure 4.2); open arrows represent antecedent pathways prior to the present day; charcoal grey arrows represent contemporaneous incremental adaptation pathways; light grey arrows represent pathways leading to maladaptive dead ends (x); and dashed arrows represent transformative pathways from maladaptive space back into adaptive space (simplified from Wise et al., 2014.)

and opportunities that may arise with future climate change (Allen et al., 2018). Both incremental and transformative pathways require intervention from higher levels of governance (Box 4.1). They will most likely be driven by government or private sector responses to natural disasters or other catastrophic environmental, social or economic events (Wise et al., 2014). Transformative adaptation strategies (Box 4.1) generally require systemic (or structural) changes in natural, economic and social systems, i.e. deliberate government or private sector policy decisions to avoid maladaptation pathways (Figure 4.3).

Climate adaptation pathways must identify any drivers and barriers (see Box 4.2). Barriers to climate adaptation pathways include governments at all levels (e.g. lack of consistent policy direction), the private sector (e.g. perceived unaffordability of adaptation) and communities (e.g. lack of resourcing of adaptation). Enablers to adaptation include appropriate governance frameworks (e.g. a clear climate change mandate), reducing systemic vulnerabilities, building capacity for adaptation (e.g. accessible adaptation tools and information),

dynamic and adaptive decision making, sensible funding mechanisms, and community partnership and collaborative engagement (e.g. community awareness and network building).

Figure 4.4 provides a simple model showing an adaptation pathway that may guide communities or organisations in their adaptation planning for climate change. The model identifies factors that will drive adaptation planning along the pathway and barriers that are likely to be encountered along the way, and enablers (drivers) to strengthen the engagement process (see Box 4.2). Before embarking on an adaptation pathway exercise, it is essential to secure specialised knowledge, leadership and prior experience to bring to any stakeholder engagement (Stafford Smith and Ash, 2011). For example, a local government planning department might follow this process in their local climate adaptation plans. Similarly, private industries such as tourism and agriculture may adopt a similar approach in adapting their businesses to climate change or risk.

Enablers and barriers will differ among countries and regions within countries (Box 4.2).

BOX 4.1: Five strategies for active adaptation of natural, economic and social systems to climate change

AVOID [*A priori* transformative adaptation]: Identify future 'no use' and/or 'no build' public and private land areas and utilise planning tools to prevent new land use and/or infrastructure development in areas at risk now or in the future. Examples: housing development areas exposed to sea-level rise; commercial development areas exposed to sea-level rise or greater riverine flood risk; areas likely to be unsuitable for agriculture in the future due to rising temperatures or saltwater intrusion; prevent degradation of undeveloped areas that may provide future climate 'refugia' for biodiversity.

ACCOMMODATE [Incremental adaptation]: Allow for continued land use and occupation of an area, but make changes to develop resilience to adverse impacts. Examples: raising road beds, building houses on piles, relocating essential infrastructure, converting agriculture to fish farming, growing flood and salt-tolerant crops, precision agriculture.

PROTECT [Incremental adaptation]: Implement defensive actions and/or structures to prevent exposure to adverse impacts and enable continued land use or occupation of an area. Examples: application of 'soft' engineering solutions (such as re-vegetation) to coastal sand dunes and river riparian areas; application of 'hard' engineering solutions (such as sea walls and levees) to protect land from sea and river floods. Hard structures raise environmental concerns and are likely to be cost-effective only in areas of very high land value or population density. May be prohibitively expensive in the long term.

RETREAT [Transformative adaptation]: Cease use and/or occupation of an area, abandon assets that are at risk and relocate to a less vulnerable area that is not exposed to adverse impacts. Examples: relocating wine growing and dairy farming to more suitable climate zones; abandonment of assets in low-lying coastal areas and re-building elsewhere; where feasible ecosystems (such as mangroves) should be allowed to retreat landward as sea level rises. This retreat may be a problem if urban infrastructure is in the way.

ATTACK [Transformative adaptation]: Reclaim land. Examples: mostly used in areas with very high land values, such as the Netherlands, Hong Kong and Singapore. Unlikely in the Australian context. Has high economic and environmental costs.

Source: Adapted from Moran and Turton (2014), CoastAdapt, 2017, Australian Academy of Science (2021). Each adaptation strategy is classified in terms of being either incremental or transformative in nature.

Developing countries may require external funding assistance (e.g. the UN's Green Climate Fund) to facilitate the necessary drivers for climate adaptation planning while also reducing in-country obstacles to implementation, such as lack of resources and skills to facilitate the process. Similarly, regional differences exist in local capacities within developed economies like Australia to engage in successful climate adaptation planning (Figure 4.4). Remote and rural areas and Indigenous communities stand out as cases that require urgent government assistance to facilitate climate adaptation pathways through enhancing drivers and reducing inherent obstacles. Incorporation of local and Indigenous climate knowledge should be a core component of the adaptation planning process (see Chapter 7 for Australian examples).

Recently, there has been a stronger focus on climate-resilient development approaches to climate mitigation and adaptation (Allen et al., 2018, Intergovernmental Panel on Climate Change, 2022b). Climate-resilient development is 'a process of implementing greenhouse gas mitigation and adaptation options to support sustainable development for all' (Schipper et al., 2022). These options give rise to development 'pathways' that align with the SDGs, notably SDG13 (see SDGs and Box 4.5 below). Such pathways are 'trajectories that strengthen sustainable development at

BOX 4.2: Examples of barriers and enablers to adaptation action in Australia

Barriers	Enablers

Governments:
- Lack of consistent policy direction from higher levels and frequent policy reversals.
- Conflicts between community-based initiatives, city councils and business interests.
- Different framings of adaptation between local governments (risk) and community groups (vulnerability, transformation).
- Competing planning objectives.
- Divergent perceptions of climate risk concepts.
- Low prioritisation of climate change adaptation among competing institutional objectives.
- Constraints in using new knowledge.
- Lack of understanding of Indigenous knowledge and practices.
- Lack of authority and political legitimacy.
- Fear of litigation.
- The upfront costs of adaptation relative to competing demands on government expenditure.

Governance frameworks:
- Clear climate change adaptation mandate.
- Measures that inform a shift from reactive to anticipatory decision making, e.g. decision tools that have long timeframes.
- Institutional frameworks integrated across all levels of government for better coordination.
- Revised design standards for buildings, infrastructure, landscape such as common land-use planning guidance and codes of practice that integrate consideration of climate risks to address existing and future exposures and vulnerability of people, physical and cultural assets.

Reducing systemic vulnerabilities:
- Economic and social policies that reduce income and wealth inequalities.
- Strengthening social capital and cohesion.
- Identifying and redressing rigid or fragmented administrative and service delivery systems.
- Review of land-use and spatial planning to reduce exposure to climate risks.
- Restoring degraded ecosystems and avoiding further environmental degradation and loss.

Private sector:
- Governance and policy uncertainty, lack of cross-sector coordination, lack of capital investment in climate solutions.
- Inconsistent hazard information and incomplete understanding of adaptation.
- Mismatch in duration of insurance cover (annual), lending (decades) and infrastructure and housing investment (50–100 years).
- Perceived unaffordability of adaptation, lack of client demand and awareness of climate change risks and limited and inconsistent climate risk regulation in the construction industry.
- Translating information into organisations to address disinterest amongst clients in the property industry.

Building capacity for adaptation:
- Provision of nationally consistent risk information through agreed methodologies for risk assessment that address non-stationarity.
- Targeted research including understanding the projected scope and scale of cascading and compounding risks.
- Education, training and professional development for adaptation under changing risk conditions.
- Accessible adaptation tools and information.

Barriers	Enablers
• Erosion of adaptive capacity and challenges of transformational adaptation in agriculture and rural communities. *Communities:* • Nature of government engagement with communities. • Lack of clarity regarding roles and responsibilities. • Lack of resourcing of adaptation. • Lack of deep engagement with climate change. • Diverging perceptions, values and goals within communities. • Inequities within and between communities. • Lack of sustained engagement, learning and trust between community, scientists and policy makers.	*Dynamic adaptive decision making:* • Increased understanding and use of decision-making tools to address uncertainties and changing risks, such as scenario planning and dynamic adaptive pathways planning to enable effective adaptation as climate risk profiles worsen. *Funding mechanisms:* • Adaptation funding framework to increase investment in adaptation actions. • New private sector financial instruments to support adaptation. *Community partnership and collaborative engagement:* • Community engagement based on principles that consider social and cultural and Indigenous peoples' contexts and an understanding of what people value and wish to protect. • Use of collaborative and learning-oriented engagement approaches tailored for the social and informed by the cultural context. • Community awareness and network building. • Building on Indigenous communities' social-cultural networks and conventions that promote collective action and mutual support.

Source: Adapted from Lawrence et al. (2022).

multiple scales and efforts to eradicate poverty through equitable societal and systems transitions and transformations while reducing the threat of climate change through ambitious mitigation, adaptation and climate resilience' (Olsson et al., 2014). Hence, climate-resilient development stops practices contributing to dangerous global warming and maladaptation (Figure 4.1). Sustainable development across and among regions of the world will help reduce shared adaptation challenges (Intergovernmental Panel on Climate Change, 2022b). Notably, over 3.3 billion people live in regions with very high and high vulnerability to climate change, while 2 billion people live in regions with low and very low vulnerability (Schipper et al., 2022). There are also intra-regional differences in climate risk and exposure that worsen disadvantage among vulnerable social groups (see Chapter 7, the 'Indigenous Peoples' section).

Effective climate adaptation avoids 'lock-in' and 'path dependency', which may lead to mal-mitigation and maladaptation (see Figures 4.1 and 4.3). Notably, alignment with the SDGs reduces vulnerabilities, increases flexibility to change, builds adaptive capacity and progresses sustainable goals and targets, thus improving intra- and inter-generational justice (Lawrence et al., 2022). On the climate mitigation side, reducing GHG emissions and structural inequalities are crucial to achieving the SDGs and contributing to resilient climate development (Schipper et al., 2022). This

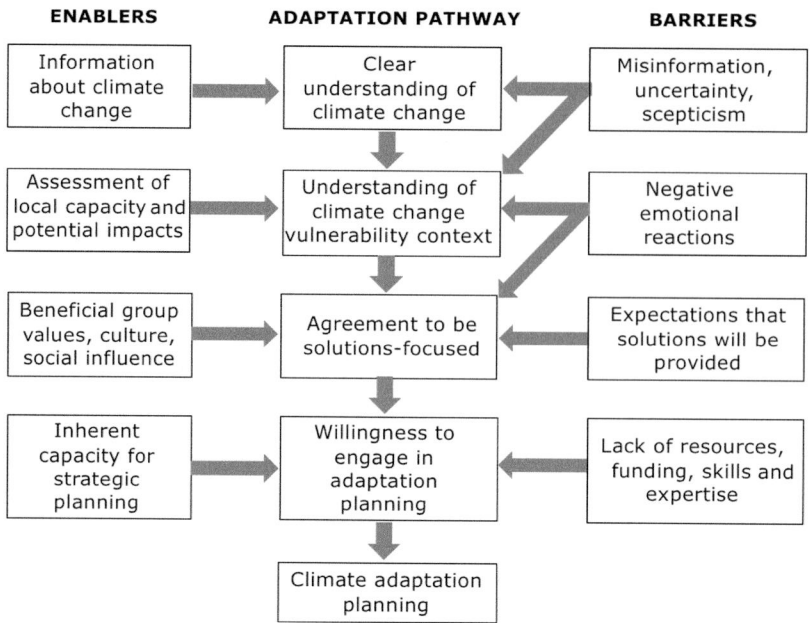

Figure 4.4 A typical climate-resilient pathway for adaptive stakeholder engagement with associated drivers and barriers (adapted from Gardner et al., 2009). For examples of other adaptation barriers and enablers, refer to Box 4.2.

SDG-based approach to climate mitigation and adaptation will also necessitate profound societal transformation to ensure the well-being of all concerned (see Chapters 5–7).

INTERNATIONAL CLIMATE CHANGE CONVENTIONS AND AGREEMENTS

United Nations Framework Convention for Climate Change

The United Nations Framework Convention for Climate Change (UNFCCC) was adopted in May 1992 and opened for signature at the 1992 Earth Summit in Rio de Janeiro (United Nations Framework Convention on Climate Change, 1992). It entered into force in March 1994 and, as of October 2021, had 197 parties (all the 196 UN countries and the European Union). The convention's ultimate objective is to 'stabilise greenhouse gas concentrations in the atmosphere at a level that would prevent dangerous anthropogenic interference with the climate system'. It also states that 'such a level should be achieved within a timeframe sufficient to allow ecosystems to adapt naturally to

climate change, ensure that food production is not threatened, and enable economic development to proceed sustainably'.

The UNFCCC fosters science and policy interaction through several mechanisms, including global climate monitoring and research. It promotes global systematic observation of the climate system as a critical prerequisite for advancing scientific knowledge on climate change and advising informed policymaking. Achieving this strategic goal requires long-term, systematic observation of the Earth's climate as the foundation for understanding climate change and its associated impacts. It also informs scientists to determine future trends. The UNFCCC also advocates climate research focused on a wide range of topics, such as earth sciences, climate processes and variability, climate modelling and prediction, climate change impacts, vulnerabilities, risks and extreme events, and research on adaptation mitigation. Such study covers various sectors, societies, economies and ecosystems and includes cross-cutting and interdisciplinary research.

The UNFCCC cooperates with the Intergovernmental Panel on Climate Change (IPCC). The IPCC is the United Nations (UN) body for assessing and synthesising the science related to climate change. The panel was established by the United Nations Environment Programme (UNEP) and the World Meteorological Organization (WMO) in 1988, with a core purpose to 'provide policymakers with regular scientific assessments concerning climate change, its implications and potential risks, and to put forward adaptation and mitigation strategies'. In the same year, the UN General Assembly endorsed the action by UNEP and WMO in jointly establishing the IPCC. As of May 2021, the IPCC had 195 member states, including Australia. Its comprehensive 5-yearly assessment reports and occasional special reports are well recognised and respected. The strong collaboration between the UNFCCC and IPCC has gathered strength and recognition through several Conference of Parties (COP) decisions.

In 2010, the COP agreed on a long-term global goal (LTGG) to reduce GHG emissions to keep the global average temperature below 2°C above pre-industrial levels. They also decided to review the adequacy of this LTGG under the convention periodically. The first review was in 2013–2015, and the outcome was captured in Article 2.1 of the Paris Agreement (see below). The second periodic review of the LTGG commenced in mid-2020 and is due to conclude in late 2022 (United Nations Framework Convention on Climate Change, 2021a, b). The provisions of the UNFCCC are pursued and implemented by two well-known international treaties: the Kyoto Protocol and the Paris Agreement.

KYOTO PROTOCOL

The Kyoto Protocol was adopted on 11 December 1997, but due to a complex ratification process, it only became official on 16 February 2005. As of October 2021, there were 192 parties (signatories) to the Kyoto Protocol, including Australia. The protocol commits industrialised (developed) countries[1] and 'economies in transition' to limit and reduce GHG emissions according to agreed individual targets. The UNFCCC requests these countries adopt policies and measures on climate mitigation and report periodically on progress towards their agreed targets. The protocol has an annex-based structure (Annex B) that binds developed countries

and consigns a heavier burden on them under the principle of common but differentiated responsibility and respective capabilities. It recognises that they are primarily responsible for the current high levels of GHG emissions in the atmosphere. Annex B sets binding targets for 37 industrialised countries, economies in transition and the European Union (as a collective). Overall, these targets add up to an average 5% emissions reduction compared to 1990 baseline levels over the 5 years from 2008 to 2012 (the first commitment period). Non-Annex B Parties in the first commitment period include most developing countries, including highly populated countries like China, India, Indonesia, Brazil and Mexico.

The Doha Amendment to the Kyoto Protocol was adopted on 8 December 2012 for a second commitment period, starting in 2013 and lasting until 2020 (United Nations Framework Convention on Climate Change, 2021a). By 28 October 2020, 147 parties deposited their instrument of acceptance and, therefore, achieved the threshold for entry into force of the Doha Amendment. The amendment entered into force on 31 December 2020 and included:

- New commitments for Annex B Parties to the Kyoto Protocol who agreed to take on commitments in a second commitment period from 1 January 2013 to 31 December 2020;
- A revised list of GHGs to be reported on by parties in the second commitment period; and
- Amendments to several articles of the Kyoto Protocol which specifically referenced issues pertaining to the first commitment period and which needed updating for the second commitment period.

During the second commitment period, parties committed to reduce GHG emissions by at least 18% below 1990 levels in the 8 years from 2013 to 2020. However, the composition of parties in the second commitment period is different from the first. Despite being a signatory to the protocol, the United States of America has never ratified the protocol. In the first commitment period, Canada had binding emissions targets but withdrew its support in December 2011. Australia was excluded from binding emissions targets in the first commitment period but officially agreed to emissions targets in the second commitment period.

---

[1] OECD stands for the Organisation for Economic Co-operation and Development. It comprises a group of 37-member countries that discuss and develop economic and social policy (OECD 2021).

BOX 4.3: Market-based mechanisms (MBMs) permitted under the Kyoto Protocol (UNFCCC: https://unfccc.int/process/the-kyoto-protocol/mechanisms)

International Emissions Trading:

Annex B Parties have accepted targets for limiting or reducing emissions. Under Article 17 of the Kyoto Protocol, countries that have emission units to spare (unused) may sell this excess capacity to other Annex B Parties that are over their targets. Given that carbon dioxide is a principal GHG (Chapter 2), carbon is now tracked and traded as a commodity that is known as the *carbon market*.

Other units that may be transferred under the Kyoto Protocol include: (1) a removal unit based on land use, land-use change and forestry (e.g. reforestation); (2) an emission reduction unit generated by a joint implementation project (see below); and (3) a certified emission reduction generated from a Clean Development Mechanism project activity (see below).

At this stage, the European Union emissions trading scheme is the largest in full operation.

Clean Development Mechanism (CDM):

The CDM is defined in Article 12 of the Kyoto Protocol and allows a country with an emission-reduction or emission-limitation commitment under the protocol (Annex B Party) to implement an emission-reduction project in developing countries These projects may earn saleable certified emission reduction (CER) credits, equivalent to 1 t of CO_2, which may be counted by the Annex B Party towards achieving its Kyoto targets.

Joint implementation (JI):

The JI is defined under Article 6 of the Kyoto Protocol and allows an Annex B Party to earn emission reduction units (ERUs) from an emission-reduction or emission-limitation commitment project in another Annex B Party, equivalent to 1 t of CO_2, which may be counted towards fulfilling its Kyoto target.

One crucial component of the Kyoto Protocol was the establishment of 'flexible market mechanisms' based on the trade of emissions permits. Under the protocol, countries must meet their targets primarily through national measures, such as emissions reductions. However, the protocol also offers them an additional means to meet their targets through three market-based mechanisms (MBMs) (Box 4.3). These policy mechanisms use markets, pricing and other economic variables to incentivise polluters to reduce or eliminate negative environmental externalities. In this context, an externality is a cost or benefit imposed on a third party who did not agree to incur that cost or benefit, e.g. carbon pollution.

Ideally, these MBMs encourage GHG abatement to commence where it is most cost-effective, e.g. in developing countries and among Annex B Parties who may exchange carbon credits under the Joint Implementation (JI) mechanisms to meet their Kyoto targets. It does not matter where emissions reductions occur, if they are removed from the global atmosphere. This mechanism has the corresponding benefits of stimulating green investment in developing countries and including the private sector in this endeavour to reduce and maintain GHG emissions at a safe level to avoid 'dangerous climate change'. It also facilitates a process called 'leap-frogging'. This process means developing and emerging economies can transition from (or avoid using) older, fossil fuel-intensive technologies for newer, cleaner infrastructure and systems like renewable energy, with obvious long-term economic and social benefits.

The Kyoto Protocol has rigorous monitoring, review, verification and compliance systems to ensure transparency and hold Annex B Parties to account. Under the protocol, Annex B Parties must monitor actual emissions and maintain accurate records of any carbon trading under various MBMs (Box 4.3). These procedures include registry systems that record and track transactions by parties under the MBMs. An international transaction log is maintained to verify that all transactions align with the strict rules of the protocol. Annex B Parties submit annual emissions inventories and national reports under the protocol at specified intervals.

The Kyoto Protocol – like the UNFCCC as a whole – is also designed to assist countries in adapting to the adverse effects of climate change. It facilitates the development and deployment of technologies that can help increase resilience to the negative impacts of climate change. For example, establishing an adaptation fund to help finance adaptation projects and programmes in developing countries that are parties to the protocol. In the first commitment period, the fund was financed mainly with a share of proceeds from CDM project activities. As part of the Doha Amendment in 2012, it was decided that international emissions trading and joint implementation (Box 4.3) would also provide the adaptation fund with a 2% share of proceeds for the second commitment period.

The Kyoto Protocol was the first global climate change treaty. Since its inception, it has received considerable criticism (e.g. Baylis et al., 2014), while other commentators are more generous, stating we should accept that the protocol was an essential first step in a long process for global action on climate change. Many small island states[2] claim that the protocol requires more ambitious emissions targets, while some economists argue its costs exceed its benefits. Others assert that the protocol is highly cumbersome and unbalanced and likely to have little impact on global GHG emissions. One of the biggest criticisms of the protocol is its focus only on emissions from Annex B Parties while not prescribing emissions commitments for rapidly growing 'emerging economies', such as China, Brazil and India. The USA has not yet ratified the protocol, and this is another glaring omission in the global fight against slowing down anthropogenic climate change. However, the latest IPCC Mitigation of Climate Change Report (Intergovernmental Panel on Climate Change, 2022b) is confident that the Kyoto Protocol has

reduced emissions in some countries and has been instrumental in building national and international capacity for GHG reporting, accounting and emissions markets. It is also highly confident that at least 18 countries with Kyoto targets for the first commitment period had sustained absolute emissions reductions for at least a decade from 2005. Two were countries with economies in transition (Intergovernmental Panel on Climate Change, 2022b).

PARIS AGREEMENT

The Paris Agreement (L'Accord de Paris) is a legally binding international treaty on climate change (United Nations Framework Convention on Climate Change, 2021b). The agreement was adopted by 196 parties at COP 21 in Paris on 12 December 2015 and entered into force on 4 November 2016. This agreement represents 97% of global GHG emissions. It builds upon the UNFCCC and – unlike the Kyoto Protocol – assembles all countries into a common cause to undertake ambitious efforts to combat climate change. It includes provisions for countries to adapt to any adverse effects of climate change, with enhanced support to assist developing countries. As of 12 October 2021, 192 parties had ratified the Paris Agreement, and 164 parties (164 countries plus the European Union) had communicated their first Nationally Determined Contributions (NDCs) to the UNFCCC Secretariat (United Nations Framework Convention on Climate Change, 2021b). Notably, both the USA and China are signatories to the agreement, representing 40% of global GHG emissions in 2019.

The primary goal of the Paris Agreement is to 'hold the increase in the global average temperature to well below 2°C above pre-industrial levels and pursue efforts to limit the temperature increase to 1.5°C above pre-industrial levels'. It does so, recognising that this would significantly reduce the risks and impacts of climate change. The global average temperature has already increased by just over 1°C since 1910 (see Chapter 2, Figure 2.6c), meaning this ambitious target requires average warming this century to be contained within 0.5°C–1.0°C from now on to avoid 'runaway climate change'. To achieve this long-term temperature goal, parties to the agreement aim to reach global peaking of GHG emissions as soon as possible to reach a 'climate neutral' world by 2050. Governments worldwide

---

[2] Small Island Developing States (SIDS) are a distinct group of 38 United Nations Member States and 20 Non-UN Members/Associate Members of United Nations regional commissions that face unique social, economic and environmental vulnerabilities (United Nations 2021).

are moving to 'net zero' to limit the impacts of climate change.

Implementation of the Paris Agreement requires both economic and social transformation – underpinned by the best available science. It works on a 5-year cycle of increasingly ambitious climate action by signatory countries. Parties were required to submit their NDC plans for climate action by late 2020 but extended due to the COVID-19 pandemic. In their NDCs, parties must communicate steps to reduce their GHG emissions to reach the Paris Agreement's goals. Parties must also communicate in the NDCs, and any actions they will take to build resilience to adapt to the impacts of climate change.

As of 17 September 2021, 86 updated, or new NDCs have been communicated by 113 parties and recorded in the interim NDC registry, covering about 59% of Parties to the Paris Agreement and accounting for about 49% of global GHG emissions (United Nations Framework Convention on Climate Change, 2021b). The implementation of most conditional elements in the NDCs depends on access to enhanced financial resources, technology transfer and technical cooperation, capacity-building support, availability of market-based mechanisms (Box 4.3) and absorptive capacity of forests and other terrestrial and marine ecosystems (United Nations Framework Convention on Climate Change, 2021b). Most of the parties that submitted new or updated NDCs have strengthened their commitment to reducing or limiting GHG emissions by 2025 and 2030, demonstrating increased ambition in addressing climate change. Within the group of 113 parties, 70 indicated carbon neutrality goals around the middle of the century. However, this current trend could lead to even more significant emissions reductions, of about 26% by 2030 compared to 2010 (United Nations Framework Convention on Climate Change, 2021b).

The total global GHG emissions level (without land use, land-use change and forestry), assuming implementation of the latest NDCs of all parties to the Paris Agreement, is estimated to be around 54.8 Gt CO_2-e in 2025 and 55.1 Gt CO_2-e in 2030 (United Nations Framework Convention on Climate Change, 2021b). However, the estimated 2025 emissions level is 58.6% higher than in 1990 (34.6 Gt CO_2-e), 15.8% higher than in 2010 (47.3 Gt CO_2-e) and 4.5% higher than in 2019 (52.4 Gt CO_2-e). The 2030 level is 59.3%

higher than in 1990, 16.3% higher than in 2010 and 5.0% higher than 2019. The Emissions Gap Report 2021 (United Nations Environment Program, 2021b) shows that new national climate pledges – combined with other mitigation measures – put the Earth on track for a global temperature rise of 2.7°C, above pre-industrial level, by 2100 (Intergovernmental Panel on Climate Change, 2022b). This track is consistent with the SSP2–4.5 emissions pathway (see Chapter 2, Table 2.6). However, a more recent study of NDC pledges argues that global warming can be maintained just below 2°C if all conditional and unconditional NDCs are implemented in full and on time (Meinshausen et al., 2022). The authors conclude that limiting warming to 'just below' and 'well below' 2°C or 1.5°C urgently requires policies and actions to bring about steep emission reductions this decade, aligned with mid-century global net-zero CO_2 emissions.

The IPCC Special Report (2018) identified mitigation options considered relevant to aligning with 1.5°C pathways (see SSP1–1.9; Chapter 2, Table 2.6) and included:

- Halting investment in unabated coal by 2030. Only a few parties communicated corresponding measures in their NDC plans, such as phasing out the use of unabated coal to produce electricity by 2025 (United Nations Framework Convention on Climate Change, 2021b);
- Phasing out of sales of fossil fuel passenger vehicles by 2035–2050. A few parties communicated corresponding measures, including banning new registration of diesel and gasoline vehicles after 2030 (United Nations Framework Convention on Climate Change, 2021b);
- Requiring newly constructed buildings to be near zero energy by 2020. Some parties communicated corresponding measures, such as requiring new buildings constructed after 1 January 2020 to consume almost zero energy (United Nations Framework Convention on Climate Change, 2021b); and
- Expanding forest cover by 2030. A few parties communicated quantitative targets for increasing national forest cover in their NDC plans (United Nations Framework Convention on Climate Change, 2021b), such as increasing forest cover to 60% of the national territory without competing for land with the agriculture sector.

The latest IPCC Assessment Report (Intergovernmental Panel on Climate Change, 2022a) examines all global modelled pathways that limit warming to 1.5°C (lower Paris target) with no or limited overshoot and those that limit warming to 2°C (upper Paris target). The report is unanimous that meeting the Paris targets can only happen with rapid and profound and, in most cases, immediate GHG emission reductions in all economic sectors (see Meinshausen et al., 2022). The report guides policy decisions that must occur on the mitigation side of climate change to limit global warming to 2°C by 2100. Recommended mitigation strategies and options at the global level focus on the energy, transportation, manufacturing (industry), construction, built environments and agriculture, forestry and land-use sectors (Box 4.4). Chapters 6 and 7 examine climate-resilient mitigation pathways for these sectors in Australia.

The Paris Agreement has a strong focus on capacity building, given many poorer countries have insufficient capacities to deal with many of the challenges brought by climate change. Consequently, the agreement 'requests all developed countries to enhance support for capacity-building actions in developing countries' (United Nations Framework Convention on Climate Change, 2021b). It provides a framework for financial, technical and capacity-building support to the countries that need it to achieve their emission targets. Many parties identified capacity building as a prerequisite for successfully implementing their NDC plans (United Nations Framework Convention on Climate Change, 2021b). Capacity building is required to formulate policy and integrate mitigation and adaptation into sectoral planning processes. This process can help parties access finance and provide the information necessary for clarity, identified transparency and understanding of NDCs. In the new or updated NDCs, compared to their previous ones, more parties expressed capacity-building needs for climate change adaptation in addition to mitigation (United Nations Framework Convention on Climate Change, 2021b).

As with the Kyoto Protocol, the Paris Agreement reaffirms that developed countries should take the lead in providing financial assistance to countries that are less endowed and more vulnerable to climate change, while for the first time, also encouraging voluntary contributions by other Parties. Climate finance is needed for a suite of mitigation programmes and projects because large-scale investments are required to reduce GHG emissions significantly. Climate finance is also crucial for climate adaptation, as significant financial resources are needed to adapt to the adverse effects and reduce vulnerability and risks associated with climate change.

In addition to financing, the Paris Agreement emphasises the importance of technology development and transfer to developing countries to improve resilience to climate change (adaptation) and reduce GHG emissions (mitigation). It institutes a technology framework to provide overarching guidance to the well-functioning technology mechanism. The mechanism comprises two bodies that work together. The first is the Technology Executive Committee which issues and provides policy recommendations that support parties' efforts to enhance climate technology development and transfer. The second is the Climate Technology Centre and Network (hosted by the UN Environment in collaboration with the United Nations Industrial Development Organization) that accelerates the development and transfer of technologies through three services:

1. Providing technical assistance at the request of developing countries on technology issues;
2. Creating access to information and knowledge on climate technologies; and
3. Fostering collaboration among climate technology stakeholders via its network of regional and sectoral experts.

There are several mechanisms in the Paris Agreement to track progress against agreed targets. The Enhanced Transparency Framework (ETF) is the primary mechanism established for this purpose. From 2024, parties to the agreement will report under the ETF on actions taken and progress in climate change mitigation, adaptation measures and support provided or received. The ETF also provides international procedures for reviewing the submitted country reports. Material gathered by the ETF feeds into the global stocktake to monitor the combined progress towards the LTGGs. It will form the basis for recommendations for countries to set more ambitious mitigation and adaptation plans for the next round of NDCs.

There is now *moderate confidence*, that the Paris Agreement has led to policy development and

BOX 4.4: Recommended climate mitigation strategies and options at the global level to limit global warming within 2°C by 2100 (Intergovernmental Panel on Climate Change, 2022a)

1. Reducing GHG emissions across the entire energy sector requires significant transitions, including a substantial reduction in overall fossil fuel use, the deployment of low-emission energy sources, switching to alternative energy carriers and energy efficiency and conservation. It notes the continued installation of unabated fossil fuel infrastructure (e.g. coal and gas-fired power stations) will 'lock in' GHG emissions.

2. Reducing industry emissions will entail coordinated action throughout value chains to promote all mitigation options, including demand management, energy and materials efficiency, circular material flows, abatement technologies and transformational changes in production processes. Moving towards net-zero GHG emissions from industry will be enabled by adopting new production processes using low and zero GHG electricity, hydrogen, fuels and carbon management.

3. Urban areas can create opportunities to increase resource efficiency and significantly reduce GHG emissions through the systemic transition of infrastructure and urban form through low-emission development pathways towards net-zero emissions. Ambitious mitigation efforts for established, rapidly growing and emerging cities will encompass (1) reducing or changing energy and material consumption, (2) electrification and (3) enhancing carbon uptake and storage in the urban environment. Cities can achieve net-zero emissions, but only if emissions are reduced within and outside their administrative boundaries through supply chains, which will have beneficial cascading effects across other sectors.

4. In modelled global scenarios, retrofitted old buildings and buildings yet to be built are projected to approach net-zero GHG emissions in 2050. This goal will occur if policy packages combine ambitious sufficiency, efficiency and renewable energy measures, are effectively implemented, and barriers to decarbonisation are removed.

5. Demand-side options and low-GHG emissions technologies can reduce transport sector emissions in developed countries and limit emissions growth in developing countries. Demand-focused interventions can reduce demand for all transport services and support the shift to more energy-efficient transport modes. Electric vehicles powered by low-emissions electricity offer the most significant decarbonisation potential for land-based transport on a life cycle basis. Sustainable biofuels can provide additional mitigation benefits in land-based transport in the short and medium terms. Sustainable biofuels, low-emissions hydrogen and derivatives (including synthetic fuels) can support mitigation of CO_2 emissions from shipping, aviation and heavy-duty land transport but require production process improvements and cost reductions. Many mitigation strategies in the transport sector would have various co-benefits, including air quality improvements, health benefits, equitable access to transportation services, reduced congestion and reduced material demand.

6. Agriculture, Forestry and Other Land Use (AFOLU) mitigation options, when sustainably implemented, can deliver large-scale GHG emission reductions and enhanced CO_2 removals but cannot fully compensate for delayed action in other sectors (e.g. energy and transportation). In addition, sustainably sourced agricultural and forest products can be used instead of more GHG-intensive products in other sectors. Barriers to implementation and trade-offs may result from the impacts of climate change, competing demands on land, conflicts with food security and livelihoods, the complexity of land ownership and management systems, and cultural aspects. There are many country-specific opportunities to provide co-benefits (such as biodiversity conservation, ecosystem services and livelihoods) and avoid risks (e.g. through adaptation to climate change).

target-setting at national and sub-national levels, in relation to mitigation, as well as enhanced transparency of climate action and support (Intergovernmental Panel on Climate Change, 2022b). There are already many examples of low-carbon solutions and new carbon markets worldwide. For instance, zero-carbon solutions are becoming competitive across economic sectors representing 25% of GHG emissions. There are now increased carbon-neutral targets among cities, regions, states/provinces, countries and critical economic sectors. Most notably, the energy and transport sectors (see Chapter 6) have moved quickly in this space, creating a boom in renewable industries. This transition is a positive sign as zero-carbon solutions could be competitive in sectors representing over 70% of global emissions by 2030.

Intergovernmental Science-Policy Platform on Biodiversity and Ecosystem Services

Over the past decade, efforts to address climate change have taken place alongside and in the context of other major environmental problems, such as biodiversity loss and degradation of essential ecosystem services (Chen et al., 2021). The Intergovernmental Science-Policy Platform on Biodiversity and Ecosystem Services (IPBES) modified the IPCC science-policy interface and assessment for its purpose (Díaz et al., 2015, see Chapter 5, Box 5.2). The platform's objective is to 'strengthen the science-policy interface for biodiversity and ecosystem services for the conservation and sustainable use of biodiversity, long-term human well-being and sustainable development' (Intergovernmental Science-Policy Platform on Biodiversity and Ecosystem Services, 2019). The Special Report on the Ocean and Cryosphere in a Changing Climate (Intergovernmental Panel on Climate Change, 2019a) and Special Report on Climate Change and Land (Intergovernmental Panel on Climate Change, 2019b) assessed the relationships between changes in biodiversity and climate change. The future work programme of IPBES (up to 2030) will address interlinkages among biodiversity, water, food and health. Working with the IPCC, the IPBES assessment will use a nexus approach to examine these interlinkages, including developing climate change mitigation and adaptation strategies.

United Nations 2030 Agenda for Sustainable Development

SUSTAINABLE DEVELOPMENT GOALS

In 2015, the United Nations General Assembly adopted the 2030 Agenda for Sustainable Development, a plan of action for people, the planet and prosperity (United Nations, 2015). Seventeen SDGs, 169 nested targets and over 200 associated indicators set the agenda for sustainable development challenges by 2030. Climate change is relevant to many of the SDGs (see Chapters 5–7 for Australian examples). However, the example given here is SDG13: 'take urgent action to combat climate change and its impacts, while acknowledging that the UNFCCC is the primary international, intergovernmental forum for negotiating the global response to climate change'. SDG13 has five targets and some associated indicators to be quantified over time (Box 4.5).

The 2020s is the 'decade of action' to deliver the SDGs by 2030. This decade is perhaps the last chance to advance a shared vision and accelerate responses to the world's gravest challenges – from eliminating poverty and hunger to reversing adverse impacts of climate change. 2020 also marked the year when the novel coronavirus created one of the most severe international crises in 100 years – with far-reaching implications for achieving the SDGs. According to the latest SDG report (United Nations, 2020a), a temporary reduction in human activities resulted in a dip in GHG emissions in 2020. However, by December 2020, emissions had fully rebounded and registered 2% higher than the same month in 2019.

Another special report on the SDGs states that financing for climate action has increased substantially, but it continues to be surpassed by investments in fossil fuels (United Nations, 2021). Specifically, climate-related financial investments increased by 17% from 2013–14 to 2015–16 (from US$584–681 billion), driven mainly by private investments in renewable energy. Nonetheless, investments in climate activities across sectors continued to be surpassed by those related to fossil fuels in the energy sector, which was US$781 billion in 2016. The report makes it clear that to 'achieve a low-carbon, climate-resilient transition, a much greater scale of annual investment is required'. Finally, climate-related financing

BOX 4.5: SDG 13, its targets and indicators (United Nations 2015)

SDG13: TAKE URGENT ACTION TO COMBAT CLIMATE CHANGE AND ITS IMPACTS

Target 13.1: Strengthen resilience and adaptive capacity to climate-related hazards and natural disasters in all countries
Indicators:

1. Number of deaths, missing persons and persons affected by disaster per 100,000 people
2. Number of countries with national and local disaster reduction strategies
3. Proportion of local governments that adopt and implement local disaster reduction strategies in line with national disaster risk reduction strategies

Target 13.2: Integrate climate change measures into national policies, strategies and planning
Indicator:

1. Number of countries that have communicated the establishment or operationalisation of an integrated policy/strategy/plan which increases their ability to adapt to the adverse impacts of climate change, and foster climate resilience and low gas emissions development in a manner that does not threaten food production (including a national plan, NDC, national communication, biennial update report or other)

Target 13.3: Improve education, awareness-raising and human and institutional capacity on climate change mitigation, adaptation, impact reduction and early warning
Indicators:

1. Number of countries that have integrated mitigation, adaptation, impact reduction and early warning into primary, secondary and tertiary curricula
2. Number of countries that have communicated the strengthening of institutional, systemic and individual capacity building to implement adaptation, mitigation and technology transfer, and development actions

Target 13a: Implement the commitment undertaken by developed-country parties to the United Nations Framework Convention on Climate Change to a goal of mobilising jointly $100 billion annually by 2020 from all sources to address the needs of developing countries in the context of meaningful mitigation actions and transparency on implementation and fully operationalise the Green Climate Fund through its capitalisation as soon as possible.
Indicator:

1. Mobilised amount of United States dollars per year starting in 2020 accountable towards the $100 billion commitment

Target 13b: Promote mechanisms for raising capacity for effective climate change-related planning and management in least developed countries and small island developing states, including focusing on women, youth, and local and marginalised communities
Indicator:

1. Number of least developed countries and small island developing states that are receiving specialised support, and amount of support, including finance, technology and capacity building, for mechanisms for raising capacities for effective climate change-related planning and management, including focusing on women, youth, and local and marginalised communities

provided by developed countries to developing countries increased by 14% in 2016, approaching US$38 billion, with 64% focused on climate mitigation and the remainder equally shared for climate adaptation and cross-cutting issues (United Nations, 2020). The financing had increased to US$48.7 billion in 2018, with a growing proportion for adaptation planning support (United Nations, 2021).

On a positive note, most developing countries have begun formulating plans to strengthen resilience and adapt to climate change (United Nations, 2020b). Such national adaptation plans (NAPs) assist developing countries in achieving the global adaptation goal under the Paris Agreement. As of May 2021, 125 of 154 developing countries formulated and implemented NAPs, and 22 countries submitted their plans to the UNFCCC Secretariat (United Nations, 2020b). The Green Climate Fund provides funding for NAPs with requests totalling US$464 million in 2020. Developed countries are stepping up their efforts to provide technical guidance and support to less-developed countries – to create and carry out such plans.

A much less positive result is the poor progress in meeting the 2020 disaster risk reduction targets (United Nations, 2020b). This result is unfortunate as climate change worsens the frequency and intensity of natural disasters each decade. The Sendai Framework for Disaster Risk Reduction (2015–2030) aims to reduce existing – and prevent new – disaster risk through clear targets and priorities for action under the 2030 Agenda for Sustainable Development (United Nations, 2020b). The framework is an ambitious – but non-binding – agreement that aims to reduce risks associated with disasters of all scales, frequencies and onset rates caused by natural or human-made hazards, including climate change. Target (e) of the Sendai Framework concerns establishing national and local disaster risk reduction strategies and has already passed the 2020 deadline. As of April 2020, 85 countries (about 40%) reported national disaster risk reduction strategies aligned in varying degrees with the Sendai Framework, with six countries reporting aligned national strategies. In 2018, 55 countries reported that at least some of their local governments had local disaster risk reduction strategies in line to some extent with national strategies (United Nations, 2020b).

Climate change is referenced in SDGs beyond SDG13, for example, in goal targets 1.5, 2.4 and 12.8.1 related to poverty, hunger and education, respectively (Liverman, 2018). Other SDGs that apply to climate mitigation and adaptation include clean energy for all (SDG7), sustainable industry (SDG9) and cities (SDG11) and the protection of life below water (SDG14) and on land (SDG15). The three chapters to follow will examine opportunities to integrate the SDGs, their targets and indicators with climate-resilient development pathways for Australian natural systems (Chapter 5), economic systems (Chapter 6) and social systems (Chapter 7).

ROLES AND RESPONSIBILITIES FOR CLIMATE CHANGE ACTION IN AUSTRALIA

Governments worldwide are moving to 'net-zero' (carbon) emissions to limit the adverse impacts of climate change on their natural environments, economies and societies. All Australian state and territory governments have agreed, intending to reach net-zero emissions by 2050. Ahead of the COP26 Glasgow Conference in November 2021, the Australian Government pledged an aspirational target to achieve net-zero emissions by 2050 (see below). At COP26, the Australian Government was heavily criticised for its low 2030 national emissions target, which remains unchanged at 26%–28% below the 2005 emissions level. In September, 2022, the newly elected Australian Labor Government has legislated a 2030 target of 43% below the 2005 level (see Chapter 6).

Australia's Long-Term Emissions Reduction Plan was released by the previous Liberal-National Coalition Government in April 2021 (Commonwealth of Australia [COA], 2021a). It aims to provide a 'whole-of-economy' plan to achieve net-zero emissions by 2050. The emissions reduction plan has four pillars:

1. Driving down technology costs;
2. Enabling deployment at scale;
3. Seizing opportunities in new and traditional markets; and
4. Fostering global collaboration. It argues that Australia needs to capture new employment and economic opportunities as the world shifts to low-emissions technologies (see examples in Chapter 6, Economic Systems).

Most aspects of the plan align with similar emissions reductions initiatives in other highly developed economies (COA, 2021a). Examples include clean hydrogen, energy storage, soil carbon, incentivising business, and voluntary carbon markets and certification. Others are building essential infrastructure, alignment of jurisdictions, regional investment, skills and training, international partnerships and offset (abatement) schemes in developing countries.

However, some core initiatives in the plan deviate significantly from other highly developed economies, e.g. carbon capture and storage (CCS), carbon capture, use and storage (CCUS), exporting natural gas and reliance on fossil fuels in manufacturing processing. The plan claims that large-scale CCUS projects can underpin new low-emissions industries (including clean hydrogen) and provide a potential 'decarbonisation' pathway for 'hard-to-abate' industries, such as cement and fertiliser production (COA, 2021a). The reliance on CCS, CCUS and fossil fuels to deliver a pathway to net zero is considered an unnecessary attempt to extend the life of high-emissions fossil fuels in Australia's energy systems (for details, see Chapter 6, Box 6.5).

The AU$4.5 billion Emissions Reduction Fund (ERF) is the main policy instrument in the Australian Government's Emissions Reduction Plan (COA, 2021a). The main purpose of the ERF is to issue Australian carbon credit units (ACCUs) and to purchase ACCUs from registered carbon offset projects by the Clean Energy Regulator (CER). The CER is the Australian Government body responsible for administering legislation that will reduce carbon emissions and increase the use of clean energy (Clean Energy Regulator, 2022). Unlike its predecessor, the Carbon Farming Initiative, an offset scheme for the agricultural, land and waste sectors, the ERF permits carbon offsets to be generated from all sectors of the economy (Macintosh et al., 2022).

Under the ERF, the primary source of demand for ACCUs is the CER, which purchases abatement on behalf of the Australian Government (Clean Energy Regulator, 2022). ACCUs are also purchased by facilities with carbon liabilities under the safeguard mechanism. The mechanism requires Australia's largest greenhouse gas emitters to keep their net emissions below an emissions limit or baseline (Department of Industry, Science,

Energy and Resources, 2022). The regulated mechanism places caps on net emissions from facilities that emit more than 100,000 t of carbon dioxide equivalent (t CO_2-e) per annum (Clean Energy Regulator, 2022). A recent analysis (Macintosh et al., 2022) has found that the caps under the safeguard mechanism are lax. The covered facilities under the mechanism have had little need to surrender ACCUs to meet their obligations. The same study found that the only other source of demand for ACCUs under the current policy settings was voluntary buyers. These buyers were entities voluntarily seeking to offset their emissions for marketing and altruistic reasons.

The Australian Government has recently released a National Climate Resilience and Adaptation Strategy (COA, 2021b). This national strategy is to support all levels of government, communities and businesses to anticipate better, manage and adapt to climate change. The adaptation strategy operates across four domains – natural, built, social and economic. It argues that effective adaptation requires the best available science and information, delivered through partnerships and investments and guided by effective governance and coordination. Key roles and responsibilities for climate change adaptation in the country are:

1. *Australian (federal) Government*: is responsible for national leadership on climate adaptation, managing Australian Government assets and services, including significant investments in public infrastructure, and providing national climate science and information. It maintains a robust and flexible economy and a well-targeted safety net to ensure that climate change does not disproportionately affect vulnerable groups.

2. *State and territory governments:* have an essential role in adaptation, with significant influence through planning laws and public infrastructure investments. This role includes vital service delivery and infrastructure areas, such as emergency services, environmental protection, health, planning and transport. They also provide science and information at local and regional scales.

3. *Local governments:* are on the 'frontline' in dealing with the impacts of climate change (see Chapter 7, Boxes 7.4 and 7.5). They have an essential role in ensuring that local

circumstances are incorporated in the overall adaptation response, and local communities are directly involved in adaptation efforts. Local governments must also be well-positioned to inform state, territory and federal governments about on-the-ground needs of local and regional communities, communicate directly with those communities and respond to local challenges.

4. *Private parties:* have an essential role in managing their climate risks, e.g. maintaining and protecting private assets and incomes (see Chapter 6, Economic Systems). This role provides a strong incentive to act on climate change. However, their response capacity will differ depending on their exposure and sensitivity to climate risk and access to resources and knowledge (see Figure 4.1).

To mitigate the risk of runaway climate change – namely limiting global heating to 1.5°–2.0°C above pre-industrial – the global GHG emissions must peak as soon as possible. Emissions must rapidly decline 45% from 2010 levels by 2030, followed by a sustained steep decline to achieve net-zero emissions by 2050 (Intergovernmental Panel on Climate Change, 2022a). There is, therefore, much work to be done to combat anthropogenic climate change, requiring structural and transformative changes to facilitate low-GHG emissions to achieve net zero by 2050. Such changes will be critical in building climate-resilient ecosystems, economies and societies. Approaches to such shifts and changes in Australia – a recognised global climate change hotspot – are discussed in more detail in the three chapters to follow.

SUMMARY

- Responding to climate change requires strategies to reduce emissions of GHGs and enhancement of carbon sinks (climate mitigation), as well as strategies to cope with climate change impacts and risks that are inevitable even with mitigation policies (climate adaptation). Adaptation may also include taking advantage of opportunities due to climate change. Both climate mitigation and adaptation are relevant in a policy context, and understanding them is central to dealing with climate change.

- Anthropogenic climate change is causing numerous impacts on the integrity, status and trend of natural, economic and social systems. Natural systems include terrestrial, freshwater, estuarine, coastal and marine (ocean) ecosystems and their biodiversity and the ecosystem services for human health and well-being (see Chapter 5). Economic systems include energy and transportation, agriculture (includes forestry, fisheries and other land use), mining (resources), manufacturing and construction and tourism

(see Chapter 6). Social systems include cities, settlements and built environments, human health and well-being, Indigenous peoples and finance and insurance (see Chapter 7).

- Planned adaptation of natural, economic and social systems to observed and potential climate change has two components: changing the fundamental attributes of systems to differing degrees. It may be incremental where the adaptation response maintains the essence and integrity or process at a given scale. Incremental adaptation is a gradual adjustment process; transformational adaptation changes the fundamental attributes of the socio-ecological system in anticipation of climate change and its impacts.

- Both incremental and transformative adaptation pathways aim to reduce the vulnerability and exposure of systems to the adverse effects of climate change. Adaptation pathways are also helpful for considering potential benefits and opportunities that may arise with future climate change. Both incremental and transformative pathways require intervention from higher levels of governance. They will most likely be driven by government or private sector responses to natural disasters or other catastrophic environmental, social or economic events. Transformative adaptation strategies generally require systemic (or structural) changes in natural, economic and social systems, i.e. deliberate government or private sector policy decisions to avoid maladaptation pathways.

- There is now a stronger focus on climate-resilient development approaches to climate mitigation and adaptation. Climate-resilient development is a process of implementing greenhouse gas mitigation and adaptation options to support sustainable development for all. These options give rise to development 'pathways' that align with the SDGs. In the near-term, progress towards climate-resilient development can be monitored by progress on the SDGs.

- The UNFCCC ultimate objective is to 'stabilise greenhouse gas concentrations in the atmosphere at a level that would prevent dangerous anthropogenic interference with the climate system'. It also states that 'such a level should be achieved within a timeframe sufficient to allow ecosystems to adapt naturally to climate change, ensure that food production is not threatened, and enable economic development to proceed sustainably'. The provisions of the UNFCCC are pursued and implemented by two well-known international treaties: the Kyoto Protocol and the Paris Agreement. The Kyoto Protocol commits industrialised (developed) countries and 'economies in transition' to limit and reduce GHG emissions according to individual targets.

- The primary goal of the Paris Agreement is to hold the increase in the global average temperature to well below 2°C above pre-industrial levels and pursue efforts to limit the temperature increase to 1.5°C above pre-industrial levels. It does so, recognising that this would significantly reduce the risks and impacts of climate change. Unlike the Kyoto Protocol, the Paris Agreement assembles all countries into a common cause to undertake ambitious efforts to combat climate change. The agreement includes provisions for countries to adapt to any adverse effects of climate change, with enhanced support to assist developing countries.

- There are already many examples of low-carbon solutions and new carbon markets worldwide. For instance, zero-carbon solutions are becoming competitive across economic sectors representing 25% of GHG emissions. There are now increased carbon-neutral targets among cities, regions, states/provinces, countries and critical economic sectors. Most notably, the energy and transport sectors have moved quickly in this space, creating a boom in renewable industries. This transition is a positive sign as zero-carbon solutions could be competitive in sectors representing over 70% of global emissions by 2030.

- Despite some promising trends, achieving the Paris targets by 2100 remains challenging according to current GHG emissions data. The current suite of NDCs places Earth's climate system on track for global warming of between 2.2°C and 3.5°C within 80 years (Intergovernmental Panel on Climate Change, 2022). Hence, to have a 50% chance of keeping global warming to 1.5°C by 2100 (lower-end Paris target), global CO_2 emissions must halve over the next 10 years, reach net zero in the 2050s and go net negative after that. More recent research shows that warming can be kept below 2°C if all conditional and unconditional NDC pledges are implemented fully and on time.

- The 2030 Agenda for Sustainable Development provides a plan of action for people, the planet and prosperity.

Seventeen SDGs, 169 nested targets and over 200 associated indicators set the agenda for sustainable development challenges by 2030. Climate change is relevant to many SDGs, notably SDG13: 'take urgent action to combat climate change and its impacts while acknowledging that the UNFCCC is the primary international, intergovernmental forum for negotiating the global response to climate change'.

- Australian state and territory governments have aligned to reach net-zero emissions by 2050. Ahead of the COP26 Glasgow Conference in November 2021, the Australian Government finally pledged to achieve net-zero emissions by 2050. The Long-Term Emissions Reduction Plan is the primary policy platform to deliver this aspirational 2050 target. Meanwhile, Australia has been criticised for its low 2030 national emissions target, which remains unchanged at 26%–28% below the 2005 emissions level. This 2030 target has recently been legislated upwards to 43% or higher by the newly elected Australian Labor Government.

5

Natural Systems

Biodiversity plays a key role in providing numerous irreplaceable services to the Australian community. Yet these remain poorly measured and demonstrated, with many changes occurring subtly on timescales that are not immediately evident to the vast majority of Australians.

State of the Environment 2016: Australia (Commonwealth of Australia 2017–2018)

INTRODUCTION

Australia is one of the world's 17 megadiverse[1] countries and is globally renowned for its distinctive ecosystems and biota (Commonwealth of Australia, 2017–2018). Its flora and fauna are highly regarded for their diversity, endemism[2] and evolutionary adaptations. Notably, they are also an inseparable part of its Indigenous culture and broader Australian identity (see Chapter 7, Social Systems). Some 46% of birds, 69% of mammals (including marine mammals), 94% of amphibians, 93% of flowering plants and 93% of reptiles are endemic to the country. Other endemic groups, such as the eucalypts, are primarily found in Australia or nearby, e.g. Papua New Guinea and eastern Indonesia. Biodiversity provides Australia, its residents and visitors with diverse economic, ecological, recreational, cultural and scientific values (Box 5.1). However, Australia has one of the worst

records of species extinctions of any continent, with about 50% of the world's known mammal extinctions in the last 200 years (Hughes, 2014; Cresswell et al., 2021). Loss of biodiversity due to climate change and other change agents (e.g. habitat loss and fragmentation, invasive species) will erode the country's critical biodiversity values and adversely affect human well-being (Cresswell et al., 2021).

Australia joined the World Heritage Convention in 1974 and, as of April 2022, has 19 world heritage sites (Valentine, 2019). Twelve are listed for natural values, three for cultural values and four for mixed (natural and cultural) values. Kakadu National Park, a mixed listing in the Northern Territory (see Figure 5.2), was the first site (among others) listed that year inscribed on the world heritage list in 1981, and the Ningaloo Coast in Western Australia (natural listing) was the most recent in 2011 (Figure 5.1). Australia's stunning biodiversity, natural beauty (e.g. sandy beaches) and world heritage sites underpin its multi-billion dollar domestic and international tourism industry. Climate change poses a significant threat to the country's ecosystems and biodiversity with knock-on implications for sustainable tourism businesses (see Chapter 6, the 'Tourism' section).

Global biodiversity hotspots are world regions that contain exceptional concentrations of endemic species (Myers, 1988, Myers et al., 2000).

[1] Together these 17 countries account for 70% of the world's biological diversity across less than 10% of the world's surface (Commonwealth of Australia, 2017–2018).

[2] Endemism is an ecological term meaning that a plant or animal lives in a defined geographical location, such as a habitat type, an island or other defined zone. For example, platypuses are found only in eastern mainland Australia and the island state of Tasmania.

DOI: 10.1201/9781003189909-5

BOX 5.1: Importance of biodiversity in an Australian context

- *Economic:* biodiversity provides humans with raw materials for consumption and production. Many livelihoods, such as those of farmers, fishers and timber workers, are dependent on biodiversity (see Chapter 6, the 'Agriculture and Land' section).
- *Ecological life support:* biodiversity provides functioning ecosystems that supply oxygen, clean air and water, pollination of plants, pest control, wastewater treatment and many ecosystem services (see Ecosystems and Ecosystem Services, Box 5.2).
- *Recreation:* many recreational pursuits rely on our unique biodiversity, such as birdwatching, hiking, camping and fishing. The Australian tourism industry also depends on the healthy status of its ecosystems and biodiversity (see Chapter 6, the 'Tourism' section).
- *Cultural:* the Australian culture is closely connected to biodiversity through the expression of identity, through spirituality and through aesthetic appreciation. Indigenous Australians have strong connections and obligations to biodiversity arising from spiritual beliefs about animals and plants (see Chapter 7, the 'Indigenous Peoples' section).
- *Scientific:* biodiversity represents a wealth of systematic ecological data that help society to understand the natural world and its origins.

Source: Adapted from Morton and Hill (2014).

These hotspots are places that are both biologically rich and highly threatened by human activities (Conservation International [CI], 2022). Hence, to be considered a biodiversity hotspot, a region must meet two strict criteria (CI, 2022):

1. It must have at least 1,500 vascular plants as endemics, i.e. the hotspot is irreplaceable.
2. It must have 30% or less of its original natural vegetation, i.e. it must be threatened. Threats include habitat loss, poaching of wildlife, invasive plants and animals, disease and pathogens and climate change.

There are currently 36 global biodiversity hotspots that meet the CI criteria (Critical Ecosystem Partnership Fund, 2022). Their intact habitats represent only 2.5% of Earth's land surface (Critical Ecosystem Partnership Fund, 2022). Before 2011, Australia only had one recognised biodiversity hotspot, the South-west Australia eco-region or 'Southwest Australia' (Myers et al., 2000). Following the efforts of Williams et al. (2011), a second hotspot was proclaimed in 2011, the Eastern Australian Temperate Forests and Queensland Tropical Rainforests or 'Forests of East Australia' (Figure 5.1).

The Southwest Australia Biodiversity Hotspot (Figure 5.1) comprises 356,700 km² of the southwestern area of the continent. More than half of the 5,570 species of the hotspot's vascular plants are endemic (Critical Ecosystem Partnership Fund, 2022). The hotspot region has a classic Mediterranean climate (see Chapter 3, Figure 3.3: *Csa* and *Csb* sub-types). With these climate sub-types, most rainfall occurs in the winter, and the summers are very dry and either warm or hot. The region's native plants are well adapted to the climate and nutrient-poor soils, as well as supporting large-scale cropping and sheep grazing in mostly cleared areas (see Chapter 6, the 'Agricultural and Land' section. The biodiversity of the hotspot region has the following characteristics (Critical Ecosystem Partnership Fund, 2022):

- Only 30% of the extant vegetation remains in relatively pristine condition.
- Dominant vegetation types are *Eucalyptus* woodlands and *Eucalyptus*-dominated 'mallee' (Kwongan) shrubland.
- The forests, woodlands and shrublands have high endemism among plants and reptiles. Its vertebrate species include the endangered numbat (*Myrmecobius fasciatus*) and honey possum (*Tarsipes rostrus*), and the red-capped parrot (*Purpureicephalus spurius*).

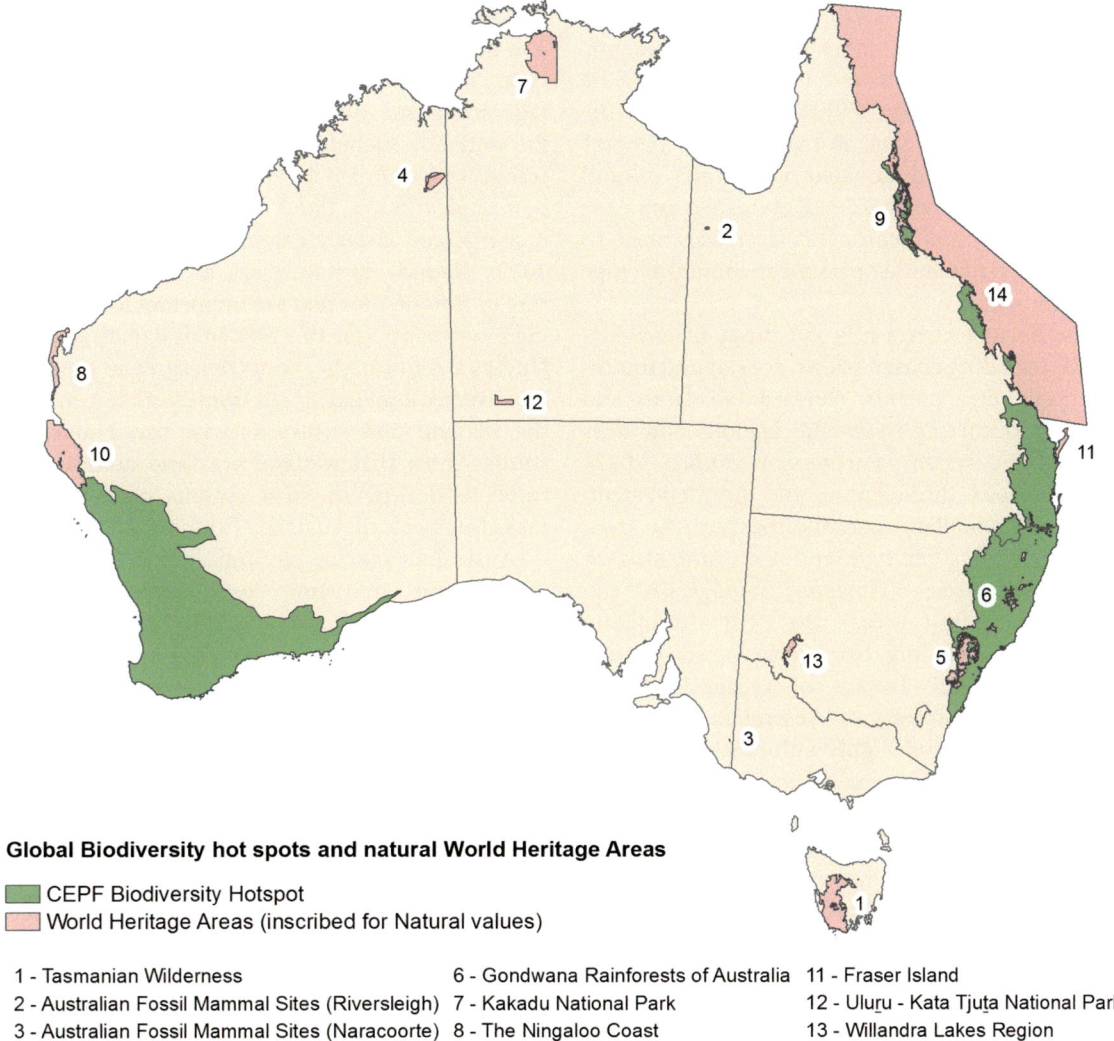

Global Biodiversity hot spots and natural World Heritage Areas

- CEPF Biodiversity Hotspot
- World Heritage Areas (inscribed for Natural values)

1 - Tasmanian Wilderness	6 - Gondwana Rainforests of Australia	11 - Fraser Island
2 - Australian Fossil Mammal Sites (Riversleigh)	7 - Kakadu National Park	12 - Uluṟu - Kata Tjuṯa National Park
3 - Australian Fossil Mammal Sites (Naracoorte)	8 - The Ningaloo Coast	13 - Willandra Lakes Region
4 - Purnululu National Park	9 - Wet Tropics of Queensland	14 - Great Barrier Reef
5 - The Greater Blue Mountains Area	10 - Shark Bay, Western Australia	

Figure 5.1 A map showing two global biodiversity hotspots and 14 natural world heritage areas in Australia. (Commonwealth of Australia, 2017–2018.)

- The region has the critically endangered western swamp turtle (*Pseudeudemydura umbrina*), which is among the world's most threatened freshwater turtle species.

The main threats to the Southwest Australia Biodiversity Hotspot include past land clearing for agriculture, bushfires (see Box 5.3) and disease caused by the root fungus *Phytophthora cinnamomi* (Laurance et al., 2011). Other threats are land clearance for bauxite mining and introduced alien species, e.g. foxes and cats (Critical Ecosystem

Partnership Fund, 2022). While not listed as a threat by Critical Ecosystem Partnership Fund (2022), climate change is a long-term pervasive threat to the region's biodiversity as the regional climate is projected to continue to warm and dry (see Chapter 3, Table 3.2). With time, the most suitable climatic environments to maintain its unique ecosystems and biodiversity will contract southwest towards the Southern Ocean (Turton, 2017).

The Forests of East Australia Biodiversity Hotspot (Figure 5.1) have a discontinuous and primarily coastal extent (about 250,000 km²) from far north Queensland to southern New South Wales. The hotspot contains more than 2,100 endemic vascular plant species and has lost nearly 80% of its original habitat (Williams et al., 2011). The hotspot comprises several tropical and temperate climate types, dominated by summer rainfall (see Chapter 3, Figure 3.3: *Af, Am, Cwa* and *Cwb* sub-types). Rainfall seasonality decreases north to south, and elevation tempers mean annual average temperatures.

The hotspot covers a broad range of environments, including coastal plains, coastal and mountain range escarpments, elevated tablelands and naturally occurring freshwater lagoons and lakes (Critical Ecosystem Partnership Fund, 2022). Sclerophyllous (fire-tolerant and fire-dependent) forest and woodland communities are the most prevalent vegetation type and are dominated by *Eucalyptus* species. However, biologically significant rainforest areas also exist throughout the region, including two world heritage areas (Figure 5.1: Wet Tropics of Queensland and Gondwana Rainforests of Australia). These two areas are considered highly vulnerable to climate change, which threatens regional tourism that capitalises on their world heritage branding (see Chapter 6, Box 6.7). Other threats to the hotspot include habitat fragmentation, invasive plants, animals and pathogens, altered fire regimes, urban encroachment, water pollution and hydro schemes for urban and irrigation use (Critical Ecosystem Partnership Fund, 2022). The hotspot also contains altitudinally restricted mountain ecosystems with many cool-adapted endemics that are highly vulnerable to regional warming and increases in heatwave events (Laurance et al., 2011, National Climate Change Adaptation Research Facility, 2013a).

Australia currently has 65 Wetlands of International Importance under the Ramsar Convention,[1] comprising more than 8.3 million hectares. Wetlands are indispensable for the countless benefits or 'ecosystem services' they provide humanity. Ecosystem services are goods and services from ecosystem structures and functions such as food, fibre and fuel and climate regulation (see Box 5.2). These various services for wetlands include the provision of freshwater resources (see Box 5.4 below), denitrification, carbon sequestration, flood mitigation, groundwater recharge and climate change regulation (Ramsar, 2022). Ramsar wetlands are also representative, rare or unique sites that are important for conserving biodiversity (see Box 5.1). In designating a wetland as a Ramsar site, countries agree to establish and oversee a management framework to conserve the wetland and ensure its wise use. Numerous studies show that wetland area and quality continue to decline in most regions of the world, including parts of Australia (Ramsar, 2022).

Australia's marine environment is the world's third-largest maritime jurisdiction, at 13.86 million km^2 (Commonwealth of Australia, 2017–2018). This vast area contains a diverse array of marine species, including many endemics. Most Australians live and work within 100 km of the coastline, and the ocean and coasts are an integral part of its culture and lifestyle (see Chapter 7, Social Systems). Australia's ocean species, directly and indirectly, support commercial fisheries and aquaculture worth AU$3 billion in 2016–2017 (Mobsby and Curtolli, 2018). The economic value of resources extracted from Australia's oceans is expected to double by 2029–2030. For example, by 2025, marine industries are expected to contribute around AU$100 billion annually to the country's overall economy (Commonwealth of Australia, 2017–2018). However, climate change is considered a significant threat to its commercial fishing and aquaculture industries, mainly due to ocean warming, the southward movement of the East Australia Current and the deoxygenation of crucial fishing grounds (see Chapter 6, Box 6.4).

Australia has globally significant marine assets, including the Great Australia Bight and the world heritage listed Great Barrier Reef, Lord Howe Island and Ningaloo Reef (Figure 5.1). In addition to their fisheries, these national icons support many regional tourism businesses and associated revenues significant to regional economies. However, many marine tourism industries are considered highly vulnerable to climate change risks and threats (see Chapter 6, Box 6.7). In addition to tourism and fisheries revenues, Australia's

[1] The Ramsar Convention's (1994) mission is the conservation and wise use of all wetlands through local and national actions and international cooperation, as a contribution towards achieving sustainable development throughout the world (Ramsar, 2022). It was originally signed in Ramsar, Pakistan, in 1971. The convention entered into force 21 December 1975 with amendments in 1982 and 1987.

BOX 5.2: Ecosystems and ecosystem services

Ecosystems are dynamic complexes of plant, animal and microorganism communities, interacting with the non-living environment (soils, water, minerals, air) in the form of functional units. These functional units occupy a diverse range of scales in the environment. Ecosystems provide the structural and hence functional basis for supply of ecosystem services for human well-being.

Ecosystem services may be defined as goods and services from ecosystem structures and functions such as food, fibre and fuel and climate regulation.

Ecosystem services are closely linked to the spatial dimension of a certain area or region in which these goods and services are provided. For example, they may be provided by natural forests ecosystems, natural and artificial wetlands, coastal and marine ecosystems and coral reefs but are also provided by human-modified environments, such as agricultural lands, tree plantations, urban forests and even urban areas.

Ecosystem services provided by natural and modified ecosystems are typically divided into four categories:

1. *Provisioning services*, such as providing food, water, timber and fibre;
2. *Regulating services*, such as the regulation of climate, floods, disease, wastes and water quality;
3. *Cultural services*, such as offering recreational, aesthetic and spiritual benefits; and
4. *Supporting services*, such as soil formation, photosynthesis and nutrient cycling.

The classification of ecosystem services has been largely superseded by the Inter-governmental Science Platform on Biodiversity and Ecosystem Services (IPBES) assessments that consider nature's contributions to people (NCP). The NCP approach recognises the pivotal and pervasive role that culture plays in outlining all links between people and nature and incorporates the role of Indigenous and local knowledge in understanding NCP (Díaz et al., 2015).

IPBES's conceptual framework includes six interlinked elements constituting a social-ecological system that operates at various scales in time and space. The six IPBES elements are:

1. *Nature* refers to the natural world with an emphasis on biodiversity, not other components of nature such as deep aquifers and fossil fuel reserves.
2. *Nature's benefits to people* refer to all the benefits that humanity obtains from nature. Ecosystem goods and services, considered separately or in bundles, are included in this category.
3. *Anthropogenic assets* refer to built-up infrastructure; health facilities; Indigenous, local and scientific knowledge systems, as well as formal and non-formal education; technology (both physical objects and procedures); and financial assets, among others.
4. *Institutions and governance systems* and other indirect drivers of change are the ways in which societies organise themselves, and the resulting influences on other components.
5. *Direct drivers of change*, both natural and anthropogenic, affect nature directly. Natural drivers are those that are not the result of human activities and are beyond human control, such as earthquakes and extreme weather events. The direct anthropogenic drivers are those that are the result of human decisions, namely, of institutions and governance systems, and other indirect drivers, such as degradation of habitats, climate change and pollution.
6. *Good quality of life* is the achievement of a fulfilled life, a concept that varies widely across different societies and groups within societies.

The IPBES framework demonstrates the main connected elements and relationships for the conservation and sustainable use of biodiversity and ecosystem services, human well-being and sustainable development.

Source: Adapted from Turton (2020).

oceans and coasts also provide a further AU$25 billion worth of essential ecosystem services (see Box 5.2), such as carbon dioxide absorption, nutrient cycling and coastal protection (Commonwealth of Australia, 2017–2018).

Since European colonisation in 1788, Australia's ecosystems, ecosystem services and biodiversity have experienced increasing pressure from human disturbances (Laurance et al., 2011, National Climate Change Adaptation Research Facility, 2013a). Many of these human-induced pressures have added to the adverse effects of climate change in more recent decades, e.g. habitat destruction, habitat fragmentation, drainage and diversion of wetlands and coastal development (Cresswell et al., 2021). When examining the impacts and risks of climate change for ecosystems, ecosystem services and biodiversity, it is essential to identify these other human change agents. In many cases, these antecedent and current human activities have compounded the effects of climate change on the integrity and functioning of ecosystems.

Australia's terrestrial, freshwater, coastal and marine ecosystems have particularly felt the brunt of climate change over the past 30 years, on top of other human-caused ecological and hydrological impacts that extend back over 200 years (Australian Academy of Science, 2021, Cresswell et al., 2021). Warming of the atmosphere and oceans and changes in rainfall patterns, notably drying across most southern continental areas, has been pervasive (see Chapter 3). This change in baseline climate has increased the severity, frequency and duration of extreme events (e.g. heatwaves, bushfires, droughts and floods), resulting in rapid and unprecedented adverse impacts on the country's ecosystems and biodiversity (Lawrence et al., 2022). For plant and animal species, differing physiological tolerances, resilience and adaptive capacity (see Chapter 4, Figure 4.1) will lead to *likely* changes in their geographical distribution, abundance and community structure with projected changes in baseline climate (National Climate Change Adaptation Research Facility, 2013a). For terrestrial ecosystems and their biota, changes in air temperature, water availability, evapotranspiration rates and the CO_2 fertiliser effect will be the most crucial drivers of ecological changes. By comparison, for freshwater and marine ecosystems and their biota, changes in water temperature and water chemistry (e.g. acidity and oxygen levels) will affect their ecological structure and function. There are concerns that some of these negative impacts on Australia's ecosystems and biodiversity may be potentially irreversible (Australian Academy of Science, 2021, Cresswell et al., 2021).

This chapter will examine ecosystems in Australia that are considered highly vulnerable to changing climate conditions and increasingly extreme climate-driven events. The two vulnerable groups of ecosystems are:

1. Terrestrial and freshwater ecosystems; and
2. Coastal and marine ecosystems.

Evaluation of each group of ecosystems will follow the conceptual framework for climate change impacts, vulnerability and climate risks (see Chapter 4, Figure 4.1). By group, the discussion will examine:

1. Any observed impacts of climate change, along with any significant issues; and
2. Potential climate impacts, threats and risks in the context of their vulnerability (i.e. exposure and sensitivity) to climate change and adaptive capacity.

The last section will examine climate-resilient development pathways for Australia's ecosystems, their ecosystem services and biodiversity and ways to integrate climate adaptation strategies with the relevant, sustainable development goals (see Chapter 4, Sustainable Development Goals, SDGs).

TERRESTRIAL AND FRESHWATER ECOSYSTEMS

Climate impacts and risks

Since European colonisation in 1788, vegetation clearance, mainly for agricultural and urban development (see Chapters 6 and 7), has been the primary threatening process for most of Australia's terrestrial ecosystems (National Climate Change Adaptation Research Facility, 2013a). This widespread process has caused loss of native species, soil degradation, ecosystem fragmentation and habitat degradation, particularly in the southern half of the continent and parts of Tasmania. Across northern Australia, changed fire regimes, the introduction of sheep and cattle grazing and

BOX 5.3: Bushfire risk and climate change in Australia

- Severe bushfire risk is increasing in Australia due to increasing high temperatures, declining rainfall in southern Australia and increasing evapotranspiration rates and the relative abundance of fuels (dead and dry foliage).
- Climate change has already accelerated bushfire risk across south-east Australia with 'Very High' fire danger days increasing in number, and fire seasons starting earlier and ending later.
- Warming and drying trends are projected to continue across southern and eastern Australia, together with an increase in the frequency, severity and duration of heatwaves (see Chapter 3, Table 3.2, Box 3.1).
- There are projected increases in pyro-convection risk for parts of southern Australia and increased dry-lightning and fire ignition for southeast Australia.
- The lengthening of the bushfire season is likely to reduce the cooler season, when controlled burns can normally be undertaken safely by bushfire authorities (includes co-management of forests with Indigenous land managers).
- The McArthur Forest Fire Danger Index (FFDI) is a tool used by emergency services to assess the risk of fire on a particular day, given prevailing conditions (see Chapter 3, the 'Fire weather' section). It was originally developed as a scale from 0 to 100, with 50–100 being categorised as 'Extreme'.
  - During the 2009 Black Saturday bushfires, the FFDI ranged from 120 to 190, prompting authorities to revise the FFDI scale to:
  - Severe (FFDI=50–75)
  - Extreme (FFDI=75–100)
  - Catastrophic (FFDI >100) or Code Red in the State of Victoria.
- Fire authorities maintain that even well-prepared homes may be impossible to defend under 'Catastrophic' conditions, with growing morbidity and mortality risks for people (see Chapter 7, Social Systems for examples).
- Yearly cumulative FFDI has increased by 42% across all bushfire regions with the greatest changes occurring in the more populated south-east of Australia, a probable consequence of elevated temperatures and a drying climate, both strongly influenced by climate change (see Chapter 3, the 'Historical Changes in Climate' section).
- FFDI ratings for the most extreme 10% of fire weather days have increased in the last 30 years, especially in southern and eastern areas where the amount of forest area in critically dry fuel states has been increasing each decade.
- Bushfires are a significant risk to Australia's terrestrial ecosystems and plants and animals, e.g. the devastating 2019–2020 'Black Summer' bushfires killed or displaced about 3 billion vertebrate animals. The same fires burnt 80% of the Blue Mountains World Heritage Area (WHA) and 50% of the Gondwana Rainforests WHA (see Chapter 6, Box 6.7).
- Although Australia's eucalypt forests and woodlands are fire adapted, increasing intensity and frequency of fires may exceed their resilience due to shorter intervals between high-severity fires.
- Carbon stocks and their stability are diminished by 'short-interval' bushfires in fire-tolerant eucalypt forests. The implications are:
  - Both single and short-interval bushfires can significantly reduce the amount of carbon stored in fire-tolerant eucalypt forests, and hence the potential for carbon-stock recovery could be compromised by projected warmer and drier future climates and by soil feedbacks to productivity.
  - Fire-tolerant eucalypt forests may not be as resilient carbon stocks as assumed, undermining their critical regulating ecosystem service role (see Box 5.2).

Sources: Commonwealth of Australia (2020), Hughes et al. (2020), Australian Academy of Science (2021), Fairman et al. (2022), Lawrence et al. (2022).

non-native predators (e.g. cats) are the main threats to ecosystems and biodiversity (National Climate Change Adaptation Research Facility, 2013a, 2016). Nationwide, pressures on terrestrial ecosystems include feral animals (e.g. pigs, cats and foxes), invasive weeds, diseases and pathogens and altered bushfire regimes (Box 5.3). Habitat clearance and fragmentation have also caused declines in populations of native plants and animals and the loss of genetic diversity.

The country's freshwater ecosystems have also been adversely affected across urban and rural environments by land clearing, habitat loss and changes to local hydrology (Capon et al., 2017). Adverse effects on rivers and wetlands include bank and bed erosion, sedimentation, pollution, eutrophication,[1] cattle encroachment, construction of barriers (dams, weirs, culverts etc.), stream diversions, channelisation and the introduction of noxious and non-native plant and animal pest species (National Climate Change Adaptation Research Facility, 2013a). All of these changes alter water quality, which is one of the most significant issues in freshwater ecosystem health in Australia. Adverse impacts of climate change on freshwater resources (Box 5.4) manifest across agriculture, industries, ecosystems and human communities, often creating challenges (and trade-offs) for managing multiple demands for a finite resource (Lawrence et al., 2022).

Globally, freshwater ecosystems are highly threatened by climate change and Australia is no different (Capon et al., 2017, Australian Academy of Science, 2021, Cresswell et al., 2021). Future climate projections indicate the likelihood of continued increases in air and water temperatures, reduction in river flows during droughts, declining water quality and changes to hydrodynamics (see Box 5.4). When climate pressures are combined with existing other stressors, such as habitat loss, invasive plant and animal species, water extraction and channelisation of creeks and rivers, there is a high risk of changes in composition, structure, function and connectivity of freshwater ecosystems (National Climate Change Adaptation Research Facility, 2013a). Additionally, saltwater

intrusion into freshwater aquifers will increase in coastal regions as sea level rises. For example, Kakadu National Park, a world heritage site (Figure 5.1), is being badly affected by rising sea levels leading to the saltwater intrusion of its globally significant Ramsar-listed freshwater wetlands (Figure 5.2). With time, mangrove ecosystems are likely to replace the freshwater wetlands (Lawrence et al., 2022). Due to regional topography, it will be impossible for the freshwater wetlands to migrate inland as the sea level rises (see Box 5.6 below).

Many terrestrial plant and animal species in Australia have responded to observed changes in key climate pressures (National Climate Change Adaptation Research Facility, 2016, Australian Academy of Science, 2021, Cresswell et al., 2021, Lawrence et al., 2022). The main climatic drivers have been increasing temperatures, drying over most of southern Australia and increasing extreme weather events, such as heatwaves, fire weather days, droughts and floods (see Chapter 3, the Historical Changes in Climate' section). Observed ecological changes include latitudinal and altitudinal shifts in the geographical location and range sizes of many species. Other changes include the timing of life cycle processes such as migration and breeding and altered behavioural, genetic and physiological traits (National Climate Change Adaptation Research Facility, 2016, Williams et al., 2017). Terrestrial species at most significant risk include many that are already endangered or rare, i.e. species that commonly have restricted or fragmented geographical distributions or habitats. They often have specialised ecological requirements (or niches) and narrow climatic tolerances (National Climate Change Adaptation Research Facility, 2016). Examples of documented climate-related pressures on terrestrial ecosystems, which include biodiversity hotspots and world heritage areas, include:

1. *Forest and woodlands of southern Australia, including the Southwest Australia Biodiversity Hotspot:* The main climate change drivers have been multiple bushfires in short succession resulting from increased fire risk conditions (see Box 5.3), including declining winter rainfall and increasing number of hot days. These climate pressures have led to drought-induced canopy dieback across several forest and woodland ecosystems (Lawrence et al., 2022), such

[1] Eutrophication refers to high concentrations of nutrients such as nitrates and phosphates leading to algal blooms and other downstream effects (e.g. inshore coral reefs).

BOX 5.4: Freshwater resources and climate change in Australia

Impacts on water resources:

- Streamflow has generally increased in northern Australia and decreased in southern Australia since the mid-1970s.
- Depending on region (e.g. southern Australia), increases in the length and severity of droughts are highly likely.
- In wetter regions (e.g. northern Australia), there may be an increase in extreme rainfall events in the short term.
- In regions with warming and drying, there will also be declines in runoff, soil moisture, streamflow, dam inflow and groundwater recharge, which will exacerbate and amplify natural variability.
- Adverse effects include: floods will damage water infrastructure; industries that require large quantities of water will be affected by droughts, e.g. coal, gas, hydropower and irrigation; bushfires will affect catchment water supplies (see Box 5.5).

Water availability (quantity):

- Decreased precipitation, increased vegetation water use and loss of soil moisture will reduce water availability and alter the water balance. This will adversely affect agriculture, urban water supplies and mining (see Chapter 6, Economic Systems), e.g. warming of 2.5°C and a rainfall decline of 20%, and associated decreases in winter runoff and soil moisture will intensify reductions in water availability in eastern Australia.
- Evaporation and increased temperatures will mean that the total water storage capacity of open water reservoirs in south-east and South-west Australia will decline greatly.
- Cities in southern Australia will have less water and will increasingly have to seek it elsewhere (e.g. desalination). Perth and Adelaide are at the highest risk, Melbourne could lose up to 35% of the water that currently flows into dams, and Sydney water supplies will decline.
- In the Murray-Darling Basin, models show that average catchment rainfall decreases of 13%–21% would reduce runoff by up to 45% in wetter catchments and up to 64% in drier and hotter ones, demonstrating that climate change would likely cause significant impacts to agriculture and land management (see Chapter 6, the 'Agriculture and Land sector').
- The drying in southern Australia, driven by declining cool season rainfall, demonstrates the urgent need for hydrological models that adequately account for climate change.
- Climate change will also result in declining ground water supplies in southeast and south-west regions.

Water quality:

- Water quality will be affected by future climate impacts. Examples include increased incidences of algal blooms (eutrophication), and contaminant mobilisation in drying of wetlands leads to the generation of acidic soils causing acidification.
- Extreme events such as bushfires affect water quality, generating ash that enters waterways and contaminates water sources, and when followed by heavy rainfall events causes flooding and nutrient runoff.
- Extreme flood events also increase turbidity, changes to water colour and incidence of organisms such as giardia and algal blooms.

Sources: Hughes et al. (2015), Australian Academy of Science (2021), Lawrence et al. (2022), Naughtin et al. (2022). For future climate projections, refer to Chapter 3, Table 3.2 and Box 3.1.
Likely impacts on water availability and water quality.

Figure 5.2 An aerial image of Kakadu National Park World Heritage Area in northern Australia during the transition from the wet to dry season. An expansion of mangroves into freshwater and brackish habitats has occurred over the last 50 years. This globally significant Ramsar Wetland site is considered highly vulnerable to climate change due to rising sea level, exacerbated by high tidal ranges (see Chapter 6, Box 6.7). (Credit: Rod Long/Unsplash/Public Domain.)

as in the northern Jarrah forest (*Eucalyptus marginate*) in Western Australia. In southeast Australia, there have also been local extirpations and replacement by woody shrubs due to 'seeders' having limited time to reach reproductive age in Alpine Ash forest (*Eucalyptus delegatensis*). A similar process occurs when vegetative regeneration capacity is exhausted, such as in Snow Gum woodlands (*Eucalyptus pauciflora*) (Lawrence et al., 2022). Warming and drying have created soil moisture and vegetation states ideal for bushfires to be ignited by lightning in areas with rarely experienced bushfires over the last few thousand years, such as the Pencil Pine forest (*Athrotaxis cupressoides*) in Tasmania (Lawrence et al., 2022).

2. *Australian Alps Bioregion and Tasmanian alpine zones:* The main climate drivers have been severe winter droughts, bushfires, overall warming and climate-induced biotic interactions. These climate pressures have resulted in shifts within and among three key alpine species related to food supply and vegetation habitat (Lawrence et al., 2022). These

threatened species are the mountain pygmy-possum *(Burramys parvus)*, the mountain plum pine *(Podocarpus lawrencei)* and the bogong moth *(Agrostis infusia)*. The retreat of the average snow line has led to an increase in species diversity in the alpine zone. In contrast, reduced snow cover has resulted in the loss of snow-related habitat for alpine zone endemic and obligate species (Lawrence et al., 2022).

3. *Wet Tropics of Queensland World Heritage Area:* The main climate drivers have been warming, increasing length of the dry season and increasing atmospheric vapour pressure deficits (see Chapter 6, Box 6.7). Regional warming has caused the decline of some endemic cool-adapted vertebrate species, including their distribution area and population size (Williams et al., 2017). These species declines have exceeded original modelling predictions. There is evidence that annual tree mortality risk across sites and species has, on average, doubled over the last 35 years, indicating a potential halving in life expectancy and carbon residence time (Bauman et al., 2022).

Results suggest thresholds involving atmospheric water stress, driven by background warming, are the primary cause of increasing tree mortality in these moist tropical forests. Another study in the Wet Tropics found that liana reproduction and abundance are likely to increase under predicted future climate regimes, with potentially important impacts on the survival, growth and reproduction of resident trees and thus the overall health of rainforests of the region (Vogado et al., 2022).

4. *Sub-Antarctic Macquarie Island World Heritage Area:* The main climate drivers have been reduced summer rainfall availability for 17 consecutive summers and increases in mean wind speed, sunshine hours and evapotranspiration over 40 years (see Chapter 6, Box 6.7). These changes have led to dieback of the critically endangered habitat-forming cushion plant *(Azorella macquariensis)* in the fellfield and herb field communities on the island (Lawrence et al., 2022).

Many of Australia's iconic native species are already showings signs of climate-related impacts. Over time, extreme climate-driven events such as heatwaves, droughts and bushfires will likely have catastrophic consequences for many species (Australian Academy of Science, 2021). For example, mass die-offs of birds (e.g. Carnaby's Cockatoo [*Calyptorhynchus latirostris*]), tree-dwelling mammals, flying foxes and freshwater fishes have already been observed following prolonged episodes of high-temperature events (National Climate Change Adaptation Research Facility, 2016, Australian Academy of Science, 2021, Lawrence et al., 2022). Koalas (*Phascolarctos cinereus*), a symbol of Australia, have been affected by increasing drought conditions and rising temperatures – exacerbated by habitat loss, bushfires and encroachment of human population on their critical habitats (Lawrence et al., 2022). These compounding effects have resulted in Koala population declines and increased risk of local extinctions.

Future climate changes (see Chapter 3, Table 3.2) will likely lead to transformative impacts on Australia's terrestrial and freshwater ecosystems and their biodiversity and ecosystem services (National Climate Change Adaptation Research Facility, 2013a, Australian Academy of Science, 2021, Lawrence et al., 2022). The main climate drivers for terrestrial ecosystems and biodiversity will be rising average day and night temperatures, increasing frequency and severity of heatwaves, declining rainfall in southern Australia, increasing bushfire risk and more extreme droughts and floods (Box 5.5). For freshwater ecosystems and biodiversity, an overall decline in runoff in southern Australian river basins, increasing evaporation rates, warming of freshwater bodies and changes in water chemistry (e.g. deoxygenation, acidification) will be the most significant climate drivers (Box 5.5). If future global warming exceeds 3°C above pre-industrial by 2100, there will be even more significant climate-driven impacts on terrestrial and freshwater ecosystems (see Chapter 3, Box 3.1).

COASTAL AND MARINE ECOSYSTEMS

Climate impacts and risks

Climate change has had significant adverse impacts on Australia's coastal and marine ecosystems (National Climate Change Adaptation Research Facility, 2013a, Australian Academy of Science, 2021, Lawrence et al., 2022). The main climatic drivers have been increasing sea levels resulting in coastal erosion and inundation during storm events, e.g. tropical cyclones and East Coast Lows (Paice and Chambers, 2016). Other significant climate drivers include oceanic warming, acidification and deoxygenation (see Chapter 3, the 'Historical Changes in Climate' section). The country's coasts occupy the interface between marine and terrestrial ecosystems, including its estuaries. Coastal ecosystems are generally highly productive and support diverse habitats and species (Paice and Chambers, 2016). Importantly, 85% of Australians live on or within 75 km of the coastline, and the population continues to grow in this narrow zone (see Chapter 7, the 'Cities, Settlements and Built Environments' section). As well as urban development, coastal ecosystems are under pressure from many non-climatic stresses, including habitat loss or disturbance, changes to nutrient and sediment dynamics, invasive species and recreational and commercial fishing (National Climate Change Adaptation Research Facility, 2013a).

Southwest Australia has experienced sustained warming and drying since the 1970s (see

BOX 5.5: Projected impacts of climate change on Australia's terrestrial and freshwater ecosystems and biodiversity

Terrestrial ecosystems:

- Increasing desertification in parts of southern Australia due to warming and drying (i.e. declining winter rainfall), e.g. some forest ecosystems (Alpine Ash, Snow Gum woodland, Pencil Pine, northern Jarrah) are projected to transition to a new state or collapse due to hotter and drier conditions with more fires.
- Woody shrub encroachment into alpine herb fields due to upward movement of the snowline and tropical savannahs due to rising CO_2 levels.
- Loss of alpine biodiversity in the south-east Australian Alps Bioregion is projected in the near term due to less snow on snow patch Feldmark (a critically endangered ecological community) and short alpine herb fields, as well as increased stress on snow-dependent plant and animal species.
- Climate change will have substantial impacts on tropical savannah birds, with Cape York species being particularly vulnerable.
- Loss of rainforests in mainland eastern Australia and Tasmania, driven by the increasing frequency, intensity and extent of droughts and bushfires (see Box 5.3).
- Rainforest habitats are expected to see changes to the structure of the rainforest canopy, damage to some species due to high-intensity cyclones and increased growth of vines associated with increasing CO_2, leading to tree mortality.
- Dramatic changes to species assemblage and species richness are predicted for rainforest vertebrates, with particular impacts to be felt among regionally endemic and restricted species that will need to move up-slope to track suitable habitat and respond to increased lowland biotic pressures.
- Forest dieback in areas affected by clearing and drought and made worse due to pest outbreaks on stressed tree species.
- Invasive plants and weeds response rates are expected to be faster than for native species due to climate change, and this could encourage the appearance of a new group of weed species, with many bioregions threatened by increased impacts from invasive plant species.
- Declining ecosystem health is likely to reduce carbon stocks in soil and vegetation, causing a positive (warming) feedback to the atmosphere (see Chapter 2, the 'Anthropogenic forcing' section).
- Some native species are likely to have potentially greater geographic ranges if they can colonise new areas, while other species may be resilient to projected climate change impacts.
- There are likely to be synergistic and compounding impacts particularly in bioregions already experiencing ecosystem degradation, threatened endemics, collapse of keystone species (e.g., Southern Cassowary), including those of value to Indigenous peoples, and high extinction rates because of other human change agents.
- Hundreds of terrestrial species are likely to suffer substantial losses of suitable habitat with a warming of 1.5°C, and these impacts are considerably greater at 2°C and above.

Freshwater ecosystems:

- Freshwater ecosystems and biodiversity are considered highly sensitive to climate change because the climatic changes to which they are exposed are typically also major drivers of their structure and function.

- Individual freshwater ecosystems can be expected to change considerably in their character, especially where hydrological regimes and/or water quality are significantly altered.
- Increases in the frequency, severity and duration of droughts in the Murray-Darling Basin will result in reduced river flow and increase the risk of mass fish kills.
- Continuous sea level rises will cause saltwater intrusion of freshwater wetland ecosystems and biota, resulting in loss of freshwater habitat and species, e.g. Kakadu National Park (see Figure 5.2).
- There is predicted to be degradation and drying of peatlands and wetlands and the loss of seasonal and ephemeral ponds.
- Rising water temperatures in freshwater ecosystems may exceed the thermal tolerances of sensitive fish species, leading to declines or local extinctions.
- Altered rainfall regimes, including droughts and lower inflows, coupled with demands for freshwater for agriculture and urban use (see Box 5.4) and threats posed by invasive aquatic species, may result in inadequate allocation of environmental flows for freshwater fishes, crayfish, turtles and frogs, e.g. there are 22 narrow range fish species in imminent risk of extinction.

Sources: National Climate Change Adaptation Research Facility (2016), Capon et al. (2017), Williams et al. (2017), Australian Academy of Science (2021), Cresswell et al. (2021), Lawrence et al. (2022).

Chapter 3, the 'Historical Changes in Climate' section). Decreased catchment runoff and higher evaporation rates have caused dramatic declines in freshwater flows of up to 70% over the past 50 years resulting in adverse impacts on the region's estuaries (Lawrence et al., 2022). Notably, there has been an increased frequency and severity of hypersaline conditions, enhanced water column stratification and hypoxia (low oxygen) and reduced flushing and greater retention of nutrients in estuaries (Paice and Chambers, 2016). Other climate change drivers in the coastal zone have caused significant adverse impacts on vulnerable fauna. For example, the increasing 'feminisation' of Green Sea Turtles (*Chelonia mydas*) in the northern GBR due to warming sand temperatures in turtle hatcheries has potential long-term population consequences (Lawrence et al., 2022). Storm surges and coastal inundation exacerbated by rising sea levels in the Torres Strait are responsible for the extinction of the Bramble Cay Melomys (*Melomys rubicola*), the first documented climate change-driven mammalian extinction globally (Australian Academy of Science, 2021).

Many of Australia's marine ecosystems have been affected by extractive industries, such as mining and fishing and contaminated runoff from land-based activities (e.g. sediments, fertilisers and pesticides). These human change agents have often exacerbated the effects of climate change in marine systems (National Climate Change Adaptation Research Facility, 2013a, Cresswell et al., 2021). The oceans around Australia are responding to regional climate change, causing serious concerns for a range of marine ecosystems, habitats and species (Australian Academy of Science, 2021). Rising ocean temperatures have worsened the frequency, severity, duration and areal extent of mass coral bleaching events on the Great Barrier Reef (GBR) over the past 25 years (Lawrence et al., 2022, see Figure 5.1). Worryingly, there have been four mass bleaching events on the GBR over the past seven years: 2016, 2017, 2020 and 2022 (Great Barrier Reef Marine Park Authority, 2022). Compared to earlier mass bleaching events, the 2022 event (a La Niña summer) was *likely* more severe than mass coral bleaching experienced in 1998 and 2002 (both El Niño summers). The 2016 strong El Niño summer remains the deadliest event so far, killing 30% of corals across the GBR and 51% in the northern section of the GBR (see Hughes et al., 2019). Coral bleaching is associated with mass mortality of corals and declines in coral recruitment (Australian Academy of Science, 2021, Lawrence et al., 2022). More frequent bleaching events bring a high risk of structural and compositional changes to complex and highly diverse shallow warm-water coral reef ecosystems (Figure 5.3). Global media outlets have reported these mass bleaching events, and a mission by UNESCO experts to Australia in

Figure 5.3 Two Pacific double-saddle butterflyfish, *Chaetodon ulietensis*, amongst a backdrop of corals at Flynn Reef, Tennis Courts dive site (Great Barrier Reef Marine Park World Heritage Area). Shallow warm-water coral reefs are highly vulnerable to climate change and associated marine heatwaves, with the GBR experiencing mass coral bleaching events in 2016, 2017, 2020 and 2022 (see Chapter 6, Box 6.7). (Credit: Wise Hok Wai Lum /Wikimedia/Public Domain.)

March 2022 is yet to report on their findings on the world heritage status of the GBR. World heritage branding is crucial to the regional marine tourism industry's viability in conducting year-round tours within the GBR marine park (see Chapter 6, Box 6.7).

The strengthening and southward movement of the East Australian Current has been directly linked to ocean warming due to climate change and observed impacts on marine ecosystems (National Climate Change Adaptation Research Facility, 2017, Cresswell et al., 2021). In addition to coral reefs, extensive changes in species' life history and distribution have been observed across Australia's marine jurisdictions. For example, seagrass meadows and kelp forests are being affected by background warming and marine heatwaves (Lawrence et al., 2022). Studies have shown that some of these adverse responses to changes in kelp forests have been caused by the southern movement of the warm-water sea urchin, facilitated by

the East Australian Current (Australian Academy of Science, 2021). Sea urchins are having negative impacts on other species via changed species interactions (e.g. predation, grazing, competition and pollination) with resulting losses of ecosystem function (National Climate Change Adaptation Research Facility, 2017).

Future changes in Australia's climate (see Chapter 3, Table 3.2) will likely lead to transformational impacts on its coastal and marine ecosystems and their biodiversity and ecosystem services (Paice and Chambers, 2016, National Climate Change Adaptation Research Facility, 2017, Capon et al., 2017, Lawrence et al., 2022). The main climate drivers for coastal ecosystems and biodiversity will be rising sea levels, increasing storm-surge risk driven by more energetic storms and more significant risks for riverine flood events (Box 5.6). For marine ecosystems and biodiversity, the most crucial climate drivers will be warming oceans, changes in water chemistry

BOX 5.6: Projected impacts of climate change on Australia's coastal and marine ecosystems and biodiversity

Coastal ecosystems:

- Saltwater inundation due to rising sea levels, storm-surge and more intense tropical cyclones and East Coast Lows will change the species composition of mangroves and coastal wetlands. Sandy beach ecosystems will likewise be impacted by these events, through erosion and inundation. Increased turbidity levels from wave activity due to storms, increased flooding events and sea-level rise will impact on coastal ecosystems such as mangroves, seagrass beds, rocky intertidal zones and estuaries.
- Rising sea levels will force colonisation of salt marsh and mangroves inland, although where there is coastal development this will prevent success (see Chapter 7, the 'Cities, Settlements and Built Environments' section). The ultimate outcome is likely to be habitat narrowing and eventual loss of these systems. Estuarine nursery habitat for many vertebrate and invertebrate species (including commercial species) will be reduced in size or destroyed, where sea level rises faster than the rate of marine plant growth and mangroves.
- Inundation of land adjacent to the coast and estuaries has implications for fringing and dune vegetation, and habitat for shorebirds and nesting turtles may be flooded.
- Complex changes to hydrology, from altered freshwater inputs and tidal exchange from the sea, will affect salinity regimes and a range of ecosystem processes and services in wave-dominated estuaries, e.g. lack of riverine flow may cause water quality issues to migrate upriver where long-term lack of flow could result in stagnation, deoxygenation and sediment nutrient release. Notably, Australian lagoons and rivers (see Box 5.4) are experiencing warming and decreasing pH at faster rates than predicted by climate models.
- CO_2 fertilisation of terrestrial vegetation, aquatic plants and phytoplankton. This effect will also enhance the growth rates of some terrestrial and aquatic weeds.

Marine ecosystems:

- Future ocean warming, together with periodic extreme marine heatwave events, is projected to lead to the continued loss of ecosystem services and ecological functions, as species further shift their distributions or decline in abundance, e.g. fish, cephalopods, crustaceans, nudibranchs, urchins and corals. Compounding climate-driven changes in the distribution of habitat-forming species, invasive macroalgae are likely to exhibit higher growth under all higher oceanic CO_2 and lower pH conditions.
- Corals and mangroves (see above) around northern Australia and kelp and seagrass around southern Australia are of critical importance for ecosystem structure and function, fisheries productivity (see Chapter 6, Box 6.4), coastal protection and carbon sequestration. These ecosystem services (see Box 5.2) are extremely likely to decline with global warming.
- Changes in marine species distribution and behaviour and survivorship have already been observed, e.g. increased occurrence of tropical species (fish and urchins) in southern seas. Ecosystems such as kelp forests will be impacted by changes to the distribution of critical predators such as sea urchins and the replacement by south-moving predator species.
- Oceanic warming and the southward movement of the East Australia Current will have impacts on a range of marine species, including commercial target species. For example, 55% of 335 fish species are projected to became smaller and 45% became larger as seas warm around Australia over the next 30 years.

- The majority (approximately 70%–90%) of the world's tropical coral reefs are projected to disappear at even low levels of warming of 1.5°C above pre-industrial. The outlook for the state of the GBR is considered 'very poor' with climate change seen as the major driver, primarily through more frequent, intense and widespread marine heatwaves.
- Ocean acidification is expected to affect the physiology and metabolism of marine organisms with carbonate body parts, such as corals and shellfish.
- The geographical range of toxic algal blooms can be expected to expand; e.g. the dinoflagellate, *Noctiluca scintillans*, has expanded its range from Sydney into southern Tasmanian waters, causing problems for the commercial salmon farm industry.
- Substantial losses in ocean productivity, ongoing ocean acidification and the increasing deterioration of coastal systems such as seagrasses are projected to occur if global warming exceeds 2°C above the pre-industrial period. The accompanying loss of ecosystem services and the release of GHGs from sediments are likely to drive positive feedbacks to the climate system (see Chapter 2, the 'Anthropogenic forcing' section).
- CO_2 fertilisation of seagrasses and phytoplankton. This effect will also enhance growth rates of some invasive marine plants.

Sources: National Climate Change Adaptation Research Facility (2013a), Paice and Chambers (2016), National Climate Change Adaptation Research Facility (2017), Australian Academy of Science (2021), Cresswell et al. (2021), Lawrence et al. (2022).

(e.g. acidity and deoxygenation) and increases in the intensity of tropical cyclones and East Coast Lows (Box 5.6). There will be more significant rises in sea level, more intense storm surges and pronounced oceanic heating and chemistry changes if global average temperatures approach or exceed 3°C above pre-industrial by 2100 (see Chapter 3, Box 3.1).

CLIMATE-RESILIENT DEVELOPMENT PATHWAYS FOR ECOSYSTEMS, ECOSYSTEMS SERVICES AND BIODIVERSITY

Mitigation and adaptation strategies

Australian ecosystems and their biodiversity may show some ecological and evolutionary (or autonomous) responses to climate change (see Chapter 4, Figure 4.1). However, the adaptive capacity of ecosystems, habitats and individual species to rapidly changing climate conditions will vary greatly and may not be enough to avoid irreversible loss of ecosystem functions and many species declines and some extinctions (Australian Academy of Science, 2021). Human planned adaptation will therefore be paramount to tackling the challenges of climate change and building ecological resilience

(National Climate Change Adaptation Research Facility, 2013a, Cresswell et al., 2021, Lawrence et al., 2022).

Based on local and Indigenous knowledge, skills and management principles, current best practices to alleviate climate pressures will be highly relevant to the future management of the country's ecosystems and their ecosystem services and biodiversity under climate change (Table 5.1). Twelve complementary strategies have been proposed that adopt current best practices and additional measures that will be required in the future (e.g. National Climate Change Adaptation Research Facility, 2013a, Australian Academy of Science, 2021, Cresswell et al., 2021), including:

1. Reducing non-climatic threats and stresses, e.g. invasive species, disease, over-fishing, over-allocation of water, tree clearing, unsustainable coastal development, and excess sediment loads, nutrient and pesticide runoff into waterways.
2. Maintaining appropriate connectivity of habitats and landscapes and seascapes, e.g. wildlife corridors and no-take zones on the GBR.
3. Identifying, protecting and planning for climate refugia and pathways, i.e. places in the landscape or seascape that are projected to

provide 'buffered' habitat for current species or where species may adaptively migrate.

4. Restoration, revegetation, reforestation and habitat creation (for terrestrial, freshwater and marine systems) to actively increase viable habitat in existing or new locations and sequester carbon into biomass or soils.

5. Managing and implementing biosecurity measures for invasive species and pathogens.

6. Management of fire regimes, including creating buffer zones around fire-sensitive vegetation and small-scale habitat alteration to reduce exposure to or impact of warming or drying.

7. Assisted evolution, i.e. large-scale genetic modification, captive breeding and release of organisms with enhanced stress tolerance, and translocation of vulnerable species to habitats projected to be more climatically suitable in the future.

8. Consolidation of information, i.e. the direction of effort towards impact assessment, spatial mapping and understanding of ecosystem services (see Box 5.2) and flow-on effects under different climate change scenarios, leading to direct prioritisation of management actions.

9. Stronger enforcement of laws and regulations and effective implementation of current management recommendations will improve ecosystem resilience, e.g. implementation of species recovery plans.

10. Low- or no-regrets actions can build resilience without committing to pathways that may become maladaptive (see Chapter 4, Figure 4.3), e.g. increasing protected areas, minimising vegetation clearing, considering cumulative impacts of major development projects in environmentally significant areas, restoring ecosystems and reducing grazing pressure.

11. Recognition that spatial scale is critical, i.e. the need for effective cross-jurisdictional management of ecosystems and threatened species.

12. Effective monitoring and evaluation of change and timely and appropriate policy and management responses.

Like all forms of planned adaptation in Australia (see Chapter 4, Box 4.2), there are also significant challenges and barriers to instigating effective ecosystem and biodiversity climate adaptation procedures, including (National Climate Change Adaptation Research Facility, 2013a):

- *Uncertainty* about site- or species-specific processes under climate change and hence management responses, e.g. what might successfully grow in an area in the future given projected warming and drying conditions.

- *Resourcing and scale* issues mean contemporary biodiversity management is typically reactive and regional and focussed on immediate threats and adverse impacts at the local scale. The challenge of climate adaptation demands a landscape-scale planning approach, e.g. building wildlife corridors and identifying climate refugia and pathways (see Table 5.1).

- *Some known solutions are unpalatable or risky*, e.g. translocation of threatened species to new areas and genetically engineered solutions (see Table 5.1). These solutions may create new problems or challenges and are often met with significant community resistance.

- *Actions in one ecosystem may add stress to another*, e.g. herbicide is an effective control mechanism for weeds but in runoff to the ocean some herbicides threaten seagrass beds and inshore coral reefs (i.e. maladaptive outcomes).

- *Limits to adaptation* is a major issue as Australia's ecosystems are already changing in response to changes in climate (see above). While passive and active ecosystem management can slow or facilitate changes (see Table 5.1), the reality is that unchecked climate change will fundamentally alter the country's ecosystems and their ecosystem services and biodiversity.

Over 30 years of research and the production of comprehensive climate adaptation plans for threatened Australian ecosystems have yielded few active adaptation strategies to reduce species loss and manage ecosystem transformation in the face of a rapidly changing climate (Hoeppner and Hughes, 2019). Plans activated so far are small-scale and relatively passive, and there are very few active ecosystem interventions, such as translocation or habitat creation, being implemented nationwide (see Table 5.1). This lack of action is very concerning, as ecosystem transformation is now inevitable across all the country's terrestrial and marine jurisdictions (see Boxes 5.5 and 5.6). This situation has led to novel 'interventionist' approaches to conservation and natural resource management

Table 5.1 Significant climate risks for highly vulnerable terrestrial, freshwater, coastal and marine ecosystems and their ecosystem services and biodiversity in Australia

Key risk	Examples of adaptation strategies
Loss of alpine biodiversity in Australia due to less snow	• Reducing pressure on alpine biodiversity from land uses that degrade vegetation and ecological condition, e.g. integrated bushfire, weed and pest animal management (M). (*Accommodate*) • Gene and seed banking, ex situ conservation. (*Retreat*)
Transition or collapse of Alpine Ash, Snow Gum woodland, Pencil Pine and northern Jarrah forests and their ecosystem services (e.g. carbon storage) in southern Australia due to hotter and drier conditions and more bushfires	• Increased capacity to extinguish bushfires during extreme fire weather events (M). (*Accommodate*) • Avoiding and reducing forest degradation from inappropriate forest management and land use (M). (*Accommodate*) • Identify, conserve and restore refugia, at locations that are protected from multiple human impacts (M). (*Accommodate*) • Promote functional connectivity at all spatial scales to aid threatened species to access resources and refugia. (*Accommodate*) • Conserve fire refugia, pathways and buffer zones within species' current distributions (M). (*Accommodate*) • Conserve or improve functional connectivity with fire refugia (M). (*Accommodate*) • Assisted colonisation to new or historical (cleared) locations (M). (*Retreat*) • Conserve fire-free refugia outside species' current distributions (M). (*Retreat*) • Active management to deliberately exclude fire from some forest types (M). (*Protect*)
Loss of cool-adapted upland endemic vertebrates in the Wet Tropics of northeast Queensland	• Minimise other stressors on upland rainforests, e.g. bushfires, feral animals, weeds, road access (M). (*Accommodate*) • Maintain functional connectivity within the entire Wet Tropics landscape, e.g. wildlife corridors (M). (*Accommodate*) • Identify and prioritise suitable refugia and pathways for rare and threatened endemics in the greater Wet Tropics region (beyond the protected areas), e.g. abandoned dairy farms in higher cooler areas (M). (*Accommodate*) • Design and manage the entire Wet Tropics landscape to increase ecosystem resilience (M). (*Accommodate*) • Assisted colonisation to new or historic (cleared) locations (M). (*Retreat*) • Assisted evolution, e.g. translocation of cool-adapted endemics further south. (*Retreat*) • Seed and gene banking, ex situ conservation. (*Retreat*)
Increase in heat-related mortality and morbidity for wildlife and known ecosystem services (e.g., pollination, seed dispersal) in Australia	• Minimise other stressors on heat-sensitive wildlife, e.g. habitat protection, fire and invasive species management (M). (*Accommodate*) • Enhance functional connectivity within species' current distributions (M). (*Accommodate*) • Conserve heatwave refugia and pathways within species' current distributions (M). (*Accommodate*) • Water misting devices installed and applied at known flying fox camps during heatwaves. (*Accommodate*) • Conserve heatwave refugia outside species' current distributions (M). (*Retreat*) • Ex situ conservation of heat-sensitive wildlife. (*Retreat*)

(Continued)

Table 5.1 (*Continued*) Significant climate risks for highly vulnerable terrestrial, freshwater, coastal and marine ecosystems and their ecosystem services and biodiversity in Australia

Key risk	Examples of adaptation strategies
Loss and degradation of freshwater, estuarine and mangrove ecosystems and their ecosystem services in low-lying coastal areas from ongoing sea-level rise	• Preserve or restore riparian vegetation cover along waterways (**M**). (*Accommodate*) • Preserve and enhance groundwater flows by minimising fine sediment input (**M**). (*Accommodate*) • Conserve freshwater refugia and pathways within species' current distributions (**M**). (*Accommodate*) • Conserve, restore or enhance vegetation to storm surges (**M**). (*Accommodate*) • Conserve or improve functional connectivity with freshwater refugia (**M**). (*Accommodate*) • Conserve landward sea-level rise refugia outside ecosystems' current distribution (**M**). (*Accommodate*) • Conserve or improve functional connectivity with sea-level rise refugia (**M**). (*Retreat*) • *Ex situ* conservation. (*Retreat*) • Sea walls, dykes and storm-surge barriers; drainage channels and tidal gates; beach renourishment and dune building. (*Protect*)
Loss and degradation of tropical shallow coral reefs and ecosystem services in Australia due to ocean warming and marine heat waves	• Minimise other stressors on coral reefs, e.g. Crown of Thorns Starfish infestations on affected reefs; reduce nutrients, pesticides and sediments from runoff from adjacent catchments (**M**). (*Accommodate*) • Identify, enhance and conserve functional connectivity of thermal refugia less exposed to marine heatwaves, e.g. outer reefs exposed to upwelling of cooler water. (*Accommodate*) • Assisted gene flow with populations located in the 'hot periphery' of their current distribution. (*Retreat*) • Trial introduction of heat-resistant corals and assisted translocation of corals to higher latitudes (**M**). (*Retreat*) • Conserve thermal refugia outside coral and other reef species' current distributions (**M**). (*Retreat*) • Coral banking and *ex situ* conservation (e.g., aquaria). (*Retreat*) • Sunshades over reefs at tourist sites. (*Protect*) • In-water fans to upwell deeper cooler seawater over reefs at tourist sites. (*Protect*)
Loss of kelp forests in southern Australia due to ocean warming, marine heatwaves and overgrazing by climate-driven geographical range extensions of herbivorous fish and sea urchins	• Maintain ecosystem functional connectivity. (*Accommodate*) • Minimise other stressors on kelp communities (**M**). (*Accommodate*) • Interventions to rehabilitate or restore degraded habitats (**M**). (*Accommodate*) • Remove locally invasive species. (*Retreat*) • Artificial habitats for displaced species (**M**). (*Retreat*) • Transplant heat-tolerant phenotypes. (**M**). (*Retreat*)

Sources: National Climate Change Adaptation Research Facility (2013a), Reside et al. (2014), National Climate Change Adaptation Research Facility (2016), Williams et al. (2017), National Climate Change Adaptation Research Facility (2017), Cresswell et al. (2021), Lawrence et al. (2022), Naughtin et al. (2022). Potential adaptation strategies are provided. Strategies that potentially mitigate GHG emissions are marked (**M**). Refer to Chapter 4 (Box 4.1) for explanations of active incremental and transformative adaptation strategies shown in italics, and Box 5.2 for ecosystem services.

Table 5.2 Integration SDGs with Australia's ecosystems, ecosystem services and biodiversity responses to climate change

SDG	Goals	Targets	Indicators
SDG6: Clean water and sanitation (*economic & environmental*)	1. Ensure availability and sustainable management of water	Protect and restore water-related ecosystems, including mountains, forests, wetlands, rivers, aquifers and lakes	• Proportion of trans-jurisdictional basin area with an operational arrangement for water cooperation • Change in extent of water-related ecosystems over time
SDG12: Responsible consumption and production (*economic and environmental*)	2. Increased natural capital	Greater stores of natural values for future conservation needs	• Biodiversity of agricultural, urban and other types of land uses
SDG13: Climate action (*environmental*)	3. Resilience and adaptive responses to natural hazards	Mitigation of climate change impacts on ecosystems, ecosystem services and biodiversity	• Reduction in native vegetation clearing GHG emissions contribution (to zero) • Climate adaptation plans for ecosystems and biodiversity, including trans-jurisdictional plans
SDG14: Life below water (*economic and environmental*)	4. Enhance the conservation and sustainable use of oceans and their resources	Sustainably manage, protect marine and restore coastal ecosystems to avoid significant adverse impacts and build resilience Minimise and address the impacts of ocean acidification Conserve at least 10% of coastal and marine areas, consistent with national and international law	• Number of jurisdictions using ecosystem-based approaches to managing marine areas • Average marine acidity (pH) measured at agreed suite of representative sampling stations • Coverage of protected areas in relation to marine areas
SDG15: Life on land (*economic and environmental*)	5. Protect, restore and promote sustainable use of terrestrial ecosystems, sustainably manage forests, and halt and reverse land degradation and biodiversity loss	Conservation, restoration and sustainable use of terrestrial and inland freshwater ecosystems and their services Ensure the conservation of mountain ecosystems, including their ecosystem services and biodiversity, to enhance their capacity to provide benefits that are essential for sustainable development Reduce loss of biodiversity due to climate change	• Forest/woodland area as a proportion of total land area • Proportion of important sites for terrestrial and freshwater biodiversity that are covered by protected areas, by ecosystem type • Coverage by protected areas of important sites for mountain biodiversity • Mountain Green Cover Index • Red List Index

Refer to Chapter 4 for an explanation of SDG goals, targets and indicators (see also Box 4.5). Triple-bottom line implications for each SDG are shown in italics.

(Australian Academy of Science, 2021). A suite of potential strategies for minimising climate threats to ecosystems can also provide effective co-benefits for human communities and well-being and climate mitigation actions, such as carbon sequestration. This ecosystem-based approach to adaptation (Scarano, 2017) aims to deliver multi-benefits for humans and nature while reducing vulnerability and building resilience to climate change (Australian Academy of Science, 2021, Cresswell et al., 2021). For example, protecting and restoring mangrove ecosystems provide coastal protection for human infrastructure, carbon sequestration and habitat provision for fisheries (see Table 5.1 for other examples).

Integration with the sustainable development goals

Five SDGs are relevant to a national strategy on climate change and Australia's ecosystems, ecosystem services and biodiversity:

1. SDG6 (Clean water and sanitation);
2. SDG12 (Responsible consumption and production);
3. SDG13 (Climate action);
4. SDG14 (Life below water); and
5. SDG15 (Life on land).

A framework (Table 5.2, opposite page) shows how to integrate ecosystems, ecosystem services, biodiversity and climate change with goals, targets and indicators for each of the five relevant SDGs. Climate-resilient development pathways for Australia's vulnerable ecosystems and biodiversity will require incremental and transformational changes in their management. Such pathways will be crucial for the country to meet (or make progress towards) the 2030 SDG targets for maintaining the health and resilience of ecosystems and ecosystem services. In that case, similar rapid changes will need to co-occur in all the economic and social systems that contribute GHGs and adversely impact nature in other ways (see Chapters 6 and 7).

SUMMARY

- Australia is a global megadiversity hotspot recognised for its distinctive ecosystems and biota. Biodiversity provides the country with economic, ecological, recreational, cultural and scientific values. Unfortunately, Australia has one of the world's worst records of species extinctions. Loss of biodiversity and essential ecosystem services due to climate change and other change agents (e.g. habitat loss and fragmentation, invasive species) are already eroding the country's critical biodiversity values, adversely affecting human well-being.

- Australia's two global biodiversity hotspots are threatened by climate change and other threats, such a habitat loss and invasive species. The Southwest Australia Biodiversity Hotspot is threatened by regional warming and drying. The Forests of East Australia Biodiversity Hotspot are exposed to regional warming, increasing heatwaves, more extreme droughts and bushfire risk. Some forest ecosystems are projected to transition to a new state or collapse due to hotter and drier conditions with more fires, e.g. Alpine Ash, Snow Gum woodland, Pencil Pine and northern Jarrah.

- Australia's globally significant wetlands (Ramsar Sites) are exposed to regional warming and drying in southern and central Australia and saltwater intrusion due to rising sea levels, e.g. Kakadu National Park. Freshwater ecosystems and biodiversity are considered highly sensitive to climate. Individual freshwater ecosystems can be expected to change considerably in their character, especially where hydrological regimes and water quality are significantly altered.

- Coastal ecosystems are threatened by saltwater intrusion due to rising sea levels, storm surges and more intense tropical cyclones and East Coast Lows. Rising sea levels force coastal ecosystems (e.g. salt marshes, mangroves) inland. However, where there is coastal development, this results in the coastal squeeze phenomenon.

- Warming oceans, deoxygenation and acidification threaten Australia's sizeable marine environment. The strengthening and southward extension of the East Australia Current have had adverse impacts on a range of marine species, including commercial target species. Ecosystem services provided by corals, mangroves and seagrasses around northern Australia and kelp and seagrasses around southern Australia (e.g. ecosystem structure and function, fisheries productivity, coastal protection and carbon storage) are highly likely to decline with global warming. The outlook for the coral reef ecosystems of the GBR is considered 'very poor', and there is a high risk for ecosystem transition with little as 1.5°C warming above pre-industrial levels.

- Australian ecosystems and biodiversity may show some ecological and evolutionary (autonomous) adaptation to climate change. However, the adaptive capacity of ecosystems, habitats and species to rapidly changing climate conditions will vary greatly and may not be enough to avoid irreversible loss of ecosystem functions and many species declines and likely extinctions. Both local and Indigenous knowledge and best management principles and practices will be highly relevant to the future management of the country's ecosystems, ecosystem services and biodiversity under climate change. For example, low- or no-regrets actions can build resilience without committing to pathways that may become maladaptive. However, the production of comprehensive climate adaptation plans for threatened ecosystems has yielded few active adaptation strategies to reduce species loss and manage ecosystem transformation in the face of a rapidly changing climate. Notably, an ecosystem-based approach to adaptation should deliver multi-benefits for humans and nature while reducing vulnerability and building resilience to climate change.

- Addressing the SDGs for Australia's ecosystems, ecosystem services and biodiversity inform incremental and transformative strategies to avoid mal-mitigation and maladaptation pathways in the future.

- Australia's ecosystems, ecosystem services and biodiversity are tightly linked to its economy and society. Observed and potential impacts of climate change on the country's economic and social systems are evaluated in the two following chapters.

6

Economic Systems

Some of the most significant risks to Australia's economic growth trajectory are the physical risks associated with a changing climate and the unplanned economic transition risk from the world's response to this changing climate.

Deloitte Access Economics (November 2020).

INTRODUCTION

Australia has a highly developed mixed economy. Such an economy uses market signals and government directives to allocate goods and resources (Schiller 2010). In 2021, the country ranked 12th in the world for gross domestic product[1] (GDP) and 18th for GDP, based on purchasing power parity[2] (PPP). Australia is home to just 0.3% of the global population but accounted for 1.7% of the global economy in 2020 (Austrade, 2021). In 2019–2020, its GDP was about AU$1.98 trillion, and its population was 25.7 million (Australian Bureau of Statistics, 2021a). Like many other highly developed countries, the Australian economy is dominated by its diversified services sector, which generated nearly 63% of its GDP and employed about 79% of its total labour force of about 13 million workers in August 2021 (Australian Bureau of Statistics, 2021a). The main services activities were

health care, finance and insurance, tourism, information media and telecommunications, education, retail, energy and transportation. COVID-19 adversely affected the economy in 2020–2021, but during 2022 there were signs of a strong recovery in most sectors.

Steffen et al. (2019) argue that climate change is a significant threat to Australia's financial stability and poses substantial systemic economic risks. Their analysis found direct macroeconomic shocks from climate change, including reduced agricultural yields, considerable damage to property and critical infrastructure, are likely to lead to painful market corrections and could trigger severe financial instability in Australia and the region. Their report notes that financial regulators also acknowledge that climate change is now a central concern for the country's economic and financial stability. This instability applies to its agricultural and property sectors and is likely to be very significant this century. However, the severity of adverse impacts on the economy (as a whole) will depend highly on global GHG emissions rates, particularly over the next three decades (see Chapter 2, Table 2.6). Even if global GHG emissions decline rapidly from now on, Australia will face many decades of climate change and associated extreme weather events (see Chapter 3, Table 3.2).

---

[1] Gross domestic product (GDP) is the standard measure of the value added created through the production of goods and services in a country during a certain period (International Monetary Fund, 2021).

[2] Purchasing power parity (PPP) allows for economists to compare economic productivity and standards of living between countries (International Monetary Fund, 2021).

The business and industry sectors are vulnerable to the significant risk generated by the compounding effects of climate change (National Climate Change Adaptation Facility, 2013b, see also Chapter 3, Table 3.2, Box 3.1). Over the past decade, Australian businesses have grown interested in engaging in climate mitigation and adaptation, recognising both the risks and opportunities. In 2021, the Business Council of Australia (BCA) published a far-reaching 'roadmap' advocating for Australian Government policies that 'reduce carbon emissions and deliver the more carbon-efficient economy Australians and our members want' (Business Council of Australia, 2021). This statement sends a strong national-level message that the BCA is willing to participate in some form of Carbon Pollution Reductions Scheme or Clean Energy Target. The BCA road map is strongly informed by the key recommendations of the King Report – prepared for the Australian Government by an expert panel – that examined additional sources for low-cost carbon abatement in the country (Department of Industry, Science, Energy and Resources, 2020).

The BCA road map asserts that 'Australia is at a crossroads: we can either embrace decarbonisation and seize a competitive advantage in developing new technologies and export industries; or be left behind and pay the price'. The BCA recommends that businesses understand the impact of 1.5°C and 3.0°C warming scenarios on their operations and develop appropriate mitigation and strategic growth plans in a warming and decarbonising world (Business Council of Australia, 2021). To achieve this ambitious goal, the BCA makes the following nine recommendations to the Australian Government:

1. *Formally commit to the net-zero emissions target by 2050*: The Australian Government has since committed to this aspirational 2050 target, bringing it in line with all its states, territories and other highly developed countries.
2. *Introduce 10-year carbon budgets with 5-year reviews*: The Climate Change Authority would advise Parliament on calibrating carbon budgets based on technology readiness and economic impacts.
3. *Lift the ambition for the 2030 interim emissions target*: Increase to the 46%–50% economy-wide range against 2005 levels (currently the national target is 26%–28%[3]).
4. *Change the safeguard mechanism to deliver a strong carbon investment signal*: Reduce the eligibility threshold for entities covered by the mechanism from 100,000 t CO_2-e per year down to 25,000 t CO_2-e per year (see Chapter 4, 'Roles and Responsibilities for Climate Change Action in Australia' section).
5. *Deepen Australia's (carbon) offsets[4] market*: Expand and deepen the domestic carbon offsets market to efficiently balance abatement activity between the easy to abate and hard to abate sectors of our economy.
6. *Invest in technology and innovation*: The Australian Government's Technology Investment Roadmap (Commonwealth of Australia, 2021c) claims to be an essential part of the pathway to net zero, as it identifies priority technologies to support the country's transition. The roadmap is the basis of the Australian Government's Long-Term Emissions Reduction Plan (Commonwealth of Australia, 2021a, see also Chapter 4, 'Roles and Responsibilities for Climate Change Action in Australia' section).
7. *Low-carbon regional road map*: Create a National Regional Transition Taskforce and establish partnerships between government and business in regional areas to proactively manage the transition, forming part of a broader regional growth strategy.
8. *Adaptation planning*: Development of the new National Climate Resilience and Adaptation Strategy as an ongoing national process, capable of keeping pace with the escalating nature of the physical risks associated with climate change. The Australian Government released the national strategy in October 2021 (Commonwealth of Australia, 2021b, see Chapter 4, 'Roles and Responsibilities for Climate Change Action in Australia' section).

---

[3] The new Australian Labor Government (elected 21 May 2022) committed to an emissions target of 43% against the 2005 level. This climate policy was formally passed into law by both houses of parliament in September 2022.

[4] A carbon offset is a reduction of emissions of greenhouse gases or an increase in carbon storage (sequestration) to compensate for emissions made elsewhere.

All states and territories also have climate change adaptation strategies acknowledging climate risks with plans to address them.

9. *Policy integration and coordination led by the Climate Change Authority*: The Climate Change Authority must be further empowered and resourced to become Australia's trusted, independent climate advisory body. It must be responsible for advising all governments on all aspects of the national overarching climate policy framework to pursue the net-zero emissions policy goal.

A study has estimated that by 2070, Australia's economy will lose AU$3 trillion from unchecked climate change (Deloitte Access Economics, 2020). However, it also argues there could be an extra AU$680 billion and 250,000 jobs added to its economy by adopting a new growth (or climate-resilient) development pathway. The study identified economic sectors that will require significant structural adjustment policies[5] to respond to climate change (see Chapter 4, 'Mitigation and adaptation pathways' section). These sectors fall into two categories: (1) those with lower demand but significant change needed, i.e., agriculture, resources (mining), tourism, energy and utilities; and (2) sectors with higher demand but significant change needed, i.e., manufacturing, construction and transport. It also identified economic sectors that will only require limited structural adjustments to deal with climate change, i.e. health care, education and other services.

The Australian Climate Roundtable (2021) of leading business, agricultural, union, social welfare and environmental sectors raised similar concerns about the impacts of climate change on the economy in their statement on Australians working together for a successful transition to net-zero emissions. They asserted adverse climate impacts are highly likely to cause economic damage, risk financial stability,

cause acute and long-lived human and community social and health implications and irreversible threats to agriculture, properties and tourism.

This chapter will examine economic sectors in Australia that are considered highly vulnerable to changing climate conditions and increasingly extreme climate-driven events. The five vulnerable groups of sectors are:

1. Energy and transportation;
2. Agriculture and land (including horticulture, forestry and fisheries);
3. Mining (resources);
4. Manufacturing and construction; and
5. Tourism.

Evaluation of each group of sectors will follow the conceptual framework for climate change impacts, vulnerability, climate risks and mitigation and adaptation options (see Chapter 4, Figure 4.1). By sectors, the discussion will examine:

1. The socio-economic context;
2. Approaches to managing risk;
3. Any observed impacts of climate change, along with any significant issues;
4. Potential climate impacts, threats and risks in the context of their vulnerability (i.e. exposure and sensitivity) to climate change and adaptive capacity; and
5. Climate-resilient development pathways and opportunities and ways to integrate these with the relevant, Sustainable Development Goals (see Chapter 4, 'Sustainable Development Goals', SDGs section). Addressing the SDGs for each economic sector will inform incremental, structural and transformative strategies to avoid mal-mitigation and maladaptation pathways in the future (see Chapter 4, Figures 4.2 and 4.3, Box 4.1).

ENERGY AND TRANSPORTATION

Socio-economic context

The ABS classifies the energy sector into (1) electricity supply industries and (2) gas supply industries. In August 2021, the electricity and gas supply industries employed about 63,900 and 18,200 persons, respectively (Australian Bureau of Statistics, 2021a). Combined, this represented about 0.62%

---

[5] Structural adjustment policies (SAPs) refer to government programs that aim to mitigate the effects of industry contractions. They are often applied to communities that are economically dependent on certain industries. SAPs can play an important role in ensuring a smooth transition process for the affected workers and business owners if they are well-designed (Adept Economics, 2022).

of the total Australian workforce. In 2019–2020, energy productivity improvement increased by 21% from 2010 to 2011 (Department of Industry, Science, Energy and Resources, 2021a). This measure is equal to the energy output divided by energy consumption. There may still be opportunities to improve energy productivity, but its rate will eventually level off. The energy sector has four components: (1) energy consumption, (2) energy production, (3) electricity generation and (4) energy trade (Department of Industry, Science, Energy and Resources, 2021a).

The country's energy consumption fell by about 3% in 2019–2020 to 6,014 petajoules due to the COVID-19 pandemic. This value is compared with an average growth of 0.7% a year over the previous decade to 2018–2019 (Department of Industry, Science, Energy and Resources, 2021a). Due to COVID-19, energy consumption fell in the transportation and commercial sectors. However, the transportation sector was the most significant energy user in 2019–2020, representing 27% of all energy use. Mining energy use (see Mining sector below) increased, supported by liquefied natural gas (LNG) and iron ore export growth – along with residential energy use due to more people working and staying at home (Department of Industry, Science, Energy and Resources, 2021a). Oil consumption fell by 7% in 2019–2020 due to lower transport use and lower crude consumption by refineries. However, oil remained Australia's largest source of primary energy consumption, at 37% of the total. Diesel fuel was the largest source of final energy consumption. The share of natural gas increased to 27% of the primary energy mix. The country now generates AU$324 million in GDP for every petajoule of energy consumed, about AU$56 million more than a decade ago (Department of Industry, Science, Energy and Resources, 2021a).

Energy production rose by 2% in 2019–2020 to 20,055 petajoules – mainly due to increased natural gas and oil production (Department of Industry, Science, Energy and Resources, 2021a). During this time, natural gas production grew by about 8%, primarily driven by increased shale gas production in the northwest of Western Australia (see map Chapter 3, Figure 3.1). Coal seam gas accounted for about 25% of gas production nationally and over 66% of gas production in eastern Australia (Department of Industry, Science, Energy and Resources, 2021a). Crude oil and condensate production increased by 18%. In comparison, naturally occurring liquid petroleum gas (LPG) production grew by 48%, primarily associated with the increased natural gas production in the northwest of the country. Black and brown coal production fell by 2% and 4% in 2019–2020, respectively (Department of Industry, Science, Energy and Resources, 2021a). The latter's decline reflected the long-term shift away from brown coal-fired electricity generation.

Total electricity generation in Australia was steady in 2019–2020 at 265 terawatt-hours (~955 petajoules). This total value included industrial, rooftop solar photovoltaic (PV) and off-grid generation (Department of Industry, Science, Energy and Resources, 2021a). Specifically, about 16% of electricity was generated outside the electricity sector by industry and households. Black and brown coal-fired electricity generation fell by 7% and 2%, respectively (Department of Industry, Science, Energy and Resources, 2021a). Coal accounted for 55% of total electricity generation and fell further to 54% in the calendar year 2020. Natural gas-fired generation rose by 5%–21% of the entire electricity generation. Renewable energy generation in 2019–2020 increased by 15%, contributing 24% to total generation (Department of Industry, Science, Energy and Resources, 2021a). That year saw a 42% increase in solar generation and a 15% increase in wind generation (Figure 6.1). Combined solar and wind contributed 8% of total generation in that year. Solar PV, especially large-scale solar PV, was the fastest growing type of electricity generation in 2019–2020 (Department of Industry, Science, Energy and Resources, 2021a).

The Australian Government's Clean Energy Regulator administers the Renewable Energy Target's (RET) two main programmes (Clean Energy Regulator, 2022): (1) the large-scale RET, which promotes investment in renewable power stations, and (2) the small-scale Renewable Energy Scheme, which supports small-scale installations, such as household solar panels and solar hot water systems. In June 2015, the Australian Parliament passed the Renewable Energy (Electricity) Amendment Bill 2015. As part of the amendment bill, the large-scale RET was reduced from 41,000 to 33,000 GWh in 2020, with interim and post-2020 targets adjusted accordingly (Clean Energy Regulator, 2018).

Figure 6.1 Wind farm near Albany, Western Australia. This image shows how a renewable energy project can co-exist with natural area conservation. (Credit: Harry Cunningham/Unsplash/Public Domain.)

The National Electricity Market (NEM) is five physically connected regions on the east coast of Australia (Department of Industry, Science, Energy and Resources, 2021b): Queensland, New South Wales (which includes the ACT), Victoria, Tasmania and South Australia (see map Chapter 3, Figure 3.1). The NEM is a wholesale market through which generators and retailers trade electricity. It interconnects the six eastern and southern states and territories and delivers around 80% of all electricity consumption in the country. Western Australia and the Northern Territory have separate electricity grids and regulatory arrangements, although the Australian Energy Market Commission also has a role in the Northern Territory grid (Department of Industry, Science, Energy and Resources, 2021b).

Australia has a long history of energy trade. In 2019–2020, net exports (exports minus imports) were equal to 70% of production (Department of Industry, Science, Energy and Resources, 2021a). Energy exports grew by 2% to 16,290 petajoules. LNG exports increased by 6% to 4,393 petajoules as new capacity came online (Department of Industry, Science, Energy and Resources, 2021a). In response to new capacity, exports of crude oil and condensate grew by 15% and LPG by 48% (Department of

Industry, Science, Energy and Resources, 2013a). Due to the COVID-19 pandemic, energy imports fell by 7% to 2,244 petajoules. This fall was primarily a response to lower transport demand and lower refinery output (Department of Industry, Science, Energy and Resources, 2021a).

Australia has a relatively small population (25.7 million in 2021) for such a large country, with coastal or near-coastal concentrations around the metropolitan cities of Brisbane, Sydney, Melbourne, Adelaide and Perth (see map Chapter 3, Figure 3.1). The highly centralised nature of most of the country's population poses challenges to its transportation sector. Many larger regional cities (e.g. Hobart, Darwin and Townsville) are distant from these major centres (see Chapter 7, 'Cities, Settlements and Built Environments' section). The country is yet to build any high-speed rail links. While there are many forms of transport, the country depends heavily on road transport to move most of its raw and manufactured goods and for business and private travel. Australia also relies on air transport for business, leisure and movement of high-value freight goods. There are over 300 airports with paved runways across the country (Australian Industry and Skills Committee, 2021). The rail network also provides urban passenger transport and intra- and interstate passenger and freight services. Maritime shipping services are central for domestic and international freight movement and export of raw materials from the mining industries to regional ports for onshore processing and overseas markets. The wider transportation sector (classified as transport, postal and warehousing) employed about 630,000 persons in August 2021, or about 4.8% of the total workforce (Australian Bureau of Statistics, 2021a). The transportation and logistics sector employed over 500,000 persons and had estimated annual revenue of AU$101.51 billion in 2019–2020, which added AU$39.91 billion to the economy that year (Australian Industry and Skills Committee, 2021).

The aviation sector – which underpins business and tourism (see Tourism sector below) – had an estimated annual revenue of AU$26.94 billion and added AU$10.73 billion to the economy in 2019–2020 (Australian Industry and Skills Committee, 2021). These revenues were significantly affected by the COVID-19 pandemic. Hence, pre-pandemic data is more representative of the past decade. For example, during 2017, the sector carried over 60 million domestic passengers, and there were over 1 million tonnes of scheduled international air freight traffic. That same year, the industry employed more than 65,000 people across various sub-sectors (Australian Industry and Skills Committee, 2021).

The rail sector is vital to business, carrying people and commodities on over 33,000 km of track across the country (Australian Industry and Skills Committee, 2021). The industry had an estimated annual revenue of AU$22.79 billion in 2019–2020. It contributed AU$8.77 billion to the economy and employed over 50,000 people across 961 companies (Australian Industry and Skills Committee, 2021). These comprised private and public operators, passenger and freight operators, track owners and managers, manufacturers (see below) and suppliers operating in urban, regional and rural areas of the country.

The maritime sector operates in a globalised shipping network and across local markets, including fishing and aquaculture, tourism, and patrol and rescue operations (Australian Industry and Skills Committee, 2021). The sector has an estimated annual revenue of AU$5.76 billion and contributed about AU$2.03 billion to the economy in 2019–2020. About 80% of Australia's imports and exports by value were carried by sea that year (Australian Industry and Skills Committee, 2021). Per capita, the country has more cruise passengers than any other country, making it the fourth-largest cruise market in the world. COVID-19 severely impacted the cruise market, but there are expectations that international cruises will begin to return in late 2022.

Managing risk

Fossil fuel-generated electricity (from coal and natural gas) is Australia's largest emitter of CO_2-e (Department of Industry, Science, Energy and Resources, 2021c). Given national and global efforts to rapidly transition to net-zero emissions by 2050, this presents a significant business risk to the energy sector from now on. New generation sources for electricity and consumer preferences are also driving transformations in the country's energy sector (Infrastructure Australia, 2019). However, this structural change is most pronounced in the electricity market (Steffen et al. 2017). Investment in energy networks multiplied from 2010 to 2019, with the value of assets in the NEM growing by around 75% (Department of Industry, Science, Energy and Resources, 2021b). The electricity generation mix is also rapidly changing (Infrastructure Australia, 2019). By 2025, over two-thirds of coal plants in the NEM will be 50 years or older – making them technically obsolete, unreliable, inefficient and costly to maintain (Stock et al., 2018a). The ageing condition of the country's fleet of coal- and gas-fired electricity plants and antiquated electricity grid are significant business risks for the energy sector now and in the future. At the same time, there have been dramatic falls in costs for solar, wind and battery storage infrastructure (Stock et al., 2018a). This fact alone makes a strong case for the electricity sector to move more quickly into renewable generation and storage.

Over the past 5 years, the dominant issue in the energy sector has been the growth in wholesale and retail energy prices (Wood et al., 2021). This growth is due to the closure of obsolete coal-fired generation assets, issues with network reliability due to ageing assets (see above) and rising costs for generation inputs such as coal and gas (Infrastructure Australia, 2019). The main impact of this steep climb in users' electricity bills has been a negative shift in user perceptions of electricity affordability. The NEM, and the institutions which support its operation, have continued in the absence of decisive Australian Government leadership and the lack of certainty on energy or emissions policy (Infrastructure Australia, 2019). The recent Australian Government's Long-Term Emissions Reduction Plan (Commonwealth of Australia, 2021a) provides some leadership and guidance on this critical issue. However, this major energy policy reform's economic and climate mitigation impacts on the NEM are yet to be thoroughly evaluated (see the 'Mitigation Pathways' section below).

Operational risks are also occurring in Australia's diverse transport sector. The sector is the country's second-largest emitter of CO_2-e (behind electricity), and its emissions are growing each year (Department of Industry, Science, Energy and Resources, 2021c). Emissions growth presents a considerable business risk to the sector, as investors are increasingly likely to divest from carbon-intensive transportation projects as time goes on (see the 'Mitigation Pathways' section below). A recent audit across the sector (Infrastructure Australia, 2019) identified some other challenges, including:

- Access to passenger transport networks and their quality is unequal across the country, e.g. there is a heavy reliance on cars to get around, and in many cases, there is a lack of suitable public transport alternatives (Stock et al., 2018b).
- Large and fast-growing cities suffer from congestion, while remote communities often have poorly utilised and maintained assets, e.g. road congestion is an AU$16 billion 'handbrake' on the productivity of Australian cities (Stock et al., 2018b).
- Passenger transport networks are at risk of becoming financially and environmentally unsustainable as the country commits to a low-carbon economy (see the 'Mitigation Pathways' section below).
- Urban and agricultural supply chains are also experiencing challenges, i.e. too often, they act as bottlenecks in national supply chains, limiting access to important markets for exporters, and increasing costs for consumers (see the 'Agriculture and Land' and 'Manufacturing and Construction' sections below).

Climate impacts and risks

The energy and transportation sectors contribute to climate change through their various activities (see the 'Greenhouse Gas Emissions' section below). For example, coal and natural gas combustion for electricity generation for manufacturing, electrified rail transportation, and business and residential electricity use. Likewise, the transportation sector relies on diesel, petroleum and LNG as energy sources for heavy machinery, trucks, diesel locomotives, ships and motor vehicles. However, energy and transportation are highly vulnerable to climate-related risks in their operations (Lawrence et al., 2022). These climate risks apply to public and private entities and infrastructure assets across both sectors.

Climate change and extreme weather events adversely impact critical national infrastructure assets, including electricity, utility and transport networks (Australian Academy of Science, 2021). These events are threatening the country's energy security and climate risks are likely to worsen in the future (see Chapter 3, Table 3.2 and Box 3.1). Critical climate risks for electricity grids and coal/gas-fired power stations are the projected increase in the frequency and intensity of extreme weather events, e.g. prolonged heatwaves, bushfires, floods, tropical cyclones and supercell storms (Stock et al., 2018a, Lawrence et al. 2022, Rice et al. 2022). Coal- and gas-fired power stations require large quantities of freshwater (Australian Academy of Science, 2021). Water supplies may become limited in some years, with future projections for more extreme and prolonged droughts (see Chapter 5, Box 5.4).

Large, 'concentrated' energy assets (e.g. large coal-fired power stations) are particularly vulnerable to extreme weather events (Lawrence et al., 2022). For example, heatwaves place enormous pressure on electricity systems due to increased demand for electricity for air conditioning and because fossil-fuelled power stations often grapple with operating effectively in extreme heat. Notably, extreme heat reduces output, and they regularly suffer mechanical failures when electricity demand is usually at its highest (Stock et al., 2018a). Critical infrastructure that transports electricity along long, narrow transmission lines from single significant sources of fossil-fuelled power is also more vulnerable to extreme weather events. This risk is much less the case with a 'distributed' grid with more minor energy sources spread over a wide range of locations (Stock et al., 2018a). For example, on 28 September 2016, South Australia experienced one of its most severe storms in many years. The supercell storm involved at least seven tornadoes, wind gusts of 190–260 km/h (equal to a Category 4 tropical cyclone), large hailstones and intense rainfall. This single but widespread event knocked down 23 transmission towers, triggering a state-wide blackout (Burns et al., 2017).

Loss of electricity supply after extreme weather events directly impacts other sectors of the economy, such as mining, manufacturing, construction and numerous service industries, largely dependent on the existing concentrated electricity generation system and its series of long-distance transmission networks. Paradoxically, more frequent extreme weather-related disruptions and increasing reliance on renewable energy (e.g. solar, wind and hydro power) could increase the short- to medium-term risks for electricity supply. This situation will lessen as new technology gets up and running (Australian Energy Market Operator, 2020, see Box 6.1). Given these deficiencies in the country's electricity infrastructure, at this stage, the energy sector has a generally low adaptive capacity to deal with anticipated changes in climate and increases in extreme weather events (see Chapter 4, Figure 4.1).

Climate change and projected increases in the frequency and intensity of extreme weather events (see Chapter 3, Table 3.2, Box 3.1) pose a significant threat to major transport infrastructure and supply chains along with Australia's transport networks. Transportation infrastructure remains exposed to a range of climate risks (Fisk, 2017), notably:

- *Risks to critical assets*: severe weather events may damage significant assets, including roads, rail, airport and seaport facilities. Support structures and buildings, drainage infrastructure, utilities, energy infrastructure and communications facilities will also be potentially affected. Damage can force owners to retire assets early, cause costly major upgrades or increase maintenance frequency.
- *Risks to operations*: more frequent or intense severe weather events may disrupt transport business operations or the useability or reliability of their services. Examples include the effect of floods on transport and electricity

supply, disruption of transport by tropical cyclones and significant storms and reduced productivity of outdoor workers due to high temperatures (see also Mining, Manufacturing and Construction sectors below).

- *Critical dependencies*: severe weather events may interrupt supply chains or services such as air, rail, sea and road transport, cargo, electricity, gas or water supply on which businesses depend (see Manufacturing and Construction sectors below). A changing climate may also affect global trading patterns, for example, by changing the supply of agricultural products or mining products (see Agricultural and Mining sectors below).
- *The national economy*: adverse climate change impacts can drag down the national economy, with a flow-on effect on individual businesses.
- *Insurance and capital markets*: climate change shocks may affect the availability of insurance and access to capital, either locally or worldwide (Chapter 7, Box 7.1).
- *Adaptive capacity*: while the transport sector relies on fossil fuels for most of its activities, its adaptive capacity to deal with climate change will remain low (see Chapter 4, Figure 4.1).

Climate-resilient development pathways and opportunities

GREENHOUSE GAS EMISSIONS

The energy and transportation sectors vary in their emissions intensity across the industries that comprise the sectors (Deloitte Access Economics 2020). Electricity supply is rated as 'extremely emissions intensive' and road transport is 'highly emissions intensive', while air and space transport and other transport are 'emissions intensive'. By comparison, rail transport and gas supply are 'moderately emissions intensive'.

Emissions from energy industries (electricity and heat production, petroleum refining and manufacture of solid fuels) were 213.8 Mt CO_2-e in 2019, or 41.2% of net national emissions (Department of Industry, Science, Energy and Resources, 2021c). By comparison, emissions from transportation industries (domestic aviation, road transport, railways, domestic navigation and other transport) were 100.5 Mt CO_2-e in 2019, or 19.4% of net national emissions. Energy and transportation produced 60.6% of the country's net national emissions in

2019 (Department of Industry, Science, Energy and Resources, 2021c). The majority of these GHG emissions were from the combustion of fossil fuels from a range of sources (i.e. coal, natural gas, diesel and petroleum products).

MITIGATION PATHWAYS

If Australia plans to achieve net-zero emissions by 2050, it will have to dramatically reduce its dependence on fossil fuels in its energy and transportation sectors (Finkel et al., 2017, Wood et al., 2021). The 2050 target presents a massive (but not impossible) challenge for energy and transportation industries and all levels of government over the next three decades. The transition away from fossil fuels to renewable energy sources and zero-carbon types of transportation must be well-executed, and there is potential for significant economic opportunities (Naughtin et al. 2022). Major structural energy reforms and transformational changes will drive both sectors towards net-zero emissions (Deloitte Access Economics, 2020). As it stands now, the Australian Government's Long-Term Emissions Reduction Plan (Commonwealth of Australia, 2021a) may be considered counterproductive to the overall process of achieving net-zero emissions as planned. For example, the plan includes the ongoing use of fossil fuels as part of the energy mix. There is also a strong dependence on expensive, largely unproven technologies for future energy production (e.g. Carbon Capture and Storage, see Box 6.5). Respected sources (e.g. Australian Academy of Science, 2021, Wood et al., 2021) argue that these technologies merely extend the life of coal and natural gas as conventional energy sources for electricity. Not only is this an economic risk, but it also has the potential to lead the country down a mal-mitigation pathway in its efforts to reach net-zero emissions by 2050 (see Chapter 4, Box 4.4). There is also a risk that such unproven technology may undermine mitigation and adaptation efforts in other energy-dependent sectors of the economy, such as agriculture, mining, manufacturing, construction and tourism (see these sectors below).

Achieving an electricity supply with zero emissions is fundamental for reducing GHG emissions across any economy (see Chapter 4, Box 4.4). This goal is a big challenge for Australia, given that 76% of its current electricity generation is from fossil-fuelled sources (Department of Industry, Science, Energy and Resources, 2021a). Over the past decade, the business sector has driven three structural and transformative changes in the electricity sector at national and global levels (Australian Academy of Science, 2021). First, wind and solar continue to be the cheapest sources of new electricity generation in Australia, including the cost of storage and new transmission network infrastructure. Second, investment in new coal power stations across the world has reduced dramatically (International Energy Agency, 2020). In fact, no new coal-fired power plants are currently under construction or planned anywhere in North America or Western Europe. Similarly, investment in new gas-fired power stations is also declining, particularly in countries with high gas prices, like Australia (International Energy Agency, 2020). Third, existing gas supplies are likely to have a diminishing role in Australia's electricity supply, including 'balancing' the variable output from wind and solar power (e.g., as in South Australia).

Improvements in energy storage are considered a long-term solution to the 'intermittency' in Australia's energy supply, with gas likely to play a decreasing role as new renewables technology is introduced (Australian Energy Market Operator, 2020). NEM's open-source platform collected recent data. It showed that renewable energy provided 31.4% of the electricity within the NEM network (eastern states and South Australia) and 32.2% for the separate West Australian grid (National Electricity Market, 2022). The national figure has rapidly overtaken the national RET, which mandated electricity retailers to sell about 23% of electricity generation from renewables (Clean Energy Regulator, 2022). Meanwhile, in 2021 gas-fired electricity declined to 5.7% of total energy generation (National Electricity Market, 2022). This decline from 12.8% in 2015 was due to the relatively high cost of gas-fired energy compared with renewable sources (National Electricity Market, 2022). The decline in demand for electricity produced by natural gas raises serious questions about the integrity of the Australian Government's 'gas-led recovery' from the COVID-19 pandemic (see below).

AEMO's draft Integrated Systems Plan (Australian Energy Market Operator, 2021) assumes that coal-fired power stations closures will occur at three times the expected rate. It is expected that all the brown coal electricity generators will be decommissioned by 2032. AEMO's 'Step Change' scenario assumes that 14 GW of coal capacity will retire by

2030 (Australian Energy Market Operator, 2021). The main reason being these ageing coal-fired generators are finding it increasingly difficult to compete financially against rooftop solar PV and large-scale renewables (Parkinson, 2021). The same scenario assumes that coal capacity in the country's primary grid will drop to about 3 GW in 2034 and zero GW by 2043. AEMO's draft plan (Australian Energy Market Operator, 2021) also includes modelling for the 'Hydrogen Superpower' scenario. This energy pathway shows that all the coal-fired generators will be offline by 2032, and wind, solar and large amounts of storage will dominate the grid. Energy experts favour the Step Change scenario at this stage but do not rule out the revolutionary Hydrogen Superpower option (Parkinson, 2021).

Meanwhile, transmission companies such as Transgrid have produced scenarios claiming that more than 90% of renewable energy is possible by 2030, which is somewhere between the Step Change and Hydrogen Superpower scenarios (Australian Energy Market Operator, 2021). The draft Integrated Systems Plan (Australian Energy Market Operator, 2021) does not support the apparent need for the Australian Government's gas-led recovery 'road map' policies (Commonwealth of Australia, 2021a, d). In regard to this assertion by AEMO, the Kurri Kurri fossil gas/diesel 'peaking' plant, under construction in the Hunter Valley, has been described by an energy engineer as a government-built white elephant (Mountain, 2022). AEMO's modelling shows that about 9 GW of gas generation may be required by 2030, but this will be limited to 'fast-start' gas generators that will rarely be needed (Parkinson, 2021).

A detailed report (Stock et al., 2018a) demonstrates how a 'modern electricity grid, powered by diverse renewable energy sources and storage can provide secure, clean and affordable power for Australians'. The report addresses many of the business and climate risks the electricity sector faces (see Risks above). Based on earlier work (Commonwealth Scientific and Industrial Research Organisation, 2017, Energy Networks Australia, 2017), the report gives a detailed road map – including crucial steps to be taken – to achieve a zero emissions, reliable and affordable electricity supply for the country (Stock et al., 2018a). However, substantial investment will be needed to strengthen the electricity grid through battery deployment and demand response measures (ClimateWorks Australia, 2021, Wood et al.,

2021). For example, for the Step Change scenario (Australian Energy Market Operator, 2021), the amount of electricity produced in the primary grid will need to double, to meet the demands of electric vehicles, the electrification of industry and households, and renewable hydrogen (Parkinson, 2021). This target will require significant structural reform and transformational changes in Australia's electricity industries going forward (Box 6.1).

On the back of the COVID-19 pandemic, O'Brien et al. (2020) propose several strategies to accelerate the country's transition to a renewable electricity supply (Box 6.1). Four stages are suggested, which should ideally be facilitated over 3 years (in order of sequence):

1. Energy efficiency incentives are needed to reduce overall load and demand response mechanisms to enable load flexibility;
2. Concurrent, fossil fuel subsidies require fiscal reform (see also Business Council of Australia, 2021);
3. The opening-up and regulatory facilitation of renewable energy zones are required, i.e. areas with high solar and wind resource potential (e.g. see Figure 6.1); and
4. Enabling higher renewable energy penetration through interconnectors, batteries and pumped solar/hydro-energy storage.

The Business Council of Australia (2021) generally supports the transition to renewable energy technologies by 2050, stating that the 'energy market policy needs to be nationally integrated, stable and durable, and technology agnostic to drive investment in the most efficient mix of technologies'. The BCA maintains that natural gas and gas infrastructure will have a critical 'enabling' role in transitioning to net zero by 2050 (Business Council of Australia, 2021). The BCA states that gas will provide critical peaking and firming capacity when wind, solar and other storage are unavailable. Other potential energy options are renewable gas alternatives to natural gas, such as syngas, biomethane and green hydrogen.[6] Others disagree (e.g., Steffen et al. 2017,

---

[6] 'Green hydrogen' is defined as hydrogen produced by splitting water into hydrogen and oxygen using renewable electricity. Not be confused with 'clean hydrogen' or 'blue hydrogen' that is made with mostly fossil fuel sources.

BOX 6.1: Five criteria that must be met to achieve a modern, efficient electricity system in Australia

---

1. *Clean*: Tackling climate change requires a rapid transition away from coal and gas to renewable-powered electricity (see Chapter 4, the 'Paris Agreement' section, Intergovernmental Panel on Climate Change 2021). The electricity system must reach a minimum of 50% renewable energy by 2030 and zero emissions well before 2050.
2. *Reliable*: Balancing demand for electricity (from households, businesses and industry) with supply from power stations, energy storage and demand flexibility (via demand management). More dispersed and diverse electricity generation. Examples: low-cost variable renewables (wind and solar PV); on-demand, or 'dispatchable' renewables (solar thermal, biomass and hydro); energy storage technologies (i.e. pumped hydro or batteries); greater grid interconnectivity; and smart grids and smart consumers, flexibility adjusting demand.
3. *Secure*: Meeting technical requirements or grid stability (described by terms such as 'frequency control' and 'inertia'), ensuring the power grid can overcome disturbances. This requirement means operating the power grid within specific technical parameters (described by terms such as frequency, voltage and fault current levels) to ensure the system maintains a stable operating state. Examples include wind turbines with grid stabilisation technologies and battery storage.
4. *Resilient*: Delivering reliable power in the face of increasingly severe weather events influenced by climate change (see Climate impacts and risks).
5. *Affordable*: Lowering electricity costs for households and businesses. Renewable energy is currently the cheapest type of new electricity generation. Commonwealth Scientific and Industrial Research Organisation's modelling found transitioning to a 'zero-emissions' electricity sector by 2050 wouldresult in AU$414 annual savings for an average household compared to business as usual.

Source: Adapted from Commonwealth Scientific and Industrial Research Organisation (2017) and Stock et al. (2017).

Australian Academy of Science, 2021, Australian Energy Market Operator, 2021, Hodges et al. 2022), arguing instead that the pathway to 100% renewables should be a short-term goal (Box 6.1).

Australia's aging coal-fired power plants will inevitably close over the next 20 years in favour of renewable energy generation (Wood et al., 2021, Australian Academy of Science, 2021, Australian Energy Market Operator, 2021). There is already a very high level of wind and solar power deployment across the country. The transition rate from fossil-fuelled energy to renewable energy affects Australia's cumulative GHG emissions until 2030 and hence its future contribution to climate change – in line with its commitments to the Paris targets (see Chapter 4, the 'Paris Agreement' section). Its recent pledge to achieve net-zero carbon emissions by 2050 means the Australian Government will have to instigate significant policy reforms to accelerate the uptake of renewable energy generation and storage technologies across the country's electricity supply networks (Box 6.1). It would be foolhardy to extend the life of existing coal plants, let alone give support to any new coal- or gas-fired power stations beyond those under construction (Australian Academy of Science, 2021, Wood et al., 2021).

Australia's transportation sector must also engage in structural reforms and transformational changes to reduce GHG emissions (see Chapter 4, Box 4.4). In this regard, battery-electric engines for cars and larger vehicles represent the most likely route for reducing GHG emissions (Australian Academy of Science, 2021). Large-scale electric vehicle (EV) adoption will require an expanded charging network and a significant expansion of a renewables-based electricity supply (see Box 6.1). Commercial and government EV fleets and private EV vehicles also provide a large, conveniently 'decentralised' energy storage capacity that can help balance electricity demand and supply on the grid (Australian Academy of Science, 2021). In fact, by 2040, 40% of vehicles in the country are likely to be electric, and these vehicles could have the potential to store electricity to a similar capacity as the proposed Snowy 2.0 hydro-electricity scheme (Infrastructure Australia 2019). A detailed study of transport solutions to climate change in Australia (Stock et al., 2018b) produced nine core recommendations to facilitate the transition of the country's transportation sector to net-zero emissions:

1. *Federal, state and territory governments to set targets for zero emissions and fossil fuel-free transport before 2050*: Develop a climate and transport policy and implementation plan to achieve these targets. The Australian Government's Long-Term Emissions Reduction Plan (Commonwealth of Australia, 2021a) discusses options to decarbonise the transport sector, focusing on EVs and 'future fuels' through its Future Fuels Fund. The Australian Renewable Energy Agency (ARENA) administers the fund and aims to support public charging infrastructure for EVs and 'clean' hydrogen and biofuel technologies.

2. *Ensure cost-benefit analyses for all transport project business cases account for the additional greenhouse gas emissions that projects will lock in over their lifetime or GHG pollution avoided (e.g. from public transport improvements).*

3. *Establish transport mode shift targets for public transport (e.g. buses, trains and ferries), cycling and walking.*

4. *Ensure that at least 50% of government (federal, state, territory) transport infrastructure expenditure is for public and active (e.g. walking and cycling) transport.*

5. *Federal, state and territory governments to introduce targets to drive uptake of electric buses, trucks, cars and bicycles powered by renewables*: Establish EV targets for specific sectors and government operations, including state and territory public transport systems, and all government vehicle fleet purchases. Victoria and New South Wales have targeted 50% of new car sales being electric by 2030, which translates to an estimated 30% of recent

car sales nationwide (ClimateWorks Australia, 2021). The Business Council of Australia (2021) recommends federal, state and local government procurement of low and zero emissions fleet vehicles and exempting EVs from luxury car taxes.

6. *State and territory governments to contract additional 100% renewable energy (e.g. battery and green hydrogen) to power public transport systems (trains, light rail and buses)*: Multiple jurisdictions address powering rail with renewable electricity and transitioning bus fleets to electric vehicles (ClimateWorks Australia, 2021). The Business Council of Australia (2021) recommends incentivising fuel switching for heavy vehicles, such as hydrogen fuel cell trucks and other vehicles in mining, construction, transport and agriculture (see sectors below). For example, green hydrogen energy for heavy goods transport on roads and mining sites (O'Brien et al., 2020).

7. *Federal Government to introduce strong vehicle GHG emissions standards (in line with most OECD partners)*: State and territory governments to advocate for vehicle emissions standards through the Council of Australian Government's Transport and Infrastructure Council. Multiple jurisdictions are addressing public transport emissions (ClimateWorks Australia, 2021). The Business Council of Australia (2021) recommends the progressive tightening of general vehicle emissions standards, starting with adopting the Euro6 Standard as soon as practicable.

8. *Federal, state and territory governments to encourage the rollout of 100% renewable-powered EV charging stations, particularly in regional areas and interstate routes*: According to the Electric Vehicle Council (Electric Vehicle Council, 2021), as of November 2021, there were about 3,000 public chargers installed across the country. The Business Council of Australia (2021) advocates that all jurisdictions adopt the New South Wales ultra-fast charging infrastructure approach as part of their EV strategies, including co-investment in EV Super-Highways and EV Commuter Corridors and EV Off-street Parking Chargers.

9. *Put a price on carbon pollution*: Consider policies or pricing that better reflect the cost of greenhouse gas pollution. Road or public

transport users bear the fee or reap economic benefits based on their chosen travel mode emissions. End government subsidies, incentives and support for fossil fuel use in the transport sector (e.g. the BCA road map, 2021).

In recent years, the aviation sector has made considerable energy efficiency improvements to its aircraft fleet and management practices (Infrastructure Australia, 2019). However, the current lack of a commercial scale low-emissions alternative fuel supply (e.g. sustainable aviation fuel) prevents the sector from securing a more rapid transition to low GHG emitting operations. In the short to medium term, the industry will rely primarily on 'offsetting' its emissions, either in-country or overseas (Business Council of Australia, 2021). However, O'Brien et al. (2020) propose three mitigation strategies for aviation industries that have the potential to reduce their high emissions:

1. Mandating of ramped emissions offset targets by aviation industries;
2. Targeted research and development funding to deliver viable solutions for low-carbon travel (e.g. battery and green hydrogen-powered aircraft); and
3. Aviation industries to engage in cross-sectoral initiatives to enable collaboration and drive innovation in low-emissions activities.

ADAPTATION PATHWAYS

Australia has all the requirements for very large-scale renewable energy production, such as abundant solar radiation, high winds, land availability, a reputation for managing large resource projects and a stable institutional and investment environment (Australian Academy of Science, 2021, see Box 6.1). Importantly, this renewable energy capacity can quickly meet domestic demand and meet needs for export industries, such as agriculture, mining and manufacturing (Garnaut, 2019). There is growing momentum in climate mitigation and adaptation actions in the energy sector. However, this has mainly been inadequate to ensure that the country's energy industries can prepare for future changes in climate and more extreme weather events (Gasbarro et al., 2016, see Climate risks above). Lawrence et al. (2022) propose a suite of climate adaptation options that may better inform climate-resilient development

pathways for the sector in the future. Strategies for the energy sector include:

- Diversify electricity supplies geographically and technically (see Box 6.1).
- Integrate planning, improved asset design and management, and disaster recovery to build resilience to more extreme weather events (see Chapter 4, Box 4.1: Accommodate and Protect strategies).
- Augment the transmission grid to support the change in generation mix using interconnectors and renewable energy zones, coupled with energy storage (see Box 6.1).
- Include climate change risks in the design, location and rating of future energy infrastructure and consideration of the implications for future transmission developments (Box 4.1: Avoid strategy).
  - *For oil and gas industries:* (1) increasing design and construction standards, (2) flood defence measures, (3) insurance, (4) improved water efficiency, (5) improved insulation of super-cooled LNG processes, (6) more efficient air conditioning and (7) creating fire breaks (see Box 4.1: Protect strategy; see also Mining sector below).
  - *For wind energy:* (1) technological developments to strengthen existing resilience (see Box 4.1: Accommodate and Protect strategies), (2) recognition that climate change projections reinforce the relative advantage of Western Australia (see Figure 6.1) and (3) consider Tasmania for new wind energy installations (see Box 4.1: Avoid and Retreat strategies).
- Energy generation diversity, demand management, pumped hydro storage and battery storage (see Box 6.1).
- Direct exports of electricity from high-solar or high-wind areas in northern Australia to countries in the Asia-Pacific region via high-voltage direct current (HVDC) cable connections (Cotts et al. 2017, Burke et al. 2022).

Like the energy sector, the transportation sector has paid limited attention to adapting to projected changes in climate and more extreme weather events. However, a recent discussion paper (Engineers Australia, 2020) provides some engineering perspectives on the building and operation of transport networks to minimise anticipated climate risks (see the 'Climate impacts and risks' section above). Transport infrastructure must be designed and built for many decades of use (see Construction sector below). For example, if society expects a bridge or railroad to last 100 years, it should be designed to operate safely in the climate range forecast in 100 years. The discussion paper proposes two climate adaptation pathways (Engineers Australia, 2020):

1. *Creating resilient transport networks*: Planning, designing and construction of new infrastructure, and maintenance of existing infrastructure (see Chapter 4, Box 4.1; Chapter 7, Table 7.4) must:
   a. Allow for higher average temperatures and a more extensive range of weather extremes in design, including thermal expansion, heat degradation and passenger comfort (see Box 4.1: Accommodate and protect strategies);
   b. Maintain or improve micro-climate in urban works by planting trees for shading and providing breeze corridors to reduce the urban heat island effect and improve liveability (see Box 4.1: Accommodate and protect strategies);
   c. Allow for sea-level rise and storm surge increases in coastal transport links (see Box 4.1: Avoid strategy); and
   d. Plan for a broader range of extreme events such as bushfires, cyclones and floods (see Box 4.1: Protect strategy).
2. *Creating less impacting transport infrastructure*: Transport networks require large amounts of energy to operate (see Emissions and Mitigation above), occupy large amounts of physical space and contribute to urban heat island effects. There is a need to adapt infrastructure planning and design processes and delivery capability to focus on infrastructure, *i.e.* the most efficient choice of mode, scale, and location. This focus may require an amendment to project planning and assessment processes, e.g. current high discount rates for economic assessment discourage long-term investment, which results in 'multiple cycles' of wasteful reconstruction and reinforces reliance on the inefficient *status quo*. Other changes require research to understand risks

and opportunities for transport infrastructure projects. Planning and design teams also need multiple skills, including climate science and urban ecology (see Chapter 7, Table 7.3).

The Australian Government's National Climate Resilience and Adaptation Strategy (Commonwealth of Australia, 2021b) attempts to provide national leadership in climate adaptation. However, business and industry's widespread adoption and implementation of this strategy are yet to be thoroughly evaluated (see Chapter 4, the 'Roles and Responsibilities for Climate Change in Australia' section). Government leadership in climate adaptation – at all levels – will be critical in building adaptive capacity and resilience to climate change in the energy and transportation sectors (Box 6.2). Notably, the transportation sector faces many barriers to climate adaptation, as many of its industries are SMEs (see Box 6.6, Chapter 4, Box 4.2: barriers). Ambiguities may hinder effective climate adaptation in both sectors in the Australian Government's gas-led recovery policies (Commonwealth of Australia, 2021a, d). These policies contradict the Australian Academy of Science (2021) and the Australian Energy Market Operator (Australian Energy Market Operator, 2021). Both sources claim that the country can achieve 100% renewable energy by the early-to-mid 2030s, without the need for government investments in new gas-generation projects.

INTEGRATION WITH THE SUSTAINABLE DEVELOPMENT GOALS

Six Sustainable Development Goals (SDGs) are relevant to a national strategy on climate change and the energy and transportation sectors:

1. SDG6 (Clean water and sanitation);
2. SDG7 (Affordable clean energy);
3. SDG9 (Industry, innovation and infrastructure);
4. SDG11 (Sustainable cities and communities);
5. SDG12 (Responsible consumption and production); and
6. SDG13 (Climate action).

A framework (Table 6.1) shows how the energy and transportation sectors can effectively integrate climate change with goals, targets and indicators for each SDG. Climate-resilient development pathways for both sectors will require structural and

---

BOX 6.2: Government-led initiatives to support climate adaptation in business and industry

---

- *Better information* wabout the potential effects of climate change on aspects of concern to business, such as adverse impacts of flooding on supply chains, effects of heatwaves on workers and clients, should be provided to business to support accurate business-led risk assessments and responses. This could be in the form of industry-specific scenarios and case studies (refer to sectors in Chapters 6 and 7).
- *Education and training programmes* can support the sector to understand its exposure to climate-associated risk and to deal with the risk in an appropriate manner.
- *Harmonisation of adaptation activities and frameworks* (e.g. sea-level rise benchmarking) across state boundaries/jurisdictions is needed. This will help to increase certainty for business and enhance the likelihood of coordinated and well-informed responses.
- The development of *relevant key performance indicators and rating tools* to support performance measurement will enable internal decision making, disclosure to shareholders and ultimately long-term improvement in performance, e.g. alignment with the United Nations Sustainable Development Goals (see Chapter 4, the 'Sustainable Development Goals' section).
- There is a need for government support for *divestment and investment strategies* and approaches (e.g. assisting industry with mechanisms to effectively disclose their climate change risks and their responses to manage risk; incentives that reward early adopters of adaptation actions), e.g. applying relevant policy changes to the Safeguard Mechanism and Emissions Reduction Fund.
- *Enabling independent peak bodies* to become trusted disseminators of information and knowledge, e.g. Business Council of Australia (2021).
- *Actions to streamline red tape*, not to reduce the requirements that need to be addressed, but to remove overlaps and apparent contradictions.
- *Provision of financial incentives for adaptation planning*, possibly funded from carbon tax revenue (or an equivalent mitigation fund), e.g. as proposed by the Business Council of Australia (2021).

Source: Adapted from NCCARF (2013b). For other examples of enablers, refer to Chapter 4 (Box 4.2).

---

transformational changes from now on. If the sectors are to meet (or make progress towards) the 2030 SDG targets, significant changes will need to co-occur in the agricultural, mining, manufacturing and construction sectors (see below). There will also need to be incremental and transformational changes in various social systems that depend on energy and transportation (see Chapter 7, Table 7.5).

AGRICULTURE AND LAND

Socio-economic context

Agriculture is a historically significant and well-established primary industry in Australia,

particularly in regional areas (Henzell, 2007). More than 90% of the food Australians consume is produced domestically, and many businesses and communities in regional towns rely on the sustainability of agriculture for their viability (McRobert et al., 2019, Howlett and Henry, 2020, see Chapter 7, Social Systems). Agriculture accounted for 55% of Australian land use (427 million ha, excluding timber production in December 2020), 25% of water extractions (3,113 GL used by agriculture in 2018–2019) and 11% of goods and services exports in 2019–2020 (Australian Bureau of Agricultural and Resource Economics and Sciences, 2021). Sustainable land management has been required over the decades to continue producing the quantity and quality of food demanded against a

Table 6.1 Integration of SDGs with the Australian energy and transportation sectors' responses to climate change

SDG	Goals	Targets	Indicators
SDG6: Clean water and sanitation (*economic and environmental*)	1. Ensure availability and sustainable management of water	Improved water quality by reducing pollution from energy and transportation operations; Substantially increased water recycling on site; Substantially increase water use efficiency	• Proportion of energy and transportation infrastructure wastewater safely treated • Change in water use efficiency over time • Level of freshwater withdrawn as a proportion of available freshwater resources
SDG7: Affordable clean energy (*economic and environmental*)	2. Support extension of renewable energy adaptation	Increased use of renewable energy; Increased energy efficiency; Lower emissions from energy use	• Percentage of renewable energy use across sector • Energy use and emissions measurements • Percentage of energy and transportation supply chain income spent on energy use
SDG9: Industry, innovation and infrastructure (*economic and environmental*)	3. Build resilient infrastructure and sustainable energy and transportation industries	Increased new, upgraded/retrofitted energy and transport machinery and infrastructure that are sustainable; Encourage new, inclusive and sustainable energy and transport industries	• Carbon emissions per unit of value added • Sustainable energy and transportation value added as a proportion of GDP and per capita • Renewable energy and transportation employment as a proportion of total employment (to 100)
SDG11: Sustainable cities and communities (*social and environmental*)	4. Make cities and settlements inclusive, safe, resilient and stable	Increase access to safe, affordable, accessible and sustainable transport systems for all	• Proportion of the population that has convenient access to sustainable public transport (using renewable energy sources), by sex, age and persons with disabilities
SDG12: Responsible consumption & production (*economic and environmental*)	5. Ensure sustainable consumption and production patterns	Increased efficiency; Increased circular economy for energy and transportation materials; Sustainable public procurement practices	• Proportion of energy and transportation infrastructure projects completed using sustainable methods, material footprint (per capita, per GDP) • Degree of public procurement practices in energy and transport sectors that are sustainable
SDG13: Climate action (*environmental*)	6. Carbon-neutral production; 7. Resilience and adaptive responses to natural hazards	Reduced global warming; Closed loop production (circular economy) and supply chain systems; Mitigation of climate change impacts on energy and transportation sectors	• Reduction in energy and transportation sector GHG emissions contributions (to zero) • Percentage of energy and transportation industries using renewable energy (to 100) • Percentage of energy and transportation industries with a climate action plan

This includes their supply chains. Refer to Chapter 4 for an explanation of SDG goals, targets and indicators (see also Box 4.5). Triple-bottom line implications for each SDG are shown in italics.

decline in suitable land (Howlett and Henry, 2020). Climate change poses a significant threat to food, fibre, timber production and sustainable land management, both now and in the future.

Agriculture, forestry and fisheries contributed 2.0% of Australia's GDP and employed about 313,700 persons (or 2.4% of the workforce) in August 2021 (Australian Bureau of Statistics, 2021a). The mix of agricultural activity depends on climate type (see Chapter 3, Figure 3.3), freshwater resources (see Chapter 5, Box 5.4), soil type and proximity to markets (Australian Bureau of Agricultural and Resource Economics and Sciences, 2021). Livestock grazing (cattle and sheep) on native vegetation (291.74 million ha) is widespread across large tracts of the country, particularly in the north and inland areas of the southeast and southwest (see map Chapter 3, Figure 3.1). Livestock grazing on modified pastures (39.81 million ha) and cropping, including crop/pasture rotation (21.78 million ha) and horticulture (0.44 million ha), are prevalent in the higher rainfall areas closer to the coast (Australian Bureau of Agricultural and Resource Economics and Sciences, 2021).

As a proportion of its economy, Australia's agricultural output remains among the highest in the 37-member countries of the Organisation for Economic Co-operation and Development (OECD) (Hughes et al. 2015). Despite this high productivity, government subsidies to Australian agriculture are meagre (2%) compared with other OECD countries, such as Norway (62%), Switzerland (55%) and Japan (46%). About 70% of Australia's primary production from agriculture, horticulture, wine production (viticulture), fisheries and forestry is exported. Export quantities vary by product (Australian Bureau of Agricultural and Resource Economics and Sciences, 2021), e.g. sugar (86%), beef and veal (75%), mutton and lamb (73%), canola (72%), wheat (71%), rice (52%), seafood (49%), dairy products (40%), fruit and nuts (31%) and pig and poultry (5%). The country is a net importer of forest products, with wood chips making up the major export component of the forestry industry (Australian Bureau of Agricultural and Resource Economics and Sciences, 2021). In 2019–2020, about 60% of all primary industry exports went to eight markets in Asia, and 40% went to other markets (Australian Bureau of Agricultural and Resource Economics and Sciences, 2021). China is the largest export market, and demand from Asia is projected to double between 2007 and 2050 (Australian Bureau of Agricultural and Resource Economics and Sciences, 2021).

Australian farms employed 326,000 workers on average during 2018–2019, including full-time, part-time, casual and contract employees (Australian Bureau of Agricultural and Resource Economics and Sciences, 2021). Broadacre farms are the largest employers in agriculture, followed by fruit, grape and nut farms, vegetable farms and dairy farms (Martin et al., 2020). Importantly, agriculture and other primary industries employ 82% of workers who live in regional and rural areas. While COVID-19 impacted the Australian economy in 2020–2021, the agricultural sector (with a few exceptions) has demonstrated an ability to adapt and transition to new opportunities (Greenville et al., 2020). The value of farm production is predicted to grow to about AU$66 billion for 2020–2021 (Australian Bureau of Agricultural and Resource Economics and Sciences, 2021).

Agriculture, horticulture and viticulture encompass supply chains from the farm, orchard and vineyard to the consumer and the industries that support growers (Howlett and Henry, 2020). The high diversity of agricultural products underpins the food and beverage sector, the largest manufacturing industry in Australia – accounting for about 32% of the country's total manufacturing turnover over the past decade (Australian Bureau of Statistics, 2021a, see Manufacturing sector below).

Managing risk

Several drivers contribute to Australia's highly variable climate (see Chapter 3, Figure 3.2). Despite very high inter-annual variability in rainfall, its agricultural sector is a world leader in managing climate risk. This risk has been a significant challenge since European colonisation in the late 18th century. Still, it has been slowly improving – even in the face of more recent accelerated climate change (Howden and Stokes, 2010, National Climate Change Adaptation Research Facility, 2013c, Howlett and Henry, 2020, Lawrence et al., 2022). Over the past 30 years, agri-businesses have responded strongly to the recent climate shifts with improvements in technology, management practices and increased farm productivity

(National Climate Change Adaptation Research Facility, 2013c, Hughes and Gooday, 2021). However, several studies have suggested that many agri-businesses are already operating close to the limits of technical efficiency (Daly et al., 2015).

In addition to climate risks, agri-businesses have experience and skills in managing significant variabilities, such as biosecurity risks and volatile commodity prices (McRobert et al., 2019, Howlett and Henry, 2020). These volatility factors generate substantial variation in farm output, more significant than that experienced by farmers in most other OECD countries and more challenging than those experienced by business owners in different sectors of the Australian economy (Keogh, 2012). Many farmers have effective strategies for managing risk, including maintaining relatively high equity levels, liquid assets and borrowing capacity, carefully using inputs, diversifying across enterprises and locations and earning off-farm income (Australian Bureau of Agricultural and Resource Economics and Sciences, 2021).

Climate impacts and risks

Australia is highly vulnerable to disruptions in its food supply due to extreme climate events, such as droughts, floods, heatwaves and bushfires (McRobert et al., 2019, Hughes and Gooday, 2021, Lawrence et al., 2022). Such disruptions are already affecting the ability of exporters to compete in tightly contested international markets (McRobert et al., 2019). Many of the observed changes in key climate indicators are projected to continue over the coming decades, even under lower GHG emission pathways (see Chapter 3, Table 3.2). However, agriculture has shown a generally high adaptive capacity in managing climate risk up until now (see Chapter 4, Figure 4.1). This capacity is mainly due to its well-developed institutional and governance structures and a strong willingness to adapt to changing conditions (Howlett and Henry, 2020). However, declining institutional and community capacity may affect future adaptive capacity resulting from high debt, absence of insurance, increasing regulatory requirements and perverse funding mechanisms that lock in ongoing exposure to climate risk (Lawrence et al., 2022).

Agriculture (includes food, fibre and ecosystem products) is now considered one of the most 'climate-vulnerable' sectors in Australia (Lawrence et al., 2022). This situation is mainly due to its high exposure and sensitivity (see Chapter 4, Figure 4.1) to changing rainfall regimes, increasing CO_2 levels, temperatures and extreme weather events (Garnaut, 2008, Stokes and Howden, 2010, Hughes et al., 2015, Howlett and Henry, 2020, Hughes and Gooday, 2021). If current rates of climate change continue – or potentially escalate – it will be challenging and expensive for many forms of agri-business to adapt to climate change (Hughes et al., 2015, McRobert et al., 2019, Howlett and Henry, 2020, Australian Academy of Science, 2021, Lawrence et al., 2022).

Since 2000, changes in rainfall patterns have cut profits across the sector by 23% compared to what could have been achieved in pre-2000 conditions (Hughes and Gooday, 2021). The most omnipresent negative impact on agriculture is drought, which disrupts cropping programmes, reduces stock numbers and erodes farm productivity and resource bases. Adding to the woes of drought, significant outbreaks of pests, weeds and pathogens often synchronise with climatically 'good years', adding financial stress to agri-businesses recovering from drought (National Climate Change Adaptation Research Facility, 2013c). Increasing risks from bushfires (see Chapter 5, Box 5.3) also affect the profitability of agri-businesses (Hughes et al., 2015). Bushfires in 1983, 2003, 2006, 2009, 2019 and 2020 burned significant areas of agricultural land in southern and eastern parts of the country. These events resulted in the loss of life, crops and pastures, livestock and critical public infrastructures, such as electricity and water supplies.

Many agri-businesses are exposed to climate change on numerous levels, including farming operations, supply chains and the provision of sustained goods and services from the rural and regional communities that depend on them for their survival (McRobert et al., 2019). Notably, across southwestern, southern and eastern Australia, many production regions are exposed to climate change, including the Murray-Darling Basin, which supports agriculture worth AU$24 billion per year (Lawrence et al., 2022). This region comprises 2.6 million people living in diverse rural communities (see Chapter 7), and critical environmental assets containing Ramsar-listed wetlands (see Chapter 5, the 'Terrestrial and Freshwater Ecosystems' section and Box 5.4). One economic modelling study estimated the cumulative costs to agricultural and labour productivity from climate change are expected to exceed AU$19 billion by 2030, AU$211 billion by 2050 and AU$4.2 trillion by 2100 (Steffen et al., 2019). This conservative study did not factor in losses from floods, bushfires, storms and tropical cyclones and the likely adverse impacts of climate change on settlements, properties and infrastructure (see Chapter 7, Table 7.4 and Box 7.1).

Businesses across the agricultural sector are strongly influenced by biophysical factors[7], government policy and regulations, and market and economic drivers (NCCARF, 2013b). Agricultural activities interact with climate change and affect the eventual impacts on agri-businesses. The average climate influences the economic viability of agricultural production at a location, together with the magnitude of climate variability. The higher the variability from 1 year to the next, the greater the risks to the agricultural sector (Crimp et al., 2014).

Most Australian soils are ancient, weathered and infertile (State of the Environment, 2016). When combined with high climate variability, this creates many challenges for maintaining sustainable agricultural systems. However, while climate change poses significant risks across the sector and most production regions, some agri-businesses and areas may benefit from climate change in the short to medium term (Howden et al., 2010, National Climate Change Adaptation Research Facility, 2013c, Hughes et al., 2015). Climate change is expected to affect natural capital, the quality and quantity of crops, biosecurity, heat stress on livestock and the geographical distribution of crops and livestock (Hughes et al., 2015, Howlett and Henry, 2020).

NATURAL CAPITAL

The economic success of Australian agriculture is highly dependent on the health and sustainability of its natural capital or natural resource base (McRobert et al., 2019, Cresswell et al., 2021). This capital includes the range of ecosystem services derived from natural capital that maintain human health and well-being (Turton, 2020, see also Chapter 5, Boxes 5.1 and 5.2). There are concerns that the high adaptive capacity of agriculture to deal with climate risks (see Chapter 4, Figure 4.1) will most likely decline in critical agricultural regions of Australia in the future. Notably, it is now widely recognised that the deleterious impacts of climate change on water availability (see Chapter 5, Box 5.4) will be the most severe challenge for agri-businesses going forward (Hughes et al., 2015).

Other environmental issues identified by various State of the Environment Reports[8] include land clearing and soil degradation due to acidification, salinisation, erosion and loss of carbon and nutrients (Hughes et al., 2015, Lawrence et al., 2022). It is now inevitable that natural capital will be increasingly stressed and degraded by more extreme climate events in the future (see Chapter 3, Table 3.2, Box 3.1). A combination of increasing temperatures, changing hydrological regimes and rising CO_2 concentrations (see Chapter 2, Figure 2.5) will have negative consequences for soil carbon (Hughes et al., 2015). Increasing temperatures will also accelerate soil decomposition and the rate of CO_2 released from soils, thereby contributing to global warming via positive radiative forcing (Heimann and Reichstein, 2008, see Chapter 2, the 'Anthropogenic forcing' section). Similarly, rangeland degradation during periods of drought may be amplified by adverse climate change impacts (Marshall, 2015), particularly in the cattle regions of northern Australia (Box 6.3).

Given that natural capital is in decline across large parts of the country, in the future, the food and agriculture value chain will have to produce more with less natural capital to ensure long-term sustainability and economic viability (McRobert et al., 2019). This decline in natural capital will create significant challenges for many regional and rural farming communities in the southern half of the country (Cresswell et al., 2021). These communities face projections for a general warming and drying climate (see Chapter 7, Social Systems).

QUALITY AND QUANTITY OF CROPS AND PASTURES

Rising atmospheric concentrations of CO_2 are having a range of effects on field and tree crops and

---

[7] In an agricultural context, biophysical factors include available arable land, fertile soils, adequate water, favourable climate conditions, pollinators, pests and weeds.

[8] Every 5 years, the state of Australia's environment is assessed at national, state and territory levels.

BOX 6.3: Climate change and the development of northern Australia

What is the Northern Australia Development Agenda?

- The rhetoric about the development of northern Australia is not new. Since Australia's federation in 1901, the 'Northern Australia Agenda' has been popularised by various federal governments at least every 20 years or so – but with little measurable impact or influence [1–5].
- The White Paper for White Paper for the Development of Northern Australia was released in June 2015 [1].
- Northern Australia (as defined in the White Paper) refers to the large region north of the Tropic of Capricorn and all the Northern Territory (see Chapter 3, Figure 3.1) [1].
- The White Paper presents a bright future for the north (rapid population growth, roads, rail, dams, mining, energy production, 'broad acre' food production, and tourism) [1,5].

What are the existing environmental and economic challenges for developing northern Australia?

- There are major constraints for agricultural development for large tracts of the region.
  - Environmental limitations for broadacre agriculture and cropping across the north include poor and easily damaged soils, highly seasonal and erratic rainfall, and complex surface and groundwater hydrology [5,7].
  - About 5%–14% of northern Australian soils are suitable for intensive agriculture [5].
  - Flooding, water and nutrient availability may make agricultural use of soils unprofitable and, in some case, even impossible [5,7].
  - Significant difficulties in capturing the huge (wet season) runoff [7].
  - While it has been estimated that 17 million ha of northern Australia has soils that could support cropping, only 1% of the area has appropriate water availability [7].
- Other constraints are long distances to markets, vulnerable transport networks (see Transportation sector) and land tenure insecurity [6].
- Climate change will exacerbate these existing environmental and economic constraints through more extreme and variable climate, e.g. severe heat waves and high rainfall variability [5,6].
- The adverse environmental, social and cultural impacts of a network of dams and irrigation schemes across one of the world's largest intact savannah regions of free-flowing rivers would be very significant [7].

What does the White Paper say about climate variability and climate change?

- There are few references to the climate of northern Australia, and no mention of what human-caused climate change might mean for proposed development of the vast region [1,5].
- The region's strong rainfall seasonality and inter-annual variability is mentioned, but it glosses over the high number of hot and very hot days that affect vast areas of the north each year under the current climate [5].
- There is no mention of official published climate change projections for the north [5,7].

What are the climate change challenges for northern Australia?

- Greatly increased number of hot days (>35° C) and very hot days (>40°C) [5,8].
- Longer fire seasons, with around 40% more very high fire danger days [8].
- More extreme rainfall events; more severe but less frequent tropical cyclones [5,8].

Climate change opportunities for northern Australia?

- Irrigation mosaics: a technique that distributes smaller areas of irrigation across a landscape – can have several benefits, such as reducing the rise of the water table (thereby decreasing the risk of salinity), reducing erosion and potentially preserving the habitats of natural wildlife [7].
- Income generation in the region includes a voluntary carbon market for landowners (including Indigenous communities), and electricity generation from geothermal, tidal, wind and solar sources that could be sold on to service grids both in northern Australia and into Asia (see Energy sector) [6,7].
- Energy-efficient buildings and communities, and climate-proof agriculture [6].

Sources: [1] Commonwealth of Australia (2015a), [2] Griffith Taylor (1924), [3] Davidson (1965), [4] Cook (2009), [5] Turton (2017), [6] Brewer (2016), [7] Hughes et al. (2015), [8] Intergovernmental Panel on Climate Change (2021).
Refer to Chapter 3 for observed and projected changes in key climate indicators for northern Australia.

wine grapes (National Climate Change Adaptation Research Facility, 2013c, Lawrence et al., 2022). Increased rates of photosynthesis[9] are causing plants to develop faster and increase their biomass in controlled greenhouse experiments. However, water availability, pests and pathogens tend to reduce this 'CO_2 fertiliser' effect under natural (field) conditions. In the case of grain crops in greenhouse experiments, yields may increase by 21% at CO_2 concentrations of 550 ppm and by 30% at 700 ppm, based on a baseline of 370 ppm (Howden et al., 2010). However, CO_2 fertilisation effects also reduce protein and micronutrient concentrations, leading to poorer nutritional quality (Howlett and Henry, 2020). In grazing regions of northern Australia, there are concerns that CO_2 fertilisation will favour the expansion of 'woody weeds' at the expense of native pastures. This phenomenon is because elevated CO_2 produces greater yields for C3 woody plants than C4 tropical grass species (Hovenden and Williams, 2010). Additionally, higher CO_2 concentrations in the future may enhance the growth rates of pervasive woody weed species (see Biosecurity below).

In an agricultural context, temperature and rainfall are strongly interconnected. Land areas have warmed by an average of 1.4°C between 1910 and 1920, with more significant warming inland

than near the coast (see Chapter 3, the 'Historical Changes in Climate' section). Rising temperatures increase evaporation from open water storages and transpiration from crops and pastures. Reduced rainfall in core cereal regions, such as southwest Western Australia and western Victoria (see map Chapter 3, Figure 3.1), has had substantial negative impacts on the yields of existing cultivars (Hughes et al., 2015). For example, the water-limited yield of wheat declined by 27% between 1990 and 2015, with a further 4% loss averted due to elevated CO_2 levels (Hochman et al., 2017).

Future climate change is likely to significantly affect cotton yields in southern Queensland and northern New South Wales (see map Chapter 3, Figure 3.1). One study showed that projected changes in key climate variables due to climate change will likely result in cotton yields declining by 17% by 2050 (Williams et al., 2015). Ameliorating these downward trends will require more water for irrigation, resulting in conflicts with downstream water users – including legislative requirements for environmental flows (see Chapter 5, Box 5.4). Similarly, rice production depends entirely on water availability (Ashton and van Dijk, 2017), and the industry has dramatically improved its water use efficiency in recent times. However, the projected climate change effects of lower rainfall, higher temperatures and higher evapotranspiration (see Chapter 3, Table 3.2) make future rice production particularly challenging (McRobert, et al. 2019).

Studies project changes in the 'frost window' that is likely to have adverse impacts on broadacre

---

[9] Photosynthesis occurs in all green plants, whereby light energy is captured and used to convert water, carbon dioxide and minerals into oxygen and energy-rich organic compounds.

cropping, viticulture and horticulture (McRobert et al., 2019). Increases in the frost season are linked to observed and projected declines in cool season rainfall over most of southern Australia (see Chapter 3, Table 3.2). Reduced cloud cover and lower humidity in winter are conducive for frosts. Over the past 20 years, frost-related production risk has increased by 30% across the country's wheat-growing region (Crimp et al., 2014). Likewise, drier and cooler winters have decreased wine grape yield and delayed budburst in southern Australia (Bonada et al., 2020).

Rising temperatures in the growing season are likely to affect the yield of several critical crop species. For example, wheat, rice and maize have reduced growth rates at temperatures greater than 30°C, particularly if soil moisture is limiting (Sánchez et al. 2014). The sucrose content of sugarcane and the flavour of wine grapes are both negatively impacted by high temperatures (Howden et al. 2014). Periods of high temperature and an expected increase in atmospheric CO_2 concentration are significant threats to wheat production (Bokshi et al., 2021). Tropical fruit crops, such as mangoes, are also exposed to increasing minimum and maximum temperatures, reducing the number of inductive days for flowering (Clonan et al., 2021). Many of these essential industries will have to relocate to areas where temperatures are more suitable (see the distribution of crops and livestock below).

About 50% of all profits from Australian agriculture are from irrigated production (Hughes et al., 2015). Observed and projected changes in rainfall and hydrology (see Chapter 3, Table 3.2 and Chapter 5, Box 5.4) will result in runoff declines in the country's central irrigated agricultural and horticultural regions, e.g. the Murray-Darling Basin. Water resources for irrigation will also be adversely impacted by rising temperatures and associated higher evapotranspiration (Lawrence et al., 2022). Other impacts include reduced connections between surface and groundwater, interceptions of runoff by farm dams and water impoundments, tree regrowth after more frequent fires and increased risk of soil erosion and loss of topsoil (Hughes et al., 2015, see Natural capital above).

BIOSECURITY

The distribution and occurrence of pests, weeds and disease can damage field and tree crops and livestock (Howlett and Henry, 2020). Complex and idiosyncratic interactions between biosecurity risks and climate change are inevitable in the future (Hughes et al., 2015). In some regions, weeds, pests and diseases already present may increase their population sizes and ranges or become more virulent (Roger et al. 2015). New biosecurity agents may become prevalent in other regions as the climate warms and creates more ideal conditions. Australia's most severe biosecurity risk will be from the southern movement of tropical and sub-tropical weed, pest and disease vectors into southern agricultural and horticultural districts. For example, cattle ticks are likely to spread into southern Queensland and northern New South Wales (Hughes et al., 2015). However, well-established weeds, pests and diseases may decline in some southern regions as the climate becomes drier and less suitable for their growing requirements, resulting in higher crop productivity.

HEAT STRESS ON LIVESTOCK

Increasing average and extremely high temperatures (see Chapter 3, Table 3.2, Box 3.1) worsen heat stress for extensive and intensive livestock industries in Australia (Hughes et al., 2015). At 3°C of warming (above pre-industrial levels), livestock in the northern third of Australia will suffer heat stress almost daily (Australian Academy of Science, 2021, see also Box 6.3). Heat stress in cattle reduces their growth rates and affects reproduction, while extreme heat waves can lead to mortality (Lawrence et al., 2022). Likewise, heat stress reduces milk yield in the dairy industry by 10%–30% and up to 40% in extreme heatwave conditions (McRobert et al., 2019). Heat stress and other climate changes (lower rainfall and droughts) are likely to affect pasture growth, quality and nutrition value, leading to declines in wool and red meat production quality and overall farm productivity (McRobert et al., 2019). There are also likely to be mixed impacts on poultry and pork agri-businesses. Increased temperatures and more frequent heatwaves will directly increase heat stress for layer hens and cause declines in overall productivity (McRobert et al., 2019). By comparison, the pork industry is likely to be indirectly affected by climate change through increased costs of inputs such as grains, electricity and fuel (McRobert et al., 2019).

DISTRIBUTION OF CROPS AND LIVESTOCK

Climate changes will continue to shift agro-ecological zones further south (Turton, 2017, McRobert et al., 2019). The most adverse impact from this southern shift will be the loss of suitable arable land for crops and livestock, although some regions may benefit in the medium term. For example, in Tasmania, wheat yields are projected to increase due to a lengthening of the growing season (Lawrence et al., 2022). In extreme cases, crops such as wheat and cotton may become unviable in areas where they are currently grown (Howlett and Henry, 2020, Australian Academy of Science, 2021). Notably, cotton production has decreased during droughts due to its firm reliance on water (Stokes and Howden, 2010). Given projections of lower water availability in the future (see Chapter 3, Table 3.2; Chapter 5, Box 5.4), it is likely that cotton production will continue to decline. Similarly, where wheat can be profitably grown, crops will diminish as growing regions experience declining seasonal rainfall, increasing temperatures and higher evaporative rates (Shabani and Kotsey, 2016, Australian Academy of Science, 2021).

Many agri-businesses will have to adapt to climate change or relocate to more suitable regions (Australian Academy of Science, 2021). For example, projected temperature increases, and the frequency and severity of heatwaves pose a severe threat to the dairy industry and likely will cause relocation to suitable regions or industry exits (McRobert et al., 2019). Australia's important viticulture sector is also highly vulnerable to climate change (Figure 6.2). By 2050, up to 70% of the country's wine-growing regions, with classic Mediterranean climates (see Chapter 3, Figure 3.3, *Csa/Csb* climate sub-types), will be much less suitable or unsuitable for grape production (Hughes et al., 2015). High-value red grape varietals are particularly sensitive to higher temperatures and lower rainfall (McRobert et al., 2019). Expansion of viticulture to cooler growing regions – such as Tasmania – is already occurring. Other compounding effects include projected increases in pests and pathogens with warming and decreases in water security for irrigated vineyards. Similarly, the diverse horticultural sector is likely to be affected across the country's growing regions, from tropical to temperate climates (see Chapter 3, Figure 3.3). However, there may be opportunities to shift some industries south to more suitable regions, but there remain concerns about horticultural production's long-term reliability and viability (Hughes et al., 2015).

The Australian sugar industry is restricted to warmer and wetter parts of coastal Queensland from just north of Cairns to as far south as northeast New South Wales (see map Chapter 3, Figure 3.1). In the short term, increasing temperatures and CO_2 levels will likely enhance sugarcane crop growth, resulting in increased yield and an extended growing season (Lawrence et al., 2022). Historically, sugarcane has been resilient to occasional dry spells, floods, tropical cyclones and market fluctuations. However, the industry is likely to become less resilient under climate change, with rising temperatures, changing rainfall patterns and declining water security for regions where crops are irrigated (McRobert et al. 2019). Sugarcane plantations on coastal flats are also vulnerable to rising sea levels, saltwater intrusion, acid sulphate soils and storm surges from tropical cyclones (Turton, 2012, Hughes et al., 2015).

Forestry (native forest products and single-species plantations) and fisheries (marine fisheries and aquaculture) are two other primary industries that are vulnerable to observed and projected changes in climate (see Box 6.4). The main climate risks include declining rainfall in native forest and plantation regions, rising air and ocean temperatures and increasing frequency and severity of extreme climate events, such as marine heatwaves and bushfires (see Chapter 3, Table 3.2; Chapter 5, Box 5.3).

Climate-resilient development pathways and opportunities

GREENHOUSE GAS EMISSIONS

Agriculture is an 'extremely' emissions-intensive sector (Deloitte Access Economics, 2020). The industry was responsible for 15% of Australia's GHG emissions in 2019, emitting 76.5 million tonnes of carbon dioxide equivalents (Mt CO_2-e) (Wood et al., 2021). Of these emissions, 69.8 Mt CO_2-e came from agricultural processes such as managing livestock or using fertilisers, while 6.7 Mt CO_2-e

Figure 6.2 Viticulture in the Barossa Valley, South Australia. This wine region produces some of the world's highest quality red wine vintages and is highly vulnerable to climate change. (Credit: Mikael Andreasson/Unsplash/Public Domain.)

came from the combustion of fuels in the agriculture, forestry and fishing sectors (Department of Industry, Science, Energy and Resources, 2021c). Cattle and sheep were responsible for 75% of emissions in the agricultural industry that year (Wood et al., 2021). However, amounts emitted each year tend to fluctuate because of the effects of drought on stocking levels. About 54% of the agricultural sector's CO_2-e emissions in 2019 was methane (CH_4) – a potent GHG (see Chapter 2, Table 2.1). About 37% of the CH_4 came from cows and other livestock due to the fermentation of plant matter in their stomachs (Howlett and Henry, 2020).

The land sector[10] includes land-based processes that are not directly related to agricultural production, such as land clearing, forestry and changes in soil carbon content (Wood et al., 2021, see also Box 6.4). This sector is the only one that currently removes more carbon from the atmosphere than it emits. This carbon sequestration process

has reduced GHG emissions by 26.3 Mt CO_2-e in 2019–2020 (Department of Industry, Science, Energy and Resources, 2021c). However, this sector produced net emissions of 87.8 Mt CO_2-e in 2005 (Department of Industry, Science, Energy and Resources, 2021c), driven mainly by deforestation for agriculture (Wood et al., 2021). Historically, it has been as high as net emissions of 191.8 Mt CO_2-e in 1990 (Department of Industry, Science, Energy and Resources, 2021c), coincidentally the baseline year for the Kyoto Protocol targets (see Chapter 4).

Australian Government modelling asserts that agricultural output could climb to AU$112 billion in 2030 (above the AU$100 billion industry goal) and AU$131 billion by 2050, while CO_2-e emissions will fall to 36% below 2005 levels (Commonwealth of Australia, 2021c). Farmers and land managers can achieve this outcome by adopting low-emissions technologies and abatement practices through the Emissions Reduction Fund (ERF) (see Chapter 4, the 'Roles and Responsibilities for Climate Change Action in Australia' section). Examples of agricultural activities under the ERF include: improving soil carbon, changing grazing practices using livestock feed to reduce methane and changing nitrogen application in cropping systems (see

---

[10]The land sector (often called 'land use, land use change and forestry' or 'LULUCF') includes processes that both add to and remove GHG emissions from the atmosphere (see Chapter 4, the 'Paris Agreement' section).

BOX 6.4: Observed and projected impacts of climate change on forestry and fisheries in Australia

Forestry

Native forests cover about 149 million ha. Over 2,000 tree species are highly vulnerable to climate change because their current distributions cover only narrow climate ranges [1,2].

Plantation forests cover 1.94 million ha [3] and are dominated by *Pinus radiata* (radiata pine) and *Eucalyptus globulus* (blue gum) that account for ~70% of the total plantation area [2].

Potential impacts of climate change on six forestry regions showed that most tree species were projected to have net reductions in growth by 2030 (varied by tree species and forestry region) [4].

Species composition could change through changes in interspecies interactions and shifts in species distributions [1].

Areas that forest types occupy could change [1].

Increasing temperatures and reduced rainfall will result in decreases in growth rates of trees [4].

Increases in the frequency and severity of bushfires will impact on tree mortality (see Chapter 5, Box 5.3) [1,2].

Some plantation forests in hot and/or dry areas may be used as 'sentinel' sites for monitoring early effects of climate change on productivity and mortality [2].

Plantation productivity may benefit from rising CO_2 levels, but growth rates may be reduced by rising temperature and increased water use (especially if rainfall is reduced) [1,2].

Other risks include increased pests and diseases, and bushfires, but some agents may decline with lower rainfall and associated dryness [2].

Climate change could compound other and existing stressors [1]. (See Chapter 3, Table 3.2)

Very little is known about the differential responses of many native forest tree species to changes in CO_2 and other climate attributes (see Chapter 5, Box 5.5) [1,2].

Fisheries

Climate change is impacting commercial, recreation and Indigenous fisheries and aquaculture [5,6,8].

The fisheries sector is seen to have a greater ability to adapt to climate change compared with land-based agriculture and horticulture. This is due to the ability of marine species to move south as oceans warm [4,7].

Ocean warming, particularly on the east coast due to the strengthening of the East Australia Current, is changing the distribution and abundance of species targeted by marine fisheries and the location of suitable environments for aquaculture species [1,7,8].

However, southward migration species will alter fisheries catch rates and target species (see Chapter 5, Box 5.6) [4,7,8].

Reproduction and development rates of target species will also be affected by rising ocean temperatures [4,5,7,8].

Future increases/decreases of commercial fish catches, depending on the species [7].

Projected increases in extreme events (e.g. marine heatwaves) will impact on fisheries, while more severe storms and cyclones will increase risks for assets (e.g. vessels and nets) [4].

Increasing adverse impacts and disruptions to seafood value chains [7].

Threats to food and nutritional security [7].

Selective breeding programmes will adapt some aquaculture species for warmer conditions, but changes in farm location may be required for some businesses [5,7].

Other climate change impacts on fisheries include ocean acidification and declines in dissolved oxygen [7].

Forestry	Fisheries
Both radiata pine and blue gum are grown over relatively wide climate ranges. This means their core growing areas should not be highly vulnerable to climate change in the short to medium term [2].	For fisheries to remain productive and sustainable (environmentally and commercially), there is a need to incorporate climate change considerations into management and planning, and to implement planned climate adaptation options [5,6,8].
With increasing warming and/or drying of some sites, alternative tree species that are better suited to such conditions should be considered.	This includes the development of decision-support tools for stakeholders and managers, together with better ways to deliver climate information to assist with adaptation [5,6,8].
Other adaptation management actions should also be applied, e.g. planting at wider spacings at drier sites [2].	

Sources: [1] Boulter (2012), [2] Booth et al. (2010), [3] Downham and Gavran (2019), [4] McRobert et al. (2019), [5] Hobday and Poloczanska (2010), [6] Fogarty et al. (2020), [7] Australian Academy of Science (2021), [8] Lawrence et al. (2022).
Climate adaptation strategies are given for both sectors. For climate projections, refer to Chapter 3 (Table 3.2).

Mitigation below). However, GHG mitigation projects are likely to be very costly to establish and maintain (Commonwealth of Australia, 2021c). There is a high reliance on carbon reduction in the agricultural and land sectors, such as incentivising improved savannah fire management to store and avoid carbon emissions. Australia's extensive tropical savannah ecosystems (see Chapter 3, Figure 3.3: *Aw* climate sub-type) will play a significant carbon storage role in the future. This role will require integrated livestock grazing and fire management systems to maximise grazing yield, carbon sequestration and GHG emissions (Scheiter et al., 2015).

Despite many challenges, the agricultural sector plans to reduce GHG emissions (Howlett and Henry, 2020). For example, the red meat sector (Meat and Livestock Australia) has an aspirational target to be carbon neutral by 2030 (MLA, 2017). At the sector-wide level, the National Farmers Federation (NFF) has approved a landmark climate change policy that supports a target of net-zero carbon emissions by 2050 (National Farmers Federation , 2020). Agri-businesses already use land management strategies that mitigate the effects of climate change (Howlett and Henry, 2020). These include no-till cultivation before sowing, which retains the previous crop stubble, reduces soil erosion and retains soil carbon. As well as offsetting (abating) GHG emissions, soil carbon retention has several co-benefits to farmers, including lower fuel costs, improved soil structure and increased water use

efficiency for crops. However, there does appear to be a strong reliance on carbon abatement schemes in the red meat sector rather than direct emissions reduction options. Similarly, there is a high expectation about potential methane emissions reductions from – yet to be fully realised and upscaled – livestock feed technologies (see Table 6.2). The Australian Government's Long-Term Emissions Reduction Plan (Commonwealth of Australia, 2021a) supports these somewhat contentious abatement pathways.

Despite numerous strong efforts by agri-businesses, projections show that combined GHG emissions from the agriculture and land sectors will rise to 82 Mt CO_2-e by 2030, 12% above 2020 levels (Department of Industry, Science, Energy and Resources, 2021c). Achieving GHG emissions cuts will require systemic and transformative changes in the management of Australia's agricultural and land sectors, including the development and application of existing and novel strategies to reduce carbon emissions and enhance carbon sinks across both sectors (see Chapter 4, Box 4.4).

MITIGATION PATHWAYS

While the agricultural sector is highly vulnerable to climate change, reducing GHG emissions from its activities is problematic (Wood et al., 2021). For example, there are no credible methods to eliminate CH_4 emissions from extensive cattle

Table 6.2 Climate-resilient mitigation pathways for Australia's agricultural and land sectors

Climate-resilient mitigation strategies

Emissions reduction: Technologies, practices or a combination of both that directly reduce GHG emissions (short to long term) [1].

The Federal Government should improve the Emissions Reduction Fund (ERF) by expanding methods related to agricultural practices (short term) [3].

Methane reductions from extensive livestock systems: Livestock and pasture management (including plant species selection) and offsetting[1] emissions through carbon sequestration (short term) [1].

Methane reductions from intensive livestock systems: Seaweed feed additive for cattle and dairy cows (medium to long term) [2].

Solutions such as alternate hydrogen sinks in the rumen of cattle and sheep, vaccines and livestock breeding (long term) [1].

Better manure and fertiliser management (short to medium term) [3].

The Australian Government should include technologies to reduce animal emissions as a priority in its Low-Emissions Technology Statements (short to long term) [3,4].

Nitrous oxide reductions: Most actions are identical to best practice management approaches to maximise the efficiency of nitrogen (N) use and its adverse environmental impacts (e.g. water quality) (short to medium term) [1,4].

Managing the availability of inorganic N relative to crop needs to slow nitrification and build-up of nitrate in soils (medium to long term) [1,4].

Managing soil water by improved drainage or better timing of irrigation to reduce water logging and enhance potential for denitrification (short to medium term) [1,4,5].

Enhancing GHG sinks: increasing the capture and storage of photosynthetically derived carbon (C) in plant biomass, soil or harvested products (short to medium term) [1,5,6].

Smarter land management can boost farm productivity and store carbon, creating carbon credits that will be in demand as the economy approaches net zero (medium to long term) [3,5,6].

Fossil fuel substitution: use of crops and residues for electricity generation or liquid fuel production (short to medium term) [1].

Land sector: Significant opportunities exist to offset GHG emissions by increasing C storage (sinks) in landscapes across the country, e.g. soil organic C and biochar (medium to long term) [1,3–6].

State and territory governments should not weaken existing land clearing laws and should aim to keep existing stocks of nature-based carbon at or above current levels (short term) [3,6].

Develop methodologies to value land sequestration, including valuation of farming practices that improve and enhance soil C storage (short term) [6].

Enable standard valuation of co-benefits as part of carbon offset schemes to provide a storefront for voluntary corporate action (short term) [6].

Climate change, notably bushfires and drought, presents significant risks to C stored in soil and vegetation – creating possibilities for positive (warming) feedbacks to the climate system. [1,3,4]

Mitigation pathways: need to be based on whole-systems emissions budgets to permit trade-offs with other factors and enterprise-scale economics (medium to long term). [1,3,5]

This requires 'best management' systems working at an integrated landscape-scale with C sequestration and GHG reductions as an integrated outcome, balanced alongside economic, social and environmental outcomes (medium to long term). [1,3,5]

Source: [1] Carlyle et al. (2010), [2] Roque et al. (2021), [3] Wood et al. (2021), [4] Battaglia (2011), [5] McRobert et al. (2019), [6] O'Brien et al. (2020).

Strategies to reduce GHG emissions and enhance carbon sinks are identified for a range of timeframes in parentheses. Short term is <10 years, medium term is 11–30 years, and long term is >30 years.

and sheep grazing. However, there are promising opportunities to reduce emissions through new sources of stock feeds, like seaweed additives, for intensive livestock (e.g., Roque et al., 2021). At the larger scale, it will take time to implement better manure and fertiliser management across the country's 50,000 broadacre farms – along with replacing diesel vehicles and machinery with low- or zero-emissions ones (Wood et al., 2021, see Transportation sector above). Nonetheless, agri-businesses are well-placed now to benefit from applying strategies to reduce GHG emissions, enhance carbon sinks and play their role in contributing to net-zero emissions by 2050 (Table 6.2). One modelling study found a net economic benefit for Australian agriculture in transitioning to a low-carbon economy, including higher productivity and the means to meet future domestic and international demand for food (Porfirio et al., 2018).

Wood et al. (2021) propose four overarching strategies to facilitate climate-resilient mitigation pathways for Australia's agricultural and land sectors. First, including agriculture and land sectors in any national net-zero target is paramount for the economy to reach net zero by 2050 (Table 6.2). This policy will reduce the risk to agricultural exporters of likely future carbon tariffs from some importing countries. Second, adoption and deployment of lower-emissions technology and practices need to happen now, not at some stage in the future. Notably, the Australian Government should improve its ERF (see Chapter 4, the 'Roles and Responsibilities for Climate Change Action in Australia' section) by expanding its methods related to agricultural practices (see also Business Council of Australia, 2021). This change includes:

- Allowing single projects to be registered under multiple methods;
- Providing a fixed-price purchasing desk for proponents of small projects;
- Developing a carbon credit exchange that differentiates between types of credits; and
- Strengthening demand signals for credits (that must have integrity).

As part of this climate mitigation strategy, the Australian Government should invest in a multi-decade outreach programme to deliver advice to farmers on practically reducing farm emissions and securing resilient income streams (see Chapter 4, Box 4.4). At the same time, it should consider alternative financing mechanisms to support the deployment of lower-emissions practices, such as income-contingent loans, to share the risk with farmers. Third, wise spending this decade will allow for more effective technologies and policies in the future. This strategy prioritises technologies to reduce animal emissions (Table 6.2). The Australian Government should expand the remit and increase funding of ARENA to support the early-stage development of low-emissions agricultural technologies that are not energy-related. In the low-emissions technology space, all governments (federal, state and territory) should consider what additional policies (subsidies, penalties or mandates) are needed to ensure the deployment of any new technologies and reduce their cost. Similarly, there should be no limits on landholders' preferences to perform credible CO_2-removing activities, e.g. the sequestration of carbon in agricultural soils under the ERF (see Business Council of Australia, 2021). Lastly, state and territory governments should not weaken existing land clearing laws. They should aim to keep existing stocks of nature-based carbon at or above current levels (Table 6.2, see also Natural capital above).

Mitigation strategies will differ in their timeframes for adoption and implementation (Table 6.2). It is also critical that actions across the agricultural and land sectors avoid mal-mitigation pathways that may have perverse economic, social and environmental outcomes (Allen et al., 2018, McRobert et al., 2019). Short-term strategies that reduce emissions from both sectors may mistakenly lock in land management practices or farm technologies that eventually contribute to higher emissions (see Chapter 4, Box 4.4). This dilemma may also counteract the effectiveness of future adaptation pathways (see Adaptation pathways below) and other forms of mitigation. Another concern is the demand to increase water security (see Chapter 5, Box 5.4) for both agricultural and urban uses may favour the development of dams and other forms of water reallocation. This decision may later have significant adverse impacts on the natural environment and downstream water users (Hughes et al., 2015, Australian Academy of Science, 2021). This predicament sends a warning to proponents of the 'Development of Northern Australia Agenda' (Box 6.3). This significant 'nation building' programme comes with considerable agronomic, environmental, institutional, social and financial challenges for agriculture. Failing to recognise these significant challenges could inadvertently create mal-mitigation and maladaptation pathways to climate change for agriculture in northern regions (Hughes et al., 2015, Turton, 2017, Lawrence et al., 2022).

ADAPTATION PATHWAYS

Future climate projections (see Chapter 3, Table 3.2, Box 3.1) suggest climate change could make conditions more challenging for Australian farmers and require significant climate adaptation responses (Hughes and Gooday, 2021, Australian Academy of Science, 2021, Lawrence et al., 2022). However, given the size of the country and the diversity of climate types (see Chapter 3, Figure 3.3), climate change impacts will vary considerably across farming regions and among agri-businesses. Some areas and agri-businesses may benefit from climate change, while others may falter and eventually fail. The magnitude and severity of any climate change will be dependent on future GHG emissions (see Chapter 2, Table 2.6). Any adaptation will likely focus on the near future rather than the long term and must address climate variability and climate change seamlessly (National Climate Change Adaptation Research Facility, 2013c). Climate change will almost certainly lead to structural changes in the agricultural and land sectors, including shifts in land use and increases in average farm sizes (Hughes and Gooday, 2021). Climate-resilient adaptation pathways will require incremental, systemic and transformative changes across both sectors (Hughes et al. 2015, see Chapter 4, Figures 4.2 and 4.3 and Box 4.1).

Building climate-resilient adaptation pathways for agriculture will need to be progressive, including resilience-focused planning for rural settlements (see Chapter 7, Social Systems), land use, industry, infrastructure and value chains (Lawrence et al., 2022). Hence, it will be essential to build resilience and reduce vulnerability in the sector to climate change (see Figure 4.1, Box 4.1: Accommodate and Protect strategies). This approach entails developing and implementing processes and actions that contribute to high adaptive capacity and resilience (National Climate Change Adaptation Research Facility, 2013c). In doing this, it needs to be recognised that some agricultural

communities, agri-businesses or regions will have a greater capacity to adapt than others (Stokes and Howden, 2011). Understanding the relevance of scale will also be important (National Climate Change Adaptation Research Facility, 2013c). For example, the scale of investment in machinery (and co-ownership across properties), the scale of supplementation with external income and the size of farm enterprises. Likewise, in this early stage, it is essential for government and non-government agencies to assist agri-businesses with understanding why adaptation is needed and how they can make it part of their management plans and strategies (Stokes and Howden, 2011). This strategy will also require education, extension activities and information (e.g. seasonal forecasts, planting dates and stock reductions) (National Climate Change Adaptation Research Facility, 2013c). Notably, many climate adaptation options in agriculture are incremental, based mainly on existing 'best practice' natural resource management. Examples include minimal or no-till practices, stubble retention, use of irrigation, drought-resistant varieties, heat-resilient varieties that can tolerate rising CO_2 concentrations, changed planting times, changed crop spacing, and nutrient and canopy management (Howden et al., 2010, Hughes et al., 2015, Bokshi et al., 2021). Such 'no regrets' practices do not require agri-businesses to make radical changes to their farming operations in the short term (Stokes and Howden, 2011).

Over time, adaptation strategies will have to involve transformative changes in agricultural practices or geographical location (National Climate Change Adaptation Research Facility, 2013c, Hughes et al., 2015, see Box 4.1: Avoid and Retreat strategies). As discussed above, the geographical relocation of agri-businesses is already occurring and is sure to continue in the future, even with lower GHG emissions scenarios. The main drivers of relocation will be declining rainfall and rising temperatures in major southern agricultural regions (see Chapter 3, Table 3.2). Some agri-businesses have purchased land in cooler and higher rainfall zones (e.g. Tasmania) with plans to relocate their industries later (National Climate Change Adaptation Research Facility, 2013c). Relocation to wetter northern regions seems unlikely due to rising average temperatures and projected increases in numbers of hot days (>35°C). Such conditions will be highly challenging for many agri-businesses experienced only in temperate-climate production systems (see Box 6.3).

Marginal lands – mainly in southern Australia – will continue to become more marginal (National Climate Change Adaptation Research Facility, 2013c). While this will eventually mean the end of some traditional farming practices, new enterprises may seize opportunities (Crimp et al., 2014). Examples include native perennial plants for forage and erosion control and carbon farming initiatives (National Climate Change Adaptation Research Facility, 2013c, McRobert et al., 2019). As time goes on, changes in land use will most likely only occur at small scales as part of incremental adaptation to reduce risk. Practices that have been standard for at least 30 years will continue to be relevant and have applications to other regions that have yet to experience significant climate change. Emerging agricultural practices will also be part of climate adaptation (National Climate Change Adaptation Research Facility, 2013c). Examples include climate change-ready crops, 'climate-sensitive' precision agriculture and agri-business diversification and risk management (Hughes et al., 2015, Lawrence et al., 2022). Investment in some significant engineering schemes to offset the adverse effects of warming and drying on productivity (e.g. dams and irrigation) may pay off in the short term. However, such schemes risk becoming maladaptive in the long term as groundwater, and large-scale water storage resources become scarce (Lawrence et al., 2022).

There has been relatively little research conducted to aid our understanding of the benefits and challenges of transformative climate adaptation in the agricultural and land sectors (Hughes et al., 2015). Such transformation will be critical in the medium to long term if agri-businesses aim to survive and thrive under a changing climate. Geographical relocations of some agri-businesses, such as dairy, horticulture and viticulture, are inevitable as incremental adaptation options become unviable. Significant changes may be necessary for some vulnerable regions, such as diversification in agricultural enterprises transitioning to different land uses (Lawrence et al., 2022). Examples include carbon sequestration, renewable energy production and biodiversity conservation (see Chapter 5, Boxes 5.1 and 5.2).

In planning for transformational change in the agricultural and land sectors, there is more emphasis on sustainability, systemic thinking and risk spreading (National Climate Change Adaptation Research Facility, 2013c, McRobert et al., 2019, Lawrence et al., 2022). This change will require a whole-of-landscape and a triple-bottom-line approach to agriculture and land management (see Integration with the Sustainable Development Goals below). Avoiding maladaptation pathways (see Chapter 4, Figures 4.1 and 4.3, Box 4.1) in both sectors will require new skills, understanding new markets and many novel and expensive infrastructure (Rickards and Howden, 2012). Such transformative changes will bring high costs and risks to Australian agri-businesses in the future.

Some existing policy instruments in the agricultural and land sectors have important implications for climate mitigation and adaptation (National Climate Change Adaptation Research Facility, 2013c). These include environmental management and conservation, water trading, drought relief, irrigation and groundwater management. However, new policies and investments will be needed to support climate-resilient development pathways in the agricultural and land sectors (National Climate Change Adaptation Research Facility, 2013c, Lawrence et al., 2022). These include:

- Improved governance and collaboration to build rural resilience (see Chapter 7), including regional and basin-scale initiatives;
- Improved water policies and initiatives (such as the Murray-Darling Basin Plan) and changes in management and technologies (see Chapter 5, Box 5.4);
- Management of biosecurity issues (see Biosecurity above) will require cross-boundary (jurisdictional) decision-making powers, resourcing and capacity for rapid action;
- Policy related to infrastructure will have important implications for agricultural adaptation, including telecommunications to support agri-businesses, transport investment and energy pricing (see Energy and Transportation sectors above);
- If farmers are to be land stewards under climate change, then policies will be required to facilitate that role, e.g. payment for ecosystem services (see Chapter 5, Box 5.2);
- Carbon sequestration (Table 6.2) and carbon pricing are likely to provide farmers with additional income and so build resilience, although some land yields little potential for this;

- With the squeeze on land from competing uses (e.g. urban development), there is an urgent need for policy protection of prime agricultural land under climate change; and
- Monitoring and evaluation to understand the performance and effects of mitigation and adaptation policies that address climate change for both sectors, e.g. the Australian Government's Future Drought Fund (Department of Agriculture, Water and the Environment 2021).

Barriers to climate adaptation in the agricultural sector will need overcoming if agri-businesses plan to survive and thrive in the future (see Chapter 4, Box 4.2: barriers). As discussed above, many agri-businesses have been practising small-scale, incremental adaptation to changing climate conditions for decades. However, the substantial capital investment needed for new infrastructure and skills development to support more structural and transformational change represents a significant barrier, especially at the level of individual farms and farmers (Hughes et al., 2015). In addition, existing financial hardship – often a legacy of years of drought – will limit the capacity of farmers to make new capital investments or develop new skills and expertise. Other barriers to effective climate adaptation include confusion about the science information and data required by agri-businesses, perceptions of the uncertainty of the science and the complexity of interactions between climate change and other change agents (Hughes et al., 2015). An essential enabler for the sector is its general willingness to engage in climate mitigation and adaptation and seek new opportunities from climate change and agri-business innovation (see Chapter 4, Box 4.2: enablers).

INTEGRATION WITH THE SUSTAINABLE DEVELOPMENT GOALS

Five SDGs are relevant to a national strategy on climate change and the agricultural sector (see also McRobert et al, 2019):

1. SDG2 (Zero hunger);
2. SDG6 (Clean water and sanitation);
3. SDG7 (Affordable and clean energy);
4. SDG12 (Responsible consumption and production); and
5. SDG13 (Climate action).

In the case of fisheries and forestry (Box 6.4) and the broader land sector, two other SDGs are also relevant:

1. SDG14 (Life below water); and
2. SDG15 (Life on land).

Table 6.3 provides a framework for how the agricultural and land sectors can effectively integrate climate change with goals, targets and indicators for each SDG.

Climate-resilient development pathways for the agricultural and land sectors will require structural and transformational changes in both sectors over the next 20 years, beyond the largely incremental changes that have happened so far (see above). Addressing the seven SDGs (Table 6.3) will contribute to lower carbon emissions for both sectors and build resilience to climate change through planned climate adaptation (see Chapter 4, Figure 4.1). If these sectors are to meet (or make progress towards) the 2030 SDG targets, structural and transformational changes will also need to commence in other 'climate-sensitive' economic and social sectors. These sectors include those dependent on sustainable agriculture and land management (see the 'Manufacturing and Construction' section below, and Chapter 7, Table 7.5).

MINING

Socio-economic context

The Australian mining sector has a long post-Colonial history (Blainey, 2013) and is a pillar of today's economy (Granwal, 2022). It is a significant industry in regional Australia, where it underpins the local economy of many small and remote communities. It also generates a critical minerals and metals processing sector (see Manufacturing sector below). Over 350 operating mines are in the country, with one-third located in Western Australia, one-quarter in Queensland and one-fifth in New South Wales (see map Chapter 3, Figure 3.1), making them the three main mining states in the country (Geoscience Australia, 2021). The sector produces 19 valuable minerals and metals in significant amounts, including bauxite (aluminium ore), iron ore, black coal, lithium, gold, lead, diamonds, rare earth elements, critical minerals, uranium and zinc. Mining is

Table 6.3 Integration of SDGs with the Australian agricultural and land sectors' responses to climate change

SDG	Goals	Targets	Indicators
SDG2: Zero hunger (*social*)	1. Produce enough food/fibre for needs (domestic and export) 2. Fair and just distribution of food/fibre 3. Improved nutritional value of produce	Reduced food/fibre waste Improved productivity Increased food security Healthier global population	• Prevalence of undernourishment/malnutrition (trading partners) • Prevalence of obesity (domestic) • Prevalence of food insecurity based on United Nations Food Security Experience Scale
SDG6: Clean water and sanitation (*economic and environmental*)	4. Ensure availability and sustainable management of water	Improved water quality entering waterways Substantially increase water use efficiency Integrated water resources management	• Proportion of bodies of water with good ambient water quality • Change in water use efficiency over time • Level of freshwater withdraw as a proportion of available freshwater resources • Degree of integrated water resources management in agriculture (0–100)
SDG7: Affordable clean energy (*economic and environmental*)	5. Support extension of renewable energy adaptation	Better farmer/grower/business resilience to energy price shocks Lower emissions from energy use	• Percentage of renewable energy use across sector • Energy emissions measurements • Percentage of farmer/grower/production supply chain income spent on energy use
SDG12: Responsible consumption and production (*economic and environmental*)	6. Sustainably productive environments 7. Increased natural capital	Increased efficiency Improved environmental health Greater stores of values for future production needs	• Proportion of food/fibre/wood production area under sustainable growing methods • Volume of production per labour unit by enterprise size • Biodiversity of agricultural and other land production • Increased soil health (0–100)

(Continued)

Table 6.3 (*Continued*) Integration of SDGs with the Australian agricultural and land sectors' responses to climate change

SDG	Goals	Targets	Indicators
SDG13: Climate action (*environmental*)	8. Carbon-neutral production 9. Resilience and adaptive responses to natural hazards	Reduced global warming Closed loop production (circular economy) and supply chain systems Mitigation of climate change impacts on food/fibre production	• Reduction in food/fibre/wood production • GHG emissions contribution (to zero) • Percentage of producers with a drought plan • Percentage of producers using renewable energy • Percentage of farmers using carbon farming or sequestration
SDG14: Life below water (*economic and environmental*)	10. Enhance the conservation and sustainable use of oceans and their resources	Effectively regulate harvesting Increase science and Indigenous knowledge about sustainable fisheries under climate change	• Proportion of fish stocks within biologically sustainable levels • Proportion of total budget for research on marine science and technology
SDG15: Life on land (*economic and environmental*)	11. Sustainable land management, reduced land degradation and biodiversity loss	Restore degraded land and soil, including land affected by droughts and floods Implement sustainable management of all types of forests, halt deforestation and restore degraded forests Increase science and Indigenous knowledge about sustainable land management under climate change	• Proportion of forest/land degraded over total land area • Progress towards sustainable forest management • Proportion of land that is degraded over total land area • Proportion of total budget for research on agricultural technology and sustainable natural resource management

Source: Expanded from McRobert et al. (2019).

This includes horticulture, viticulture, forestry and fisheries and their supply chains. Refer to Chapter 4 for an explanation of SDG goals, targets and indicators (see also Box 4.5). Triple-bottom line implications for each SDG are shown in italics.

strongly export-oriented, with minimal processing onshore. Major exports include iron ore, thermal and metallurgical coal, natural gas, crude oil, manganese, antimony, nickel, silver, cobalt, copper and tin. In 2021, the country was the world's largest exporter of coal and liquified natural gas. Most mineral exports go to China, followed by Japan and the Korean Republic, while coal exports go to India, China and Japan.

Unlike agriculture, mining has a tiny national 'footprint'. Specifically, granted mining leases account for around 0.6% of the landmass and mining's current footprint (including its waste) occupies less than 0.1% of the country (Minerals Council of Australia, 2021). By volume, the two most important mineral commodities are iron ore (29 mines), of which 97% comes from Western Australia, and coal (over 90 mines) originates from the east coast, in the states of Queensland and New South Wales (Minerals Council of Australia, 2021). In contrast to most global production, most (around 75%) of black coal in Australia is from open-cut mines (Figure 6.3). This ratio of 3:1 open-cut/surface to underground mines also applies to the broader (i.e. non-coal) local mining sector (Minerals Council of Australia, 2021).

The mining sector contributes to the Australian economy through export income, royalty payments and employment. The gross value added by the mining sector was in the hundreds of billions of Australian dollars in the past decade alone (Granwal, 2022). In 2020, the sector contributed 10.4% of the country's GDP and resources exports reached AU$222.1 billion in that year (Australian Bureau of Statistics, 2021a). Mining directly employed about 267,800 persons (or 2.0% of the workforce) in August 2021 (Australian Bureau of Statistics, 2021a). The industry also employed over 300,000 persons through various mining equipment, technology and services businesses (Commonwealth of Australia, 2021d). However, the sector has a strong multiplier effect, so the number of people employed in the more extensive resources sector is closer to 1.1 million (Commonwealth of Australia, 2021d). Many of these jobs are in metropolitan centres. During the COVID-19 pandemic, the mining sector has been described as a 'pillar of stability' for the Australian economy. However, the Australian Government provides fossil fuel subsidies (~ AU$12 billion annually) and exploration rights, which drives more extensive extraction and exports, ultimately increasing global GHG emissions. This situation is a problem for the Australian Government and fossil fuels industries, as international efforts unite on the pathway to net-zero emissions by 2050 – to

Figure 6.3 Open-cut coal mine in Hunter Valley. (Credit: Max Phillips/Wikimedia/Public Domain.)

be driven by more rapid decarbonisation after 2030 (see Chapter 4, the 'Paris Agreement' section).

Managing risk

The Australian mining sector has considerable experience managing a range of risks across its domestic and international operations. These risks include workplace health and safety, environmental, public health and safety, regulatory, competition, production, reputation and conflict minerals (Hodgkinson et al., 2014). This status has evolved due to a combination of a long history of implementing effective environmental and safety regulations and adopting voluntary codes of practice and standards (Austrade, 2013).

Energy costs and carbon risks are ongoing issues for the sector in a business environment of declining ore grade concentrations (Smith, 2013). Operationally, the energy cost is an upfront input to assessing mine suitability for potential or continued extraction. Current energy costs comprise 7%–12% of total mining costs, but this could be as high as 20%–30% for some mining sub-sectors by the 2030s, particularly for energy-intensive gold and copper mine operations (Smith, 2013). Another risk is diesel price exposure, as this fuel is essential to power generators, machinery and mining haul trucks. The sector currently receives generous taxation deductions on its use of diesel fuel. This tax exemption is a business risk, as the G20[11] and OECD (both include Australia) have committed to phasing out fossil fuel subsidies by 2030. A price on carbon is another risk to the mining sector because of its energy-intensive operations (Smith, 2013, Deloitte Access Economics, 2020). If carbon mitigation strategies are not instigated (see Mitigation pathways below), any increases in carbon prices will inflate energy costs for the sector, reducing profits and making some mine operations marginal or unsustainable.

Climate impacts and risks

The mining sector contributes to global climate change through its various activities (e.g. the combustion of fossil fuels in-country and elsewhere). There are also significant climate risks (e.g. floods, heatwaves, bushfires, tropical cyclones) for mine operations, supply chains and distribution networks (Hodgkinson et al., 2010, Nelson and Schuchard, 2011, Smith, 2013, Mason and Giurco, 2013). Fundamentally, the fossil fuels extractive industries (coal and natural gas) have limited adaptive capacity to deal with climate change due to their significant contributions to global GHG emissions. The global push for net-zero carbon emissions by 2050 (see Chapter 4, the 'Paris Agreement' section) and the associated rise of 'carbon-constrained' economics will negatively impact future thermal coal, natural gas and crude oil exports. With the development of 'green steel',[12] there is also likely to be a fall in the global demand for metallurgical coal over time as new technology for making steel evolves (see Manufacturing sector below). However, as discussed later, the mining sector is well-placed to utilise renewable energy options for its existing (and future) mine operations. There are also opportunities to engage in carbon offsets and growing new export markets for many of its products, e.g. rare earth elements and critical minerals (see Hodgkinson and Smith, 2021).

Despite the sector's high exposure to climate change and its inherent low adaptive capacity to deal with it (see Chapter 4, Figure 4.1), there is limited research on climate change impacts on mining in Australia and many other countries (Hodgkinson et al., 2014, Odell et al., 2018, Mavrommatis et al., 2019). This lack of research is somewhat perplexing, as climate change is highly likely to increase the sector's exposure to extreme weather events, energy costs and carbon price risks (Smith 2013, Lawrence et al. 2022). The mining industry is known for its many long-life assets, extended supply chains and significant freshwater requirements for its operations (Nelson and Schuchard, 2011, Smith, 2013, Mason and Giurco, 2013). Observed and potential impacts of climate change on the mining sector are likely across four key areas (Nelson and Schuchard, 2011):

1. Disturbance to mine infrastructure and operations;
2. Changing access to supply chains and distribution routes;
3. Challenges to worker health and safety conditions, and community relations; and
4. Challenges to environmental management and mitigation.

DISTURBANCE TO MINE INFRASTRUCTURE AND OPERATIONS

Depending on location, natural disasters and rising sea levels threaten mining operations and their supply chains (see Chapter 3, Table 3.2, Box 3.1). More intense (but less frequent) tropical cyclones in northern regions may damage mine infrastructure and equipment, leading to construction and operational disruptions (Smith, 2013, Mason and Giurco, 2013). Heavier rain events will likely exacerbate soil erosion near opencast mines, and rising sea levels may hinder access to coastal installations. For example, the 2010–2011 floods in Queensland disrupted commodity exports for an extended period, with a sizeable impact on gross state product (Steffen et al., 2019). With projections for more extreme droughts and floods for all key mining regions, there will be increased climate risk for mine site infrastructure and operations. During prolonged droughts, less water will be available for mining and mine site processing activities, bringing the increased risk for mine waste management and potential conflict with other water users for surface and groundwater supplies (see Chapter 5, Box 5.4). Hence, water costs are highly likely to increase for pre- and post-use mine water treatment (Nelson and Schuchard, 2011).

Projections for hotter and drier conditions in southern and central Australia (see Chapter 3, Table 3.2) will likely worsen bushfire risk for mining infrastructure and operations, increasing the risk of coal mine fires (Smith, 2013). At the same time, rising temperatures and more intense heatwaves will increase energy demand to cool underground mines and surface processing facilities (Nelson and Schuchard, 2011, Mason and Giurco, 2013). Increased heat will result in rising energy costs and a higher risk for self-combustion of coal

mining spoil piles (Smith, 2013). Heatwaves also increase energy demand by affecting the capacity of power transmission and distribution facilities, which then causes energy disruptions to mining operations (Nelson and Schuchard 2011). Energy rationing may lead to permanent decreases in mining production, affecting profits and commodity prices (see Energy sector above).

Greater intensity and damage from extreme weather events have already led to an exponential increase in weather-related insurance and reinsurance costs over the last 50 years (Smith, 2013). With the projected future increases in the intensity and frequency of extreme weather events (see Chapter 3, Table 3.2, Box 3.1), the cost of insurance will continue to rise, affecting mining company profits (see Chapter 7, Box 7.1). Several primary international insurance, reinsurance and finance companies have signalled their intent to divest their business from new thermal coal mines. This signal is a significant business risk to the thermal coal mining industry, as they may have to consider mechanisms to self-insure any extended or new mine operations.

CHANGING ACCESS TO SUPPLY CHAINS AND DISTRIBUTION ROUTES

Mining mostly occurs in isolated parts of Australia (Hodgkinson et al., 2014). Therefore, servicing the sector requires extensive transportation networks to allow providers to supply goods and services to mine sites, transport workers and move ore to facilities for processing and eventually to ports for export (Nelson and Schuchard, 2011, see the 'Energy amd Transportation' section above, and the 'Manufacturing and Construction' section below). Tropical cyclones, heavy rain events and extreme heat are likely to disrupt land transportation routes, degrade road and rail networks and damage processing plants and ports (Smith, 2013). Supply chains will also be affected by these extreme climate events. Disruptions are likely in the delivery of input materials for mine sites, such as steel, timber, cement, hydrochloric acid and cyanide, and consumables such as diesel, tyres and reagents (Nelson and Schuchard, 2011). Major disruptions cause curtailed mine production, reduced efficiencies and lower mine profits. Over time, sea-level rise and more frequent storms may affect port availability and timely transport to overseas markets.

CHALLENGES TO WORKER HEALTH AND SAFETY CONDITIONS AND COMMUNITY RELATIONS

Severe climate events present immediate health and safety risks to mine sites (Smith, 2013, Mason and Giurco, 2013). Long-term changes, such as rising average temperatures, may negatively impact worker recruitment, retention, safety and productivity (Nelson and Schuchard, 2011, McTernan et al., 2016). Increasing temperatures and associated heatwaves increase the risk of heat-related morbidity, enhancing the likelihood of accidents, injuries and fatalities, hence decreasing productivity (Mason and Giurco, 2013). In some cases, underground cooling systems, water and energy supplies may not cope with extreme heatwaves. Chronic heat-induced illness, irritation and absenteeism are likely to increase among the mining workforce (McTernan et al., 2016), to a point where it may prove difficult to live in some regions of northern Australia (Turton, 2017, see Box 6.3). Flooding and bushfires are also significant issues for mine sites and their supply chains and transportation networks (Nelson and Schuchard, 2011, Mason and Giurco, 2013). These may affect workers' safety on site, and along road and rail routes.

Mining companies face challenging physical environments (Hodgkinson et al., 2010). Adjacent human communities – beyond the actual mine sites – are also exposed to the same range of climate stressors (Nelson and Schuchard, 2011, Mason and Giurco, 2013). Some remote communities play an essential role in mining operations and supply chains and may host mineworkers and their families (see Chapter 7, the 'Cities, Settlements and Built Environment' section). Mining employs some Indigenous people whose communities are often near mine sites. These communities are socially marginalised and vulnerable to extreme climate events, which adversely impact their health, well-being and cultural values (see Chapter 7, Table 7.2). Mining companies may compete with Indigenous communities for scarce resources, such as water and energy, and this presents an increasing future risk to their social licence to operate. Lastly, labour deficits are a potential risk to the mining sector in the future. Local communities may move away from mine sites due to lack of water, power and more frequent isolation after extreme weather events (Hodgkinson et al., 2010).

CHALLENGES TO ENVIRONMENTAL MANAGEMENT

Australian mining industries are highly regulated, and their operations must comply with various federal, state and territory laws to minimise and palliate any adverse impacts on the natural environment. Projected increases in water scarcity and higher temperatures will make it more difficult to re-establish vegetation cover at open-cut mine sites in some regions (Nelson and Schuchard, 2011, Wardell-Johnson et al., 2015). However, elevated CO_2 levels may favour vegetation growth rates in some mining areas, provided water is not limited (see Chapter 5, Box 5.4). Projected increases in rainfall events will elevate the risk for tailings dam failure and the discharge of contaminated water into the wider natural environment (Metcalfe and Bui, 2016). However, higher evaporation rates may reduce the need for water treatment and disposal by reducing volumes and associated costs (Nelson and Schuchard, 2011). Another risk is accidental spillages along transportation routes which will worsen the impacts of mining operations on the ecology, hydrology and air quality of areas away from the actual mine site (Grech et al., 2016, Phillips, 2016, Ali et al., 2018).

For most mining companies, increases in extreme climate events pose a significant risk to business profits and viability. These include increased mine site remediation costs, increases in environmental liability, adverse impacts on community health and safety and considerable potential for reputational damage (Nelson and Schuchard, 2011, Hodgkinson et al., 2014). Decommissioning costs after mine closure are likely to increase due to more stringent site mitigation requirements to accommodate future extreme weather events, such as floods and bushfires (Smith, 2013, see also Chapter, Box 5.3). Another issue is legacy mine sites rehabilitated under antecedent climate conditions (Mason and Giurco, 2013). These may require supplemental protection measures to ensure they can cope with more extreme weather events in the future. This issue, in turn, will impose costs and risks of environmental liability and monitoring responsibilities on mine companies (Hodgkinson et al., 2014). In some cases, state and territory governments will eventually have to bear these costs of remediation and monitoring legacy mine sites.

Climate-resilient development pathways and opportunities

GREENHOUSE GAS EMISSIONS

Mining is mostly an emissions-intensive sector (Deloitte Access Economics, 2020). From an emissions perspective, coal mining and oil and gas extraction are 'extremely emissions intensive', while mining and quarrying (metals and minerals) are 'emissions intensive'. Australia's national GHG inventory identifies several categories that apply to emissions from the Australian mining and minerals processing sector (Department of Industry, Science, Energy and Resources, 2021c). These categories combine emissions from several sectors, so it is impossible to extract a single GHG emission value for the mining sector. Electricity, stationary energy and transport emissions – including those associated with mining operations – were described earlier (see the 'Energy and Transportation' section above). Emissions related to minerals and metals processing occur under industrial processes and product use (see the 'Manufacturing and Construction' section below). Fugitive emissions from fuel extraction occur during the production, processing, transport, storage, transmission and distribution of fossil fuels such as coal, crude oil and natural gas. These emissions amounted to 51.0 Mt CO_2-e in 2019, representing about 10% of the country's GHG emissions (Department of Industry, Science, Energy and Resources, 2021c).

A recent independent study – utilising remote sensing to detect fugitive methane (CH_4) emissions from coal mining basins in eastern Australia – has challenged the accuracy of the Australian Government's reporting of GHG emissions to international bodies (Kayrros, 2021). Their study found fugitive CH_4 emissions from 50-odd coal mines in the Bowen Basin in Central Queensland added up to about 1.5 Mt of CH_4 per year in 2018–2019. Government data for the same period was only one-third of that amount, raising concerns about the accuracy of the original data – as reported by the mining companies. Another peer-reviewed Dutch study (Sadavarte et al., 2021) has raised similar concerns about Australia's publicly released CH_4 data from mine sites. The authors call for increased monitoring and investment in methane recovery technologies for surface and underground mines. Given that CH_4 is

responsible for more than 30% of global warming to date (Intergovernmental Panel on Climate Change, 2021), any under-reporting of fugitive CH_4 emissions is a significant global climate risk.

Compared to other primary production sectors (e.g. agriculture), the mining sector has been relatively unconcerned about climate change until recently (Loechel et al., 2013). In 2020, the peak body for the mining sector, the Minerals Council of Australia (MCA), stated publicly – for the first time – its continued support for action on climate change (Minerals Council of Australia, 2020). MCA acknowledges that sustained global effort is required to reduce the risks of human-induced climate change. Their Climate Action Plan (2020–2023) supports a measured transition to a low carbon emissions global economy in line with the Paris Agreement and its goal of net-zero emissions by 2050 (Minerals Council of Australia, 2020, see also Chapter 4, the 'Paris Agreement' section). The plan outlines a series of actions around three key themes:

1. Support developing technology pathways to achieve significant reductions in Australia's greenhouse gas emissions;
2. Increased transparency on climate change-related reporting and informed advocacy; and
3. Knowledge sharing of the sector's responses to addressing climate change.

These three key themes attempt to address climate-resilient mitigation and adaptation pathways for the sector, which have importance to the future of regional Australia, where most mining activities occur.

MITIGATION PATHWAYS

The mining sector now recognises that climate change is a significant threat to its various sub-sectors and has instigated an action plan to meet future challenges and opportunities (Minerals Council of Australia, 2020). Given the sector's diversity, this presents significant challenges as the global economy divests from fossil fuels and transitions to renewables for energy sources for manufacturing, energy and transportation (see Chapter 4, Box 4.4). The low-carbon energy transition poses substantial risks for Australia's fossil fuel extractive industries, such as black coal and natural gas (Smith, 2013). On the other hand, growing demands for many other minerals and metals, such as iron ore, rare earth metals and critical minerals, are likely to bring increased economic benefits to the sector in the future (Hodgkinson and Smith, 2021).

The mining sector has identified some climate-resilient mitigation pathways for adoption over the coming decades (Minerals Council of Australia, 2020). Many of these are aspirational and include adopting unproven and expensive technologies (e.g. carbon capture and storage, see Box 6.5). There is also a firm reliance on 'offsetting' emissions from the sector, both within-country and overseas. The industry encourages substantial investment across a broad range of low-emissions technologies in emissions reductions. This strategy includes defining a more significant role for the minerals processing sector in the global and national transformation to lower-emissions technologies and export products (see the 'Manufacturing and Construction' section below).

The Business Council of Australia (2021) supports many MCA initiatives. Notably, they call for an expansion and tightening of the Safeguard Mechanism (see Introduction) as the primary policy lever to reduce emissions in the mining, energy and manufacturing sectors. Recommendations include:

1. A shorter timeframe under the Emissions Reduction Fund (ERF) (see Chapter 4, the 'Roles and Responsibilities for Climate Change Action in Australia' section) to reduce the barriers faced by projects with high upfront capital costs;
2. Establishing a pilot method programme to test new method ideas and expedite method preparation;
3. Implementation of a National Emissions Intensity Reduction Programme, to support businesses to transition their plant, equipment and processes to low-emissions alternatives; and
4. To fast-track major project approvals to accelerate new investment in low or zero-emissions projects and to accelerate the economic transition to a net-zero economy (Business Council of Australia, 2021).

In response to some of these recommendations, the Australian Government (Commonwealth of Australia, 2021a) has recently introduced a Safeguard Crediting Mechanism (SCM) to help incentivise new low-emissions technologies in mining, manufacturing, transport and gas sectors. This mechanism arose from the King Review (Department of Industry, Science, Energy and Resources, 2020). Until 2030, the SCM scheme will encourage projects that:

- Significantly reduce the emissions intensity of facilities covered by the Safeguard Mechanism; and
- Help develop and deploy emerging low-emissions technologies. However, the eligibility threshold for entities covered by the Safeguard Mechanism is currently 100,000 t CO_2-e per year. However, this high threshold will limit the number of entities that may actively participate in the scheme (Business Council of Australia, 2021).

The mining sector also supports innovative solutions for zero-emissions energy production at mine sites. For example, the uptake of electric and 'green' hydrogen-powered vehicles and machinery in mining activities is essential to reduce significant emissions from current technology, which largely relies on fossil fuels (Smith, 2013). This uptake could include using the same technologies for their supply chains (see Transportation sector above). In addition, the uptake of renewable energy sources at mine camps and mining towns will reduce their carbon emissions. In this regard, the sector has indicated its support of ARENA to continue accelerating the uptake of clean energy in the country. However, the reference to 'clean' energy implies a mixed-use of traditional fossil fuels and renewables for energy production.

In terms of enhancing carbon sinks, the mining sector advocates for access to low-cost international abatement for voluntary and compliance purposes (Minerals Council of Australia, 2020). In practice, this means offsetting carbon-equivalent emissions (CO_2-e) from mining and minerals processing in abatement schemes in other countries. These could include reforestation, avoided deforestation and soil and blue carbon[13] initiatives. The most controversial mitigation strategy advocated by the mining sector

---

[13] 'Blue carbon' is the carbon stored in coastal and marine ecosystems, e.g. mangroves, seagrasses, kelp forests.

and the Australian Government (Commonwealth of Australia, 2021a) is the development and upscaling of carbon capture and storage (CCS) and carbon capture, use and storage (CCUS). CCS and CCUS technologies claim to capture CO_2 emissions from fossil fuel-fired power stations and other industrial processes and store them deep beneath the ground. Both technologies are promoted in the Australian Government's highly touted gas-fed recovery (Commonwealth of Australia, 2021a). A demonstration CCUS site is in the Surat Basin in Queensland, where the deep geological conditions are considered suitable for storing large quantities of CO_2. Both technologies generate potential benefits and costs that must be weighed up by industry and Government (see Box 6.5).

O'Brien et al. (2020) suggest several climate mitigation strategies for mining industries and their supply chains and distribution networks. These include electric charging infrastructure and green hydrogen energy for heavy goods support. They also advocate for energy efficiency incentives to reduce load and demand response mechanisms to enable power load flexibility (see Box 6.1). In the case of remote mine sites and mining communities, there are considerable carbon mitigation benefits in the regulatory facilitation of renewable energy zones, particularly in areas with high wind or solar resource potential (see Figure 6.1). Australia has many mine sites in high solar radiation zones (Hodgkinson et al., 2014). Several renewable energy initiatives are underway across the sector, including the use of solar photovoltaic (PV) arrays with battery storage, pumped solar/hydropower, utilising existing and abandoned mine water reservoirs and innovative green solar/hydrogen energy schemes (see Chapter 4, Box 4.4).

ADAPTATION PATHWAYS

Future climate changes (see Chapter 3, Table 3.2) will adversely impact Australia's mining sector. Therefore, the industry and its supply chains must develop effective climate adaptation strategies to deal with future climate risks. In an international review of the hydrocarbons, metals and minerals extraction industries, Loginova and Batterbury (2019) argue that regions hosting mining activities are strategically important in the global response to climate change. Their global review suggests that adapting to climate change must take three forms that have applicability to Australia:

1. Incrementally improving the resilience of mining operations;
2. Transitioning to more inclusive governance through institutional and policy innovations; and
3. More profound transformations that shift the balance of power, including profit-sharing, localised control or cessation of mining entirely.

Finally, they state that clarifying adaptation pathways helps identify priorities and inform policies for a fairer and more sustainable future for mining and the regions where it takes place (Loginova and Batterbury, 2019).

The Australian mining sector will also need to develop and implement adaptation strategies to reduce any adverse climate impacts and risks while also seeking opportunities to benefit from future changes in climate (Hodgkinson and Smith, 2021). The Minerals Council of Australia's Climate Action Plan (Minerals Council of Australia, 2020) makes specific references supporting climate adaptation. It refers to the need to understand the types of adaptation investments needed in the sector, especially concerning operations, employee health, supply chains, water use, energy resources and local communities, to help minimise the adverse impacts of a changing climate. Other adaptation strategies in the plan include assessing and managing the physical effects of climate change on mine sites to build operational resilience. In addition, there are strategies to articulate how the sector can transform towards a circular economy[14] and how it could enhance value-adding activities. Examples include climate-smart minerals and metals, and 'non-energy' uses for oil, gas and coal in a low-carbon economy (Minerals Council of Australia, 2020, see Hodgkinson and Smith, 2021).

Adaptation to specific climate hazards and risks will be required across the mining and minerals and processing sector (Smith, 2013, Mason and Giurco, 2013). Some adaptations are already happening in the industry and are mainly incremental. In contrast, other systemic and transformative adaptations will need to occur in the future (Loginova and Batterbury, 2019, see also

[14] The circular economy is defined as an economy that balances economic development with environmental and resources protection.

BOX 6.5: Carbon capture and storage (CCS) and carbon capture, use and storage (CCUS) in Australia

What are CCS and CCUS? Carbon capture and storage (CCS) involves capturing, transporting and storing GHG emissions from fossil fuel power stations, energy-intensive industries and gas fields by injecting the captured GHGs back into the ground. Carbon capture, use and storage (CCUS) involves capturing emissions through various technological options and then injecting the captured emissions deep underground [1].

Globally, about 40 Mt of CO_2 are captured and stored annually [2].

The Global Carbon Capture Storage Institute concedes CCS needs to increase at least 100-fold by 2050 to be viable [2].

How much has been invested in CCUS technologies and hubs? Australian Government has committed AU$250 million over 10 years to fund CCUS projects, establish CCUS hubs and support research, development and commercialisation of CCUS technologies. This includes AU$50 million CCUS Development Fund supporting technologies, such as direct air capture, CCS from power stations and carbon-negative construction materials [1].

Which industrial processes can use large-scale CCUS technologies? Natural gas processing, cement production, steel production, fertiliser production, power generation and hydrogen production from fossil feedstocks [1].

Where will CCUS be deployed? Advanced stage of development in the Surat, Cooper and Gippsland Basins (onshore) and Barrow and Petrel sub-basins (offshore). The combined storage capacity is 20 billion tonnes of CO_2. There are plans to expand these five storage basins by 2050 [1].

What are the potential benefits of CCUS technologies? They may play an important long-term role in negative GHG emissions projects that store CO_2 drawn down from the atmosphere [2].

Australia has abundant, world-class (stable) geological storage basins [1].

Storage sites are close to industries producing highly concentrated streams of CO_2 emissions [1].

What are the potential costs of CCUS technologies? Leakage of CO_2, either gradual or in a catastrophic leakage, could negate the initial environmental benefits of capturing and storing CO_2 emissions and may also have harmful effects on human health.

CCUS produces extra oil, which when combusted puts 74 million tonnes of CO_2 a year (globally) back into the atmosphere (a net positive technology at this stage) [2].

Globally, CCUS has not been trialled and tested at the scale required to tackle the climate crisis [2,3].

When attached to fossil fuel developments (e.g. coal, oil and gas), CCUS is not a climate solution, as mining and combusting fossil fuels only add more CO_2 to the atmosphere [3].

Chevron's Gorgon Gas Plant in Western Australia (one of the biggest in the world) is described as a big, expensive failure [3].

The technology is incredibly expensive, i.e. at least six times more expensive than electricity generated from wind power backed by battery storage (see Box 6.5) [3].

CCS for coal-burning plants is no longer seen as a cost competitive option, at least not in locations where renewable energy is plentiful and relatively cheap as in Australia [4].

Sources: [1] Commonwealth of Australia (2021a), [2] Global Carbon Capture Storage Institute (2021), [3] Climate Council (2021a), [4] Wood et al. (2021).
Definitions, investments, uses, locations and their potential benefits and costs.

Chapter 4, Box 4.1: Accommodate, Protect and Retreat strategies). With a projected increase in the frequency and intensity of extreme weather events, adaptation strategies will need to be developed and applied to reduce risks of costs of disruptions to supply chains and risks for asset operation and performance (Smith, 2013, Hodgkinson et al., 2014). Examples of adaptation strategies include:

- Diversify the supplier base to ensure the resilience of infrastructure, working with government and transport providers to explore private investment in transport for vulnerable routes and locations;
- Consider geographical location (or relocation) as a strategy to manage potentially severe regional climate impacts;
- Consider the provision of emergency power and water supplies, pumping and wastewater treatment;
- Apply flood defence measures;
- Develop a bushfire management plan;
- Explore and develop alternative water supplies in partnership with nearby local and Indigenous communities (see Chapter 7, Tables 7.2 and 7.4);
- Assess health and safety risks from climate change and their potential impact on assets and operational processes;
- Develop robust climate-proof business continuity plans;
- Review and increase insurance policies to cover risks;
- Transition to mine site automation and remote operations; and
- Ensure that future climate change considerations occur in mine site decommissioning plans.

The mining sector has only recently started engaging in climate change. Therefore, much less is known about potential barriers and enablers to climate adaptation across the industry (see Chapter 4, Box 4.2). Earlier work (Mason and Guiro, 2013) found a lack of consensus about the foundation and need for climate change adaptation in the sector. This poor communication also included the extent of transmission of climate information between experienced professionals and less experienced professionals and public availability of information on risk assessment, preparedness and

management for climate change at mine sites and their supply chains. Other barriers identified by researchers (Loechel et al., 2013, Hodgkinson et al. 2014) include:

- Adaptation advantage does not outweigh the cost of assessment or adaptation;
- Perception of or insufficient evidence of low or no impacts of climate change in the near term;
- Climate threat not the most immediate issue;
- Uncertainty of impacts and adaptation effectiveness;
- Political and regulatory change uncertainty;
- Lack of risk management tools (see Box 6.2); and
- Existing mining infrastructure restricts change options.

There are several enablers (drivers) to raise awareness of climate adaptation issues and options for the mining sector (Mason and Giurco, 2013, Hodgkinson et al., 2014, see Chapter 4, Box 4.2: enablers). There is also a significant role for government-led initiatives in driving climate adaptation (see Box 6.2). The main adaptation enablers for the mining sector include:

- Existing knowledge that climate events can impact heavily on mining operations and communities (see Climate Risks and Impacts above);
- Given the high risks of doing nothing, there is a strong business case for improving planning and preparation to incorporate climate adaptation (see the Minerals Council of Australia Climate Action Plan above);
- National and international case studies exist that provide examples of the adverse impact of extreme weather events and leading practices to improve responses (e.g. Mason and Giurco, 2013, Sharma et al., 2013);
- High-quality data and information exist to support a greater understanding of historical and future climate trends (see Chapters 2 and 3); and
- Existing or modified planning processes can be used to support risk assessment and adaptive planning (see Chapter 4, Figure 4.4, Box 4.2: enablers). Notably, managing changes to the risk profiles of mining industries are already embedded within their existing management processes (see the 'Managing risk' section above).

INTEGRATION WITH THE SUSTAINABLE DEVELOPMENT GOALS

Five SDGs are relevant to a national strategy on climate change and the mining sector:

1. SDG6 (Clean water and sanitation);
2. SDG7 (Affordable clean energy);
3. SDG9 (Industry, innovation and infrastructure);
4. SDG12 (Responsible consumption and production); and
5. SDG13 (Climate action).

A framework (Table 6.4) shows how the Australian mining sector can effectively integrate climate change with goals, targets and indicators for each SDG. The sector's climate-resilient development pathways will require incremental, structural and transformational changes over the next 20 years. If the mining industry is to meet (or make progress towards) the 2030 SDG targets, structural and transformational changes will need to co-occur in other economic sectors (see the 'Energy and Transportation' sections above, and the 'Manufacturing and Construction 'sections below). Similarly, there will need to be incremental and transformational changes in social systems that depend on mining, particularly those relevant to remote mining towns and mining-dependent Indigenous communities (see Chapter 7, Table 7.5).

MANUFACTURING AND CONSTRUCTION

Socio-economic context

Manufacturing has been historically significant in Australia and contributes to the economy and society. The manufacturing sector is diverse and encompasses 141 industries that employed over 1 million persons (or 7.7% of the workforce) in August 2021 (Australian Bureau of Statistics, 2021a). Most manufacturing industries are in the country's largest cities, but manufacturing is vital in regional cities and towns. As in many advanced economies, over the past 30 years manufacturing output and employment have fallen steadily as a share of Australia's GDP. In 1990, the sector accounted for nearly 15% of the country's GDP, compared with 6% in 2020 (World Bank, 2020). Revenue for the manufacturing sector has declined about 0.5% each year between 2015 and 2020 to AU$397.3 billion, making it the 12th fastest growing sector in the country (Australian Bureau of Statistics, 2021a). Over the same 5-year period, the industry directly supported about 79,000 businesses (Australian Bureau of Statistics, 2021a). In 2020, the food and beverage, grocery and fresh produce sector (the most prominent manufacturing sub-sector) were worth AU$132 billion (Australian Food and Grocery Council, 2021). It entailed 16,000 businesses of all sizes that employed over 270,000 people. The closely allied construction industry employed over 1.1 million persons in August 2021, which accounted for 8.6% of the total workforce (Australian Bureau of Statistics, 2021a). This diverse industry involves nearly 400,000 small and medium enterprises (SMEs) (Australian Bureau of Statistics, 2021a).

Manufacturing has several sub-sectors, including food and beverage industries (see above), coal and chemicals, and metal products (Langcake, 2016). These sub-sectors comprise about 80% of manufacturing output and employment. The food and beverage industries and metal products are highly dependent on Australian agriculture and mining inputs, respectively. The primary products for these industries are in abundant supply (see Agriculture and Mining sectors above). The manufacturers in this first group use a relatively low share of intermediate components that are imported (Langcake, 2016). They also have below-average exposure to competition from imports and export more processed products than other manufacturing industries. By comparison, the machinery and equipment and petroleum, coal and chemicals industries use relatively few inputs from primary industries in Australia and have a relatively high share of imported intermediate components (Langcake, 2016). Consequently, this second group experience intense competition from imports, and their final goods are geared primarily to the domestic market.

Managing risk

Over many decades, the Australian business (private) sector (including manufacturing and construction) has become well-versed in managing various risks. A recent study identified many business and industry risks that apply to other economic sectors covered in this chapter (Allianz

Table 6.4 Integration of SDGs with the Australian mining sector's responses to climate change

SDG	Goals	Targets	Indicators
SDG6: Clean water and sanitation (*economic and environmental*)	1. Ensure availability and sustainable management of water	Improved water quality by reducing pollution from mine operations Substantially increased water recycling on site Substantially increase water use efficiency	• Proportion of mine site wastewater safely treated • Change in water use efficiency over time • Level of freshwater withdrawn as a proportion of available freshwater resources
SDG7: Affordable clean energy (*economic and environmental*)	2. Support extension of renewable energy adaptation	Increased use of renewable energy Increased energy efficiency Lower emissions from energy use	• Percentage of renewable energy use across sector • Energy use and emissions measurements • Percentage of mining, onsite processing and supply chain income spent on energy use
SDG9: Industry, innovation and infrastructure (*economic and environmental*)	3. Build resilient infrastructure and sustainable mining industries	Increased new, upgraded/retrofitted mining machinery and infrastructure that are sustainable	• Carbon emissions per unit of value added
SDG12: Responsible consumption and production (*economic and environmental*)	4. Ensure sustainable consumption and production patterns	Increased efficiency Increased circular economy for certain mined materials Sustainable public procurement practices	• Proportion of minerals produced under sustainable mining methods, material footprint (per capita, per GDP) • Degree of public procurement practices at mine sites that are sustainable
SDG13: Climate action (*environmental*)	5. Carbon-neutral production 6. Resilience and adaptive responses to natural hazards	Reduced global warming Closed loop production (circular economy) and supply chain systems Mitigation of climate change impacts on mining	• Reduction in mining and onsite processing production of GHG emissions contributions (to zero) • Percentage of mine sites using renewable energy • Percentage of mine sites with a drought and flood hazards plan

This includes hydrocarbons, metals and minerals extraction, mine site processing industries and their supply chains. Refer to Chapter 4 for an explanation of SDG goals, targets and indicators (see also Box 4.5). Triple-bottom line implications for each SDG are shown in italics.

Global Corporate and Specialty, 2021). These business risks included (in order of severity):

1. The COVID-19 pandemic outbreak;
2. Business interruption, including supply chain disruption;
3. Changes in legislation and regulation, i.e. trade wars and tariffs, economic sanctions, protectionism, Brexit and Euro-zone disintegration; cyber incidents;
4. Natural catastrophes, e.g. 2019–2020 bushfires (see Chapter 5, Box 5.3);
5. Climate change and increasing weather extremes (see below);
6. Market volatility, competition, market stagnation and fluctuation;
7. Macroeconomic developments; and
8. New technologies.

Two other significant challenges are forecast to impact business going forward:

9. Sustainability measures in response to a transition to a net-zero carbon economy; and
10. Data management and digital security (Allianz Global Corporate and Specialty, 2021). The Business Council of Australia has recently addressed the former in their high-level strategic road map towards a net-zero emissions economy by 2050 (Business Council of Australia, 2021).

A recent survey of Australian CEOs (Price Waterhouse Coopers Australia, 2019) confirmed this rising awareness that climate change threatens business growth and sustainability. The survey identified three main reasons for their business concerns:

1. Increased consumer awareness around environmental considerations;
2. Increased community activism on climate issues; and
3. Increased focus from investors and analysts on companies' environmental, social and governance reporting.

Climate impacts and risks

Manufacturing and construction businesses contribute to climate change through their processing and construction activities and supply chains

emissions. However, they are also exposed to climate risks in their business operations and generally have a low capacity to deal with many risks. The most significant climate threats to the manufacturing sector are flow-on effects from adverse climate change impacts on the Australian energy, transportation, agricultural and mining sectors (see above). For example, extreme weather events in agricultural and mining production areas can cause supply disruptions to food and beverage and metal and mineral processing industries. This disruption will affect volumes and prices of processed goods supplied to domestic and international markets, including any processed materials required for construction businesses (National Climate Change Adaptation Research Facility, 2013b).

Climatic disruptions to manufacturing industries in other countries that supply intermediate components is another business risk to Australian manufacturers and construction businesses. This disruption will affect the sustainable supply and price of products like petroleum, chemicals and building materials to the domestic market. Any disruption to manufacturing processes and construction activities will impact business profits and supply chains (or customer base), and their viability in often tightly contested domestic and international markets (National Climate Change Adaptation Research Facility, 2013b). The rising cost of insurance, driven by increases in the frequency and intensity of extreme weather events, is another business risk to both sectors (see Chapter 7, Box 7.1).

The construction sector is vulnerable to physical risks associated with changes in the workplace environment caused by climate change, such as projections for more extreme weather events. These events include heat waves, floods, intense East Coast Lows and tropical cyclones that will adversely impact future construction sites (see Chapter 3, Table 3.2, Box 3.1). There may also be potential water shortages in drought years, such as what occurred in Cape Town, South Africa during the 2015-2018 water crisis (Müller, 2020). Extreme weather events can affect worker health and safety, supply chain reliability, project delays and business profits (Hurlimann et al., 2019, see Chapter 7, Box 7.1). Therefore, the construction sector has a low adaptive capacity to deal with climate change. Notably, climate conditions may become particularly challenging for the industry in northern Australia in future decades. Climate risk and adaptation are essential considerations for the 'Northern Australia Development' Agenda (see Box 6.3).

The recent Business Council of Australia Climate Action Plan (Business Council of Australia, 2020) demonstrates that the business and industry sector – at its highest level – is now more focused on climate risk than in the past. It will take time for the raft of actions in the plan to be adopted and fully implemented by the 140-odd manufacturing companies and the 400,000-odd construction companies in the country. Many of these actions will require implementing structural changes and transformative climate adaptation strategies across both sectors over time (see Chapter 4, Figure 4.3, Box 4.1). Regarding climate mitigation, the development and deployment of emissions reduction technologies into manufacturing production and construction systems and their supply chains will be required this decade and beyond. In driving these significant structural changes, governments have a role to play by building engagement through education and recognising and promoting good practice (National Climate Change Adaptation Research Facility, 2013b).

Climate-resilient development pathways and opportunities

GREENHOUSE GAS EMISSIONS

Manufacturing varies greatly in its emissions intensity across the various industries that comprise the sector (Deloitte Access Economics, 2020). Manufacturing (primary metal/metal) is 'highly emissions intensive'; manufacturing (basic chemical/ polymer/rubber, non-metallic mineral product, petroleum and coal) is 'emissions intensive'; manufacturing (food product/beverage/tobacco) is 'moderately emissions intensive'; and manufacturing (fabricated metal; textile/leather/clothing/footwear; transport/machinery/equipment; furniture) is 'marginally emissions intensive'. The construction sector also varies in emissions intensity (Deloitte Access Economics, 2020). Construction services (e.g. supply chains) are 'emissions intensive' while building construction is 'moderately emissions intensive'.

The manufacturing and construction industries emissions were 40.8 Mt CO_2-e in 2019, or

7.9% of net national emissions (Department of Industry, Science, Energy and Resources, 2021c). This quantity appears in emissions produced from fuel combustion in manufacturing industries, ferrous and non-ferrous metals production, plastics production, construction and non-energy mining. Emissions generated from the electricity production used in these industries apply under electricity and heat production (see Energy sector above). In addition, total emissions estimated from petroleum refining and manufacture of solid fuels were 34.4 Mt CO_2-e in 2019, or 6.9% of net national emissions (Department of Industry, Science, Energy and Resources, 2021c). However, the built environment, which includes the construction, operation and maintenance of buildings, accounts for about 25% of Australia's GHG emissions (Van der Heijden, 2018).

MITIGATION PATHWAYS

Like agriculture and mining of fossil fuels (see above), many industrial manufacturing processes are often called 'hard to abate' industries, as there are few opportunities to substitute emission-free inputs or processes (Australian Academy of Science, 2021). Examples include the production of cement, chemicals, steel, fertiliser and other products (Butler et al., 2020). There remains a strong dependence on fossil fuel energy sources in the manufacturing sector (see Energy sector above). For instance, in 2019–2020, natural gas contributed 42% of manufacturing final energy use (Department of Industry, Science, Energy and Resources, 2021a). Nonetheless, Australia's business sector is well-placed to reduce GHG emissions by transitioning to renewable energy sources for its manufacturing and construction industries and their supply chains (Australian Academy of Science, 2021). The decarbonisation of the electricity grid (see Box 6.1) is considered a key enabler of fuel switching across the economy and within critical industries (see Chapter 4, Box 4.4). Businesses and households can then substitute away from emissions-intensive fuel sources towards renewable electricity (see the report by Deloitte Access Economics, 2020). The proposition is that technology drives the emissions efficiency required for a net-zero (carbon) economy through technological innovations. To achieve these low-emissions innovations requires incentives created by a whole-of-government commitment to support a net-zero carbon transition (Australian Academy of Science, 2021). The Australian Government's Long-Term Emissions Reduction Plan (Commonwealth of Australia, 2021a) attempts to provide this leadership. However, significant concerns have been raised about its reliance on fossil fuels and CCUS as crucial components in its 'critical pathways' to net zero by 2050 (Australian Academy of Science, 2021). In particular, the frequently quoted term 'clean hydrogen' (or blue hydrogen[15]) can be misleading, as proponents of fossil-fuelled hydrogen have used this to describe hydrogen production linked to CCUS (Box 6.5), as well as renewably sourced green hydrogen (Climate Council, 2021a, Hodges et al., 2022). Given the highly restricted GHG emissions budget that all countries must comply with from now on, the exploration and development of new gas fields (including coal seam gas) are counterproductive to dealing with climate change (Australian Academy of Science, 2021).

Building on this requirement for government incentives, O'Brien et al. (2020) propose three strategies to assist manufacturing industries with reducing their carbon footprint:

1. Provide tax incentives for the uptake of efficient appliances, lighting and energy demand response capabilities;
2. Underwrite and facilitate corporate Power Purchase Agreements; and
3. Support cross-sectoral initiatives in 'hard to abate' sectors such as cement, steel and chemicals.

In addition, there is much to be gained by expanding and tightening the Safeguard Mechanism as the primary policy driver to reduce emissions across the manufacturing and construction sectors (Business Council of Australia, 2021, see the 'Mining' section above). The aim is to shift energy export industries to net-zero emissions as quickly as possible and produce energy-intensive products using renewable energy (Australian Academy of Science, 2021).

The Australian Government's Resources Technology and Critical Minerals Processing Roadmap (Commonwealth of Australia, 2021d) discusses the enormous economic opportunities from enhanced processing of rare earth and critical minerals needed for batteries, solar panels and wind turbines. Australia has significant reserves of the critical minerals and metals (see Mining sector above), which are required to drive the modern global economy. For example, Australia is the largest lithium producer globally, supplying just over half of the worldwide supply (Commonwealth of Australia, 2021d). Critical minerals and metals are the basis for manufacturing advanced technologies such as electric vehicles, mobile phones and renewable energy systems (Hodgkinson and Smith, 2021, Commonwealth of Australia, 2021d). Currently, there is little or no processing of these ores in Australia, and the road map proposes undertaking further value-adding and manufacturing domestically to reverse this situation (Commonwealth of Australia, 2021d). Examples of critical minerals processing include batteries, solar cells, magnets for traction motors, lightweight alloys for aerospace and automotive industries, wind turbine components and fuel cells. Opportunities for climate mitigation identified in the road map include:

- Construction of renewable energy microgrids (see the 'Energy and Transportation' section above);
- The use of renewable energy sources to cost-effectively process and value-add to critical minerals and metals;
- Carbon capture, use and storage technologies (see benefits and costs in Box 6.5); and
- Green hydrogen technologies: i.e. hydrogen made through water electrolysis from renewable energy (see Hodges et al., 2022).

Australia also has significant opportunities to develop a 'green steel' industry, utilising vast supplies of local iron ore, solar resources and renewable energy-generated hydrogen sources (Australian Academy of Science, 2021, Venkataraman et al., 2022). This innovative process for making steel has the potential for upscaling and would take advantage of substantial amounts of locally produced green hydrogen (Wood et al., 2021, Hodges et al.,

[15] 'Blue hydrogen' is hydrogen (H_2) produced from natural gas with a process of steam methane reforming. This process creates hydrogen and carbon monoxide (CO). Water is added to the mixture, which turns the CO into CO_2 and more H_2. The remaining CO_2 emissions are then captured and stored underground (see Box 6.5).

2022, Venkataraman et al., 2022). There may also be opportunities to use renewable energy for manufacturing (without hydrogen production), such as through a rejuvenation of Australia's aluminium smelting industry powered entirely by renewable energy (Australian Academy of Science, 2021).

The construction sector must also play its role in climate action. The sector's industries must seek pathways to reduce their carbon emissions and effectively manage any 'net-zero' transition risks (see Chapter 4, Box 4.4). For example, electricity can easily replace natural gas for heating, cooling and cooking in buildings, and by using electric heat pump technology (Madeddu et al., 2020). The sector must implement strategies and actions to introduce decarbonisation throughout the construction value chain. Müller (2020) propose three measures to reduce GHG emissions from the construction sector and its supply chains:

1. The lowering of carbon intensity of building materials in the upstream production process of materials (e.g. zero-carbon buildings by 2050);
2. The implementation of 'climate-smart', low and clean energy consumption in the use phase of real estate and infrastructure (see Chapter 7, Table 7.3); and
3. The design of more recyclable materials and closed materials flow in the refurbishment and demolition phases (i.e. the circularity of building materials).

The National Construction Code of Australia (NCC, 2019) sets minimal obligatory requirements for energy efficiency, while National Housing Rating Scheme assesses compliance (Martek and Hosseini, 2019). The NCC only stipulates minimal mandatory requirements for energy efficiency, while two other schemes provide aspirational (voluntary) measures: (1) National Australian Built Environment Rating System and (2) Green Star. Martek and Hosseini (2019) assert that the combined compulsory and voluntary performance rating measures are inadequate as schemes to significantly reduce emissions from the construction sector. They argue that an 'effective strategy to cut emissions must encompass the whole lifecycle of planning, designing, constructing, operating and even decommissioning and disposal of buildings' (see Chapter 7, Table 7.3 and Box 7.2).

ADAPTATION PATHWAYS

For the broader business sector, climate change adaptation presents three key challenges (National Climate Change Adaptation Research Facility, 2013b):

1. Maintaining economic viability;
2. Managing legal climate risk and responding to adaptation regulations; and
3. Positioning business to identify opportunities.

The business sector has not historically engaged with climate adaptation (National Climate Change Adaptation Research Facility, 2013b). However, it has long been aware of the risks and opportunities associated with GHG mitigation and climate change policies (West and Brereton, 2013). Its willingness to engage in adaptation planning is slowly changing – driven mainly by the strategic intent that businesses now need to consider climate risks in their operations and supply chains (Business Council of Australia, 2021). While the BCA's position is on climate mitigation, there is an acknowledgement of the need for business and industry to engage in climate adaptation planning. This direction seems contingent on leadership and guidance by the Australian Government. This leadership is now progressing with the recent release of the National Climate Resilience and Adaptation Strategy (Commonwealth of Australia, 2021b). However, it is too early to assess the efficacy of this high-level strategy in preparing businesses and industries for dealing with climate change.

Manufacturing industries (e.g. factories) are particularly exposed to existing and projected extreme weather events that may threaten critical infrastructure and adversely impact production activities (see Chapter 3, Table 3.2, Box 3.1). Climate adaptation in the manufacturing sector is likely to be incremental in the early stages, e.g. retrofitting factories to withstand increasing ambient temperatures, heatwaves, bushfires and floods (see Chapter 4, Box 4.1: Accommodate and Protect strategies). Over time, as climate risks increase in vulnerable areas, transformative adaptation will need to be considered to ensure long-term business viability (see Chapter 7, Boxes 7.4 and 7.5). For example, some factories may have to relocate to new sites for their operations (see Box 4.1: Retreat strategy). Hence, the incorporation of climate change projections (see Chapter 3, Table 3.2) in the design, location and investment decisions of future (yet to be built) long-life assets, such as factories, is paramount (Commonwealth of Australia, 2021c). This approach will mean avoiding any identified 'no build' zones in developing new manufacturing precincts (see Box 4.1: Avoid strategy).

The construction sector is already being affected by extreme weather events (see Chapter 7, Box 7.1). The frequency and intensity of such events are highly likely to increase in the future (see Chapter 3, Table 3.2, Box 3.1). However, the construction sector can facilitate a leadership role in climate adaptation by preparing its employees, products, clients and the general population for significant changes ahead (Röck et al., 2020). The sector must start planning now to ensure the durability and sustainability of any long-term building assets that will need to tolerate future climate conditions (Steffen et al., 2019, see also Chapter 4, Box 4.1, Accommodate and Protect strategies). Adaptation strategies include (Müller, 2020):

- Increasing the durability of building materials against extreme weather conditions;
- Overhauling heating/cooling and insulation concepts and energy self-sufficiency in new buildings (e.g. 'green buildings', 'biodiversity enablement');
- Revising water management towards more 'climate-smart' water management systems during the construction and use phases of buildings; and
- Lowering the potential adverse effects of the construction sector on the natural environment from soil sealing (i.e. change in water flows from heavy rain) or land use change (i.e. land that would otherwise store carbon).

Across Australian business and industry more widely, a study (National Climate Change Adaptation Research Facility, 2013b) has collated a list of government-led initiatives that may also support climate adaptation in the manufacturing and construction sectors (see Box 6.2). Unlocking any barriers to these initiatives will be necessary, along with using them as enablers to assist businesses and industries with implementing climate adaptation (see Chapter 4, Box 4.2: barriers). Of concern, is that many companies in the manufacturing and construction sectors rely on short planning cycles for their operations (National Climate Change Adaptation Research Facility, 2013b). Climate adaptation is a medium- to long-term risk for many companies, so CEOs and company directors do not recognise its urgency. Additionally, most companies in both sectors are small and medium enterprises (SMEs), which limit their capacity to incorporate climate adaptation strategies in their strategic plans. While this is a challenge for SMEs, the procedures required to strengthen the business continuity under climate change can be largely integrated into existing planning and operational processes and through industry networks. Nonetheless, barriers remain for SMEs to fully engage in adaptation planning (see also Chapter 4, Box 4.2: barriers). Kuruppu et al. (2013) found that several critical socio-economic and political factors in which SMEs operate are likely to constrain and influence their adaptive capacity to respond to climate change in Australia (Box 6.6).

INTEGRATION WITH THE SUSTAINABLE DEVELOPMENT GOALS

Five SDGs are relevant to a national strategy on climate change and the manufacturing and construction sectors:

1. SDG6 (Clean water and sanitation);
2. SDG7 (Affordable clean energy);
3. SDG9 (Industry, innovation and infrastructure);
4. SDG12 (Responsible consumption and production); and
5. SDG13 (Climate action).

A framework (Table 6.5) shows how the manufacturing and construction sectors can effectively integrate climate change with goals, targets and indicators for each SDG. Climate-resilient development pathways for both sectors will require structural and transformational changes from now on. If the sectors are to meet (or make progress towards) the 2030 SDG targets, significant changes will need to co-occur in the agricultural, mining, energy and transportation sectors (see above). A whole-of-economy perspective will be critical in driving innovation and change towards a net-zero world. There will also need to be incremental and transformational changes in various social systems that depend on manufacturing and construction (see Chapter 7, Table 7.5).

BOX 6.6: Barriers to climate change adaptation in small and medium enterprises (SMEs) in Australia

---

- Limited access to funding and limited human resources for delivering climate action programmes for SMEs.
- The absence of dedicated climate adaptation-related programmes or business continuity planning targeting SMEs within local government (see Chapter 7, Tables 7.3 and 7.3).
- Linkages between SMEs and local government is often weak.
- Poor coordination between government, non-government and private sector stakeholders who provide support to SMEs. This leads to limited information sharing and missed opportunities for joint learning and reflection amongst stakeholders. Underpinning these failures are struggles over power due to the desire of organisations to protect their niches in the wider business landscape supporting SMEs.
- A lack of urgency amongst stakeholders about the need to develop climate risk reduction initiatives for SMEs. Such initiatives were not seen as a priority issue for SMEs in the short term (i.e. the next 5 years).
- The limited formal mechanisms for monitoring and evaluating current risk reduction initiatives and thus the constrained opportunities to learn and improve upon them.
- The adaptive capacities of SMEs are to a large extent shaped by the adaptive capacity of the organisations that support them. This limits the agency of SMEs in securing business continuity.

Source: After: Kuruppu et al. (2013).

Note: In Australia, 99.8% of all enterprises (businesses) are SMEs (<200 employees), 97.4% are small businesses (0–19 employees), 2% are medium (20–199 employees) (Australian Bureau of Statistics, 2021a). Refer also to Chapter 4 (Box 4.2: barriers).

TOURISM

Socio-economic context

Tourism is a significant industry in Australia and, compared to the rest of the economy, was disproportionately affected by the COVID-19 pandemic. While impacts on employment in the sector varied considerably across the country, in 2020–2021, persons employed in tourism fell by 20.3% to 507,000 people (Australian Bureau of Statistics, 2021a). International arrivals and domestic tourism numbers declined significantly in response to the pandemic, primarily due to travel restrictions imposed within Australia and overseas. These included lockdowns and closed international and domestic borders in 2020, 2021 and early 2022. Consequently, in 2020–2021, domestic tourism consumption fell 12.1% and international fell 94.9% (Australian Bureau of Statistics, 2021a). The tourism industry is in recovery mode, with the resumption of domestic travel between most states and territories, as of February 2022. Inbound international travel will resume from March 2022. However, it may take several years for international tourism arrivals to return to pre-pandemic levels, and some markets may take even longer.

Due to the devastating effects of the pandemic, the 2018–2019 fiscal year provides a better indicator of the economic importance of the country's tourism sector over the past decade (Australian Bureau of Statistics, 2019). In that year, the combined value of total international and domestic visitor spending increased by 11.2% from the previous year to AU$122 billion (Tourism Research Australia, 2020). From this spending, the direct tourism GDP increased by 6% to AU$60.8 billion, but its share of total GDP remained unchanged at 3.4%. The sector was the country's fourth-largest exporting industry, accounting for 8.2% of its export earnings or AU$39.1 billion (Tourism Research Australia, 2020). In the same fiscal year, 9.3 million international visitors arrived in Australia, an increase of 3.0% compared to

Table 6.5 Integration of SDGs with the Australian manufacturing and construction sectors' responses to climate change

SDG	Goals	Targets	Indicators
SDG6: Clean water and sanitation (*economic and environmental*)	1. Ensure availability and sustainable management of water	Improved water quality by reducing pollution from manufacturing and construction operations Substantially increased water recycling on site Substantially increase water use efficiency	• Proportion of manufacturing and construction wastewater safely treated • Change in water use efficiency over time • Level of freshwater withdrawn as a proportion of available freshwater resources
SDG7: Affordable clean energy (*economic and environmental*)	2. Support extension of renewable energy adaptation	Increased use of renewable energy Increased energy efficiency Lower emissions from energy use	• Percentage of renewable energy use across sectors • Energy use and emissions measurements • Percentage of manufacturing and construction supply chain income spent on energy use
SDG9: Industry, innovation and infrastructure (*economic and environmental*)	3. Build resilient infrastructure and sustainable manufacturing and construction industries	Increased new, upgraded/retrofitted manufacturing and construction machinery that are sustainable Encourage new, inclusive and sustainable manufacturing and construction industries	• Carbon emissions per unit of value added • Sustainable manufacturing and construction value added as a proportion of GDP and per capita • Sustainable manufacturing and construction employment as a proportion of total employment (to 100)
SDG12: Responsible consumption and production (*economic and environmental*)	4. Ensure sustainable consumption and production patterns	Increased efficiency Increased circular economy for manufactured and construction materials Sustainable public procurement practices	• Proportion of products made using sustainable manufacturing methods and proportion of construction projects completed using sustainable methods, material footprint (per capita, per GDP) • Degree of public procurement practices in manufacturing and construction sectors that are sustainable
SDG13: Climate action (*environmental*)	5. Carbon-neutral production 6. Resilience and adaptive responses to natural hazards	Reduced global warming Closed loop production (circular economy) and supply chain systems Mitigation of climate change impacts on manufacturing and construction sectors	• Reduction in manufacturing and construction sector GHG emissions contributions (to zero) • Percentage of manufacturing and construction industries using renewable energy • Percentage of manufacturing and construction industries with a climate action plan

This includes their supply chains. Refer to Chapter 4 for an explanation of SDG goals, targets and indicators (see also Box 4.5). Triple-bottom line implications for each SDG are shown in italics.

the previous year (Tourism Research Australia, 2020). In 2018–2019, persons employed in tourism increased by about 21,500 employees to over 666,000 employed persons (Australian Bureau of Statistics, 2019). However, tourism's share of total employment remained stable at 5.2%. In 2018–2019, there were 302,520 tourism-related businesses in the country (Tourism Research Australia, 2020), and the sector is vital in many regional areas of the country. In 2018–2019, it accounted for 4.1% of regional GDP and 8.1% of the regional workforce (Tourism Research Australia, 2020). Tourism also can help create economic sustainability by dispersing spending across the regions. In 2018–2019, almost 44% of visitor spend was in regional areas that contain some of the most popular tourist attractions, such as the Great Barrier Reef (Tourism Research Australia , 2020).

Before the pandemic, tourism was one of the most extensive and fastest growing economic sectors in Australia and the world (Hughes et al., 2018). Tourism businesses across Australia are highly dependent on its natural and cultural assets (Turton et al., 2010, Turton 2014, see Chapter 5, Natural Systems). Many of the highly popular 'natural' places tourists visit are also culturally significant to First Nations Peoples, such as Uluṟu -Kata Tjuṯa and Kakadu National Parks in the Northern Territory and the Great Barrier Reef and Daintree Rainforest in Queensland. Other popular tourist attractions include the Sydney Opera House and Harbour Bridge in New South Wales and Melbourne's 'cultural scene' in Victoria. Not surprisingly, nature-based tourism (which often emphasises Indigenous culture) dominates the country's market for domestic and international visitors. Surveys of international visitors (Tourism Australia, 2017) list the following top five tourist attractions: (1) beaches (50%), (2) wildlife (43%), (3) the Great Barrier Reef (40%), (4) unspoilt natural wilderness (36%) and (5) rainforests/forests/national parks (36%).

Managing risk

Over many decades, Australian tourism businesses have demonstrated a generally high resilience to market shocks. These shocks have adversely impacted both domestic and international markets. Pre-COVID-19 examples include:

- The 1989 Australian pilots' strike seriously affected domestic flights, including inbound international connections to tourist destinations in the country.
- The 1997 Asian financial crisis particularly affected inbound tourism from Japan, which was a primary international market at that time.
- The 2002–2004 SARS-CoV-1 virus outbreak mainly affected inbound international tourism.
- The global financial crisis severely disrupted international inbound and some domestic tourism.

In all these cases, some form of government financial assistance saved many tourism businesses, particularly in deeply affected tourist destinations, such as Cairns in far north Queensland (see map Chapter 3, Figure 3.1).

Like all Australian businesses, tourism is exposed to various risks in its operations. Tourism businesses are affected by pandemics, supply chain disruptions, natural catastrophes (e.g. bushfires, tropical cyclones, coral bleaching), market volatility and competition, and the tyranny of distance. Notably, Australia has long been an aspirational, once-in-a-lifetime destination for international visitors. This travel attribute requires long-haul travel, high costs and extensive pre-travel planning (Tourism Research Australia, 2020). At present, carbon offsetting is the only practical global solution to dealing with emissions from air travel. Nonetheless, the aviation sector is working on battery electric-powered aircraft for short-haul and green hydrogen-powered aircraft for long-haul flights (see the 'Energy and Transportation' sections above).

Climate change policy at the international level (see Chapter 4, the 'Paris Agreement' section) affects tourists' decision making about travel destinations and modes of travel. As the world increasingly divests from fossil fuels, more people become concerned about their 'individual' or 'business' carbon footprints. Discerning travellers and companies may defer from long-haul leisure and business travel in favour of regional travel via electrified high-speed rail (e.g. within Europe). When compared to 22 OECD economies and Russia, selected due to comparable wealth, population and development, it becomes clear that Australia is significantly behind in the energy transition (Saddler,

2021). The country's renewables electrification performance (see the 'Energy and Transportation' section above) is the worst of the 24 OECD countries. Additionally, its decarbonisation performance has also been among the worst. On this basis, Australia's overall energy transition performance has been worse than any of the other 23 highly developed countries (Saddler, 2021).

Climate impacts and risks

The Australian tourism sector has demonstrated its remarkable resilience to a range of adverse impacts, and this suggests it has a high adaptive capacity to most change agents (see Chapter 4, Figure 4.1). But can the sector and its diversity of industries respond effectively to the projected changes in climate and the increased risk of more extreme weather events (see Chapter 3, Table 3.2)? Tourism also contributes directly to climate change through its GHG emissions from aviation and land and sea travel to and from destinations (see the 'Energy and Transportation' section above). Likewise, there are emissions from energy use in significant tourism infrastructures, such as hotels and resorts (see Energy sector above). Climate change also presents a considerable threat to many of the country's key tourism destinations and outdoor attractions (Turton et al., 2010, Hughes et al., 2018, Cresswell et al., 2021, Lawrence et al., 2022). These threats include adverse impacts on vulnerable natural and cultural sites (Turton et al., 2009). There are also potential risks for weather-related effects on the health and well-being of tourists visiting popular tourist sites during the hotter months of the year, particularly those in northern and central Australia (see Box 6.3).

IMPACTS ON TOURIST TRAVEL DECISIONS

Knowledge about the climate at tourism destinations is crucial for tourism travel, both within and between countries. A few researchers have examined the relationship between climate attributes – such as temperature and humidity – and the comfort level for tourist activities. The best-known example of this was the tourism climatic index (TCI), developed by Mieczkowski (1985) and applied by others. The TCI uses monthly means of routinely measured weather data (e.g. maximum daily temperature, relative humidity and wind speed), primarily available at key tourism destinations (or at nearby sites). The TCI methodology requires mean monthly climate data and assumptions about everyday tourist activities, such as sightseeing, shopping and other outdoor activities. Scores range from low to moderate levels of physical exertion (Hughes et al., 2018). A TCI score, from below 9 (impossible comfort level) to 90–100 (ideal comfort level), is then calculated for the tourism destination of interest for each month of the year.

Amelung and Nicholls (2014) applied the TCI methodology to continental Australia and utilised a future high emissions scenario (see Chapter 2, Table 2.6: SSP5–8.5). They produced spatial TCI projections for the 2020s, 2050s and 2080s. Projections for the 2080s show many tourist destinations in northern Australia, such as Darwin and Kakadu National Park, and Cairns, Townsville and Airlie Beach (access points to the GBR) could become 'unfavourable' during the hottest months of the year (see Chapter 3, Figure 3.1, and Box 6.7). By the 2080s, the unfavourable TCI region will also extend south into the sub-tropics, affecting key summer tourist destinations, such as the Gold and Sunshine Coasts and Brisbane (Amelung and Nicholls, 2014, Hughes et al., 2018). At the same time, the far southwest of Western Australia, far south of South Australia, southern Victoria and alpine areas will likely experience ideal human comfort conditions (Amelung and Nicholls, 2014). Their study excluded the island state of Tasmania, a popular tourist destination. However, ideal human comfort conditions will apply for the 2080s based on projections for other mainland regions with similar climates (e.g., southern Victoria).

Australia's iconic 'Red Centre', which includes the world heritage-listed Uluṟu-Kata Tjuṯa National Park in the Northern Territory, is a top-rated destination for domestic and international tourists. Pre-pandemic, the national park had more than 300,000 visitors per year (Hughes et al., 2018), generating vital income for the An̲angu First Nations People, who are the traditional custodians of the area and share joint management of the park with the Australian Government. The most significant adverse impact of climate change on future tourism in the Red Centre will be increasing extreme heat (Hughes et al., 2018). As early as 2030, large parts of central Australia are projected to experience over 113 days per year over 35°C. Under a high emissions scenario (see Chapter 2, Table 2.6:

Figure 6.4 A lone visitor inspects the landscape, Uluṟu-Kata Tjuṯa National Park, Northern Territory, Australia. The Red Centre is projected to experience a significant increase in the number of hot days (>35°C) over coming decades. Under a high emissions scenario (see Chapter 2, Table 2.6: SSP5–8.5), iconic tourism destinations, like Uluṟu-Kata Tjuṯa National Park, may exceed 250 hot days per year by the 2080s. (Credit: Ian Cochrane/Flickr/Public Domain.)

SSP5–8.5), this may exceed 250 days per year by 2080 (see Figure 6.4). By then, plus 50°C hot days will be expected during the hotter months. This thermal regime will present visitors and residents with highly unfavourable climate conditions for much of the year (see TCI above). There are also concerns about future groundwater resources in the park, which are vulnerable to projected reductions in runoff, driven by higher evaporation (Hughes et al., 2018).

IMPACTS ON TOURISM DESTINATIONS

Climate-related events are already having adverse impacts on tourism destinations in Australia (see global review by Scott et al., 2019). Examples include mass coral bleaching events affecting tourist sites on the GBR in 2016 and 2017 (Ma and Kirilenko, 2019, see Box 6.7) and the devastating 2019–2020 summer bushfires (see Chapter 5, Box 5.3) adversely affecting visitor numbers, critical tourism infrastructure and popular destinations in

New South Wales, Victoria, Tasmania and South Australia (Schweinsberg et al., 2020). Climate change impacts at tourist destinations include those affecting biodiversity and ecosystem services (see Chapter 5, Boxes 5.1 and 5.2), water security (see Chapter 5, Box 5.4), accessibility to beaches, the incidence and geographical distribution of vector-borne diseases and toxic jellyfish (Hughes et al., 2018). Adverse effects are also likely on the wine industry in popular tourist destinations, like the Barossa Valley in South Australia (Turton et al., 2009, see Agricultural sector above).

Sea-level rise and associated coastal inundation events threaten the country's iconic beaches. For example, the beaches of the Gold Coast (see Chapter 7, Figure 7.2) in southern Queensland are world famous and underpin the AU\$4.7 billion local economy and directly employed over 30,000 jobs in 2014 (Hughes et al., 2018). The beaches are highly vulnerable to rising sea levels, and storm surges from East Coast Lows and southward moving tropical cyclones (see Chapter 3, Table 3.2). Under current climate conditions, the beaches require 400,000–700,000 m³ of sand transported each year to repair storm damage (Hughes et al., 2018). One study estimated the cost of beach renourishment to the Gold Coast City Council would be AU$11–54 million per year over the next century, varying according to which GHG emissions scenario was used (Cooper and Lemckert, 2012).

Australia's alpine areas and ski tourism in New South Wales, Victoria and Tasmania are also highly vulnerable to climate change (Lawrence et al., 2022). A tiny portion of Australia receives accumulating snow each year (see Chapter 3, Figure 3.3: *Dfb* climate sub-type). The snow season is very marginal compared with other ski tourism regions (e.g. Europe, North America, Japan, New Zealand and Chile), with a strong dependence on artificial snowmaking in most years. With projected rises in average temperatures, declines in winter snowfall and increasing scarcity of freshwater for artificial snowmaking, the future security of the ski industry is restricted to a few decades at most. Rising energy costs to power the snow guns is another risk for the alpine ski industry (Hughes et al., 2018).

CLIMATE CHANGE THREATS TO WORLD HERITAGE SITES

Valentine (2019) notes that world heritage listing affords considerable international standing to countries with such global status and provides significant in-country economic benefits, mainly through tourism. However, world heritage status also comes with a responsibility for wise and ongoing management of the Outstanding Universal Value (OUV) of all listed sites. Sites that can no longer maintain OUV are often placed on the world heritage 'in-danger' list. There are currently 53 world heritage sites on this list (International Union for Conservation of Nature, 2020). Reasons for listing sites as 'in danger' worldwide include climate change, armed conflict, inappropriate development, poaching and other illegal activities such as logging, fishing, mining, and invasive plants and animals (International Union for Conservation of Nature, 2020). Australia's world heritage sites are not immune to being included on the in-danger list in the future, with the Great Barrier Reef now being considered a potential candidate for inclusion due to adverse climate change impacts and land-based water quality runoff concerns (see map Chapter 5, Figure 5.1).

Observed and projected changes in key climate indicators (see Chapter 3, Table 3.2, Box 3.1) pose significant threats to Australia's world heritage sites and the tourism industries that depend on them (Australian National University, 2009, Turton et al., 2009, Turton, 2014, Hughes et al., 2018). These threats include rising atmospheric and oceanic CO_2 levels, increasing average temperatures and land and ocean heatwave events, declining rainfall in southern Australia, increasing frequency and severity of bushfire weather events, rising sea levels and rising cloud bases (see Chapter 5, Boxes 5.3, 5.5 and 5.6). At least 13 of the 19 world heritage sites in the country (see map Chapter 5, Figure 5.1) are vulnerable to climate change, with some places facing multiple climate threats (Box 6.7).

Climate-resilient development pathways and opportunities

GREENHOUSE GAS EMISSIONS

Before the COVID-19 pandemic, an international survey of 160 countries – between 2009 and 2013 – calculated the quantity of carbon emissions produced by the tourism sector globally (Lenzen et al., 2018). Their study evaluated more than a billion supply chains to track tourists' goods and services. They found the sector's carbon footprint around the planet grew by 15%, over that period, from 3.9 to 4.5 Gt CO_2-e. This amount is four times more than previously thought and accounted for about 8% of global GHG emissions in the survey period. The carbon footprint of Australian tourism has likely grown at a similar rate. An earlier paper (Dwyer et al., 2010) estimated that depending on the approach, tourism contributes between 3.9% and 5.3% of the total industry GHG emissions in Australia. GHG emissions from tourism contribute to the energy and transportation sectors (see above). However, manufacturing and agricultural supply chains to tourism businesses, and the construction of tourism infrastructures, such as hotels and resorts are other sources of GHG emissions (see the 'Agricultural and Land' and the 'Manufacturing and Constructions' sections above).

BOX 6.7: Impacts of climate change on Australia's world heritage sites

Climate attribute	Climate-related impacts and threats to outstanding universal value (OUV)	Vulnerable world heritage areas/properties and jurisdiction responsible for management (in parentheses)
Rising levels of CO_2	Ocean acidification, adverse effects on calcite lifeforms, e.g. corals and shell fish (see Chapter 5, Box 5.6)	Great Barrier Reef (AG/QLD) Ningaloo Coast (WA) Lord Howe Island Group (NSW)
Rising air and ocean temperatures, increases in frequency and severity of land and marine heatwaves	Bleaching of shallow-water coral reefs, loss of coral cover, changes in reef structure and composition, loss of sea grass beds (see Chapter 5, Box 5.6) Decline of cool adapted endemic vertebrates, shrinking areas of suitable habitat for mountain top endemic flora and fauna, loss of keystone species during heat waves (see Chapter 5, Box 5.5)	Great Barrier Reef (AG/QLD) Ningaloo Coast (WA) Lord Howe Island Group (NSW) Shark Bay (WA) Wet Tropics of Queensland (QLD) Gondwana Rainforests (QLD/NSW) Greater Blue Mountains Area (NSW)
Declines in rainfall	Increased drought risk, changes in forest structure and composition, extinction risk, changes in catchment hydrology and ecology (see Chapter 5, Box 5.4), more extreme events	Gondwana Rainforests (QLD/NSW) Greater Blue Mountains Area (NSW) Fraser Island (QLD)
Reduced summer rainfall, increasing temperatures, evaporation, wind and sunshine hours	Effects on food availability and vegetation distribution, effects on fauna, e.g. four penguin species, extinction risk	Macquarie Island (TAS)
Increases in frequency and severity of bushfire weather events	Changes in forest structure and composition, loss of fire-sensitive species and habitats, changes in catchment hydrology and ecology, damage and loss of infrastructure, properties and Indigenous cultural landscapes and sites (see Chapter 5, Boxes 5.3 and 5.4; Chapter 7, Table 7.2)	Wet Tropics of Queensland (QLD) Gondwana Rainforests (NSW) Greater Blue Mountains Area (NSW) Tasmanian Wilderness (TAS) Fraser Island (QLD) Kakadu National Park (AG)

Climate attribute	Climate-related impacts and threats to outstanding universal value (OUV)	Vulnerable world heritage areas/properties and jurisdiction responsible for management (in parentheses)
Rising sea levels	Loss of freshwater wetlands due to nascent saltwater intrusion, conversion to saltwater ecosystems (see Chapter 5, Box 5.4), loss of habitats and species, damage and loss of Indigenous cultural sites, increased wave energy and dune erosion, damage to building foundations	Kakadu National Park (AG) Ningaloo Coast (WA) Fraser Island (QLD) Lord Howe Island Group (NSW) Tasmanian Wilderness (TAS) Sydney Opera House (NSW) Australian Convict Sites, i.e. Port Arthur (TAS)
Rising cloud base, increasing atmospheric water vapour deficits	Reduction of upland forest 'cloud stripping' mechanism in the drier months, reduced base flows in perennial water courses, adverse impacts on arboreal and in-stream fauna, extinction risk	Wet Tropics of Queensland (QLD) Gondwana Rainforests (QLD/NSW) Lord Howe Island Group (NSW)
Projected increases in intensity of more severe storms, but decreases in frequency and forward motion of all tropical cyclones	Changes in coral reef structure and composition, interactions with coral bleaching events and ocean acidification, loss of seagrass beds, changes in forest structure and composition, interactions with bushfire events, storm surge risk for littoral forests and coastal wetlands	Great Barrier Reef (AG/QLD) Ningaloo Coast (WA) Wet Tropics of Queensland (QLD) Fraser Island (QLD) Gondwana Rainforests (QLD/NSW) Kakadu National Park (AG)

Source: Adapted from Turton (2014).
Refer to Chapter 5 (Figure 5.1) for the location and areal extent of the 14 world heritage areas inscribed for natural values.
AG, Australian Government; NSW, New South Wales; QLD, Queensland; TAS, Tasmania; WA, Western Australia.

Research shows that the rapid increase in tourism demand is effectively outstripping the decarbonisation of tourism-related technology (Lenzen et al., 2018). With a return to pre-pandemic travel levels and its high carbon intensity and continuous growth, the tourism sector will increase its overall and relative contributions to global GHG emissions in the future. Some factors contribute to the exceptionally high carbon intensity of Australian tourism. First, getting to and from Australia requires long-haul flights or long-distance cruise ship travel. Both forms of transportation are carbon-intensive and directly contribute to climate change (see Transportation sector above). In addition, the country's domestic tourism travel includes air travel from major cities to regional tourist destinations, as the country lacks any high-speed electric train services. The self-drive tourist market is also primarily by fossil-fuelled vehicles, as Australia has a poor uptake of electric cars and camper vans (EVs). Most land- and water-based transport modes in the tourism sector rely on diesel-powered coaches and ocean-going vessels (see below). Second, most tourism infrastructures (e.g. hotels, resorts) rely on electricity from one of the national grids, supplied mainly by coal- and gas-fired power stations (see Energy sector above).

MITIGATION PATHWAYS

The Australian tourism sector has not had a high level of interaction with climate change agendas (Turton et al., 2010), but there are signs of change across the industry (Hughes et al., 2018). Some of the reasons for this lack of engagement will be examined below (see adaptation pathways). The most significant climate mitigation challenge to the sector is its reliance on air travel, which at this stage is entirely dependent on fossil fuel energy for aircraft (Becken and Mackey, 2017). This problem also extends to many land and ocean transport forms that carry tourists to and from destinations, e.g. fast-speed catamarans accessing outer GBR pontoon sites. Until new forms of renewable energy become available (e.g. green hydrogen- and electric-powered aircraft and marine vessels), the sector will have to rely on offsetting its GHG emissions (see the 'Energy and Transportation' sections above). Examples of offsetting schemes include protecting and restoring forests to increase carbon sequestration and renewable energy projects (Hughes et al., 2018). However, Becken and Mackey (2017) assert that while forest protection and restoration can offset biomass-based emissions, they cannot fully compensate for fossil fuel emissions because such emissions are a separate and additional source of CO_2 to the atmosphere (see Chapter 2, Table 2.2). There are similar issues with energy consumption by tourism infrastructures attached to national electricity grids (e.g. hotels and resorts). However, as the share of renewable energy into the national grids increases, the carbon footprint from tourism infrastructure energy use will also decline. In the meantime, the industry needs to do more to reduces its carbon emissions.

Tourism Australia, the Australian Government Agency responsible for the growing demand for Australia as a tourism destination, both domestically and internationally, is mandated under law to foster a 'competitive and sustainable tourism industry' (Tourism Australia, 2022). In the climate mitigation space, the agency refers to two sources for Australian Government support for tourism businesses:

1. Green Initiatives Grants (GIG); and
2. Emissions Reduction Fund (ERF) (see Chapter 4, the 'Roles and Responsibilities for Climate Change Action in Australia' section).

Tourism industry grants under the GIG vary between AU$50,000 and AU$200,000. The GIG provides an opportunity for tourism businesses to secure grants to reduce emissions and utilise clean energy alternatives (e.g. solar PV, wind and hydropower).

The Clean Energy Regulator (2022) must publish and maintain details of projects registered under the ERF. As of 10 February 2022, there are 1,316 projects listed under the ERF, including those revoked (Clean Energy Regulator, 2022). Projects are categorised by method type: agriculture, waste, carbon capture, energy efficiency, facilities, industrial fugitives, savannah burning, transport, vegetation and waste. A closer inspection of the list of projects, allowing for the removal of revoked projects shows no projects related to the tourism sector. This absence is surprising because the industry aligns with many categories permitted under the ERF. This omission also questions why Tourism Australia considers the ERF a mechanism for tourism businesses to mitigate their carbon emissions. The lack of uptake of the ERF by tourism businesses is likely to be related to the strict conditions of the fund, particularly the barriers faced by tourism projects in the form of resource outlays or foregone profits. Many of these upfront outlays are not offset by carbon revenues and secondary benefits (e.g. reduced energy costs) in the project's early years. (King Report, Department of Industry, Science, Energy and Resources, 2020). Another barrier is that the tourism sector comprises SMEs (see Box 6.6). They may lack the skills, size of business enterprise (i.e. critical mass) and time to devote to applications to the ERF.

Despite various challenges to obtaining significant funding assistance in climate mitigation, some leaders in the tourism sector are paving the way for sustainable practices across their business operations. Hughes et al. (2018) provide some examples in their report, and these comprise three categories:

- *Resorts and tour operators:* Examples include the Lady Elliot Island Eco-Resort on the GBR, which has installed a 28 kW hybrid solar and power station; other GBR tourism operators who are utilising offset schemes or new technologies to reduce their carbon footprints; Uluṟu Resort in the Red Centre (see Figure 6.4), who are powering 15% of the resort's average electricity demand from a solar PV system; and the Alto Hotel in Melbourne, where all the hotel's electricity is supplied from renewable sources.
- *Tourist attractions:* Examples include Perth Zoo, with a solar PV array that supplies 30% of the zoo's energy needs, and zoos in Victoria, South Australia and New South Wales that have also installed solar PV systems.
- *Travel services:* Examples include carbon-neutral travel companies, like Intrepid Travel and airlines, like Qantas and Virgin Australia, investing in biofuels and carbon offsetting projects in the country and overseas (see the Energy and Transportation' sections above).

ADAPTATION PATHWAYS

Australia's most important tourism sites and destinations are already adversely impacted by climate change, and future impacts are highly likely (Turton et al., 2009, Turton, 2014, Hughes et al., 2018, Cresswell et al., 2021, Lawrence et al., 2022). While the tourism sector has shown a high adaptive capacity to deal with many economic shocks over the years, the industry appears reluctant to deal with climate adaptation (Turton et al., 2010, Hughes et al., 2018). There has been a systemic failure of Australian Governments to highlight and support the need for climate change adaptation and urgent action to protect the long-term future of tourism. The political dismissal of 'climate change' undoubtedly leads to hesitation within tourism peak bodies and individual tourism businesses hoping for government leadership. A large Australia-wide study examining climate adaptation in the tourism sector (Turton et al., 2009, 2010) confirms this lack of leadership. It identified three main barriers to the adoption of adaptation strategies across the industry:

1. The scale and uncertainty surrounding climate change projections;

2. Communication within and between regional and national bodies (see Chapter 4, Box 4.2: barriers); and
3. Concerns regarding the capacity of small and medium enterprises (SMEs) to adapt, relative to governments and larger operators (see below).

While some tourism businesses are active in promoting the adverse impacts of climate change on the natural assets that form the basis for their operations (e.g. selected GBR operators), some are certainly sceptical about the science of human-caused climate change (see Chapter 2). There are also concerns in the sector about the 'messaging' by national and international media about climate-driven events on crucial tourism destinations, such as the 2016 and 2017 mass bleaching events on the GBR and the 2019–2020 bushfires in southeast Australia. In the case of the former, there were reports in some media in Australia and overseas that most of the GBR was dead, when in fact, only some sections had been severely bleached. The media needs to put more effort into accurately reporting climate-driven events at tourist destinations and ensuring that their reports are based on robust science and not sensationalised to the detriment of regional tourism communities.

One of the significant limitations to climate adaptation within the tourism sector is the high proportion of SMEs that characterise regional tourism (see Box 6.6). Many small tourism businesses operate on small overheads, with little or no capital or capacity to implement effective adaptation strategies. Additionally, very few SMEs can plan on time frames more extended than a couple of years, and as a result, making changes now will cost something. Businesses are reluctant to spend limited capital on adapting to threats that may or may not eventuate for several decades (Turton et al., 2010). On the other hand, larger tourism businesses can better implement climate adaptation strategies for their enterprises. Many tourism businesses are made up of many different types of products, and the long-term impacts of climate change may be quite different from one product to another (Turton et al., 2010). Hence, some businesses (e.g. GBR island resorts) have built infrastructure and modes of transport that may be very costly or difficult to upgrade or retrofit. In contrast, others are mobile operators that have the capacity (over time) to amend tour programmes in response to destination and site changes. At the same time, other businesses have no physical assets (e.g. natural area guided walks) that could more easily transition into a new type of operation at little or no cost (Turton et al., 2009).

The tourism industry must move beyond climate mitigation and place a much greater emphasis on adapting to a changing climate (Moyle et al., 2018). Turton et al. (2010) identified a range of climate adaptation strategies and enablers for the sector:

- Instilling industry confidence that the climate is changing and that increased variability in climate is part of the process (see Chapters 2 and 3);
- Motivation to avoid risk or take up opportunities that may accompany climate change;
- Demonstration of new technologies that facilitate adaptation to changing conditions;
- Transitional and legislative support from the government (see also Moyle et al., 2018);
- Resources from the government (see Box 6.2) and tourism peak bodies and organisations; and
- Effective monitoring and evaluation of climate adaptation strategies over time.

Climate adaptation pathways for the tourism sector will need to overcome the many barriers described above (see also Chapter 4, Box 4.2: barriers). Responses to climate change and more extreme weather events (see Chapter 3, Table 3.2) will vary across the sector and geographical location. Incremental adaptation pathways may apply in the early stages, such as taking tourists to non-degraded coral reefs (see Chapter 4, Box 4.1: Accommodate strategy). Some businesses may hold on for a time by protecting their assets through soft and hard engineering solutions, such as foredune restoration, river levees and sea walls (see Box 4.1: Protect strategy). However, if the condition of crucial tourist assets declines beyond a certain tipping point, then transformative adaptation strategies will need to be considered by affected businesses. For example, some tourism businesses may have to relocate to sites and destinations that have not been adversely impacted by climate-driven events (see Box 4.1: Retreat strategy). Strategies include the relocation of GBR tourism operators from northern to southern reef sites; abandoning damaged tourism resorts on islands severely damaged by recent tropical cyclones; the shutting down of Red Centre tourist resorts during part of the year due to the escalating number of hot days each year. Lastly, future investments in tourism infrastructure (e.g. resorts) may opt to avoid locations deemed climatically unsuitable (Box 4.1: Avoid strategy). Examples include:

- Avoiding vulnerable coastal, riverine and island sites;
- Investing in summer tourism ventures in alpine areas rather than traditional winter ventures; and
- Avoiding places considered to be thermally unfavourable for much of the year due to a combination of high temperature and humidity levels, e.g. parts of northern and central Australia (see Box 6.3).

INTEGRATION WITH THE SUSTAINABLE DEVELOPMENT GOALS

Six SDGs are relevant to a national strategy on climate change and the tourism sector:

1. SDG6 (Clean water and sanitation);
2. SDG7 (Affordable clean energy);
3. SDG9 (Industry, innovation and infrastructure);
4. SDG11 (Sustainable cities and communities);
5. SDG12 (Responsible consumption and production); and
6. SDG13 (Climate action).

A framework (Table 6.6) shows how the Australian tourism sector can effectively integrate climate change with goals, targets and indicators for each of the six relevant SDGs. Climate-resilient development pathways for tourism industries will require incremental, structural and transformational changes from this decade. If the sector is to meet (or make progress towards) the 2030 SDG targets, similar changes will need to co-occur in the energy, transportation, agricultural and construction sectors (see sections above). A whole-of-economy perspective will be critical in driving innovation and change towards a net-zero world. There will also need to be incremental and transformational changes in various social systems that depend on tourism industries (see Chapter 7, Table 7.5).

Table 6.6 Integration of SDGs with the Australian tourism sector's responses to climate change

SDG	Goals	Targets	Indicators
SDG6: Clean water and sanitation (*economic and environmental*)	1. Ensure availability and sustainable management of water	Improved water quality by reducing pollution from tourism activities and operations Substantially increase water use efficiency	• Proportion of tourism infrastructure wastewater safely treated • Change in water use efficiency over time • Level of freshwater withdraw as a proportion of available freshwater resources
SDG7: Affordable clean energy (*economic and environmental*)	2. Support extension of renewable energy adaptation	Increased use of renewable energy from tourism business activities Lower emissions from energy use	• Percentage of renewable energy use across sector • Energy emissions measurements • Percentage of tourism supply chain income spent on energy use
SDG9: Industry, innovation and infrastructure (*economic and environmental*)	3. Build resilient tourism infrastructure and business operations	Increased new, upgraded/retrofitted tourism infrastructure and transport systems that are sustainable Encourage new, inclusive tourism industries	• Carbon emissions per unit of value added • Sustainable tourism value added as a proportion of GDP and per capita • Sustainable tourism employment as a proportion of total employment (to 100)
SDG11: Sustainable cities and communities (*social and environmental*)	4. Make cities and settlements inclusive, safe, resilient and stable	Strengthen efforts to protect and safeguard the country's cultural and natural heritage that underpins the tourism sector	• Total per capita expenditure on the preservation, protection and conservation of all cultural and natural heritage that is relevant to tourism businesses, by source of funding (public, private) • Proportion of tourism construction projects completed using sustainable methods, material footprint (per capita, per GDP)
SDG12: Responsible consumption and production (*economic and environmental*)	5. Ensure sustainable consumption and production patterns	Increased efficiency Increased circular economy for tourism sector materials Sustainable public procurement practices	• Degree of public procurement practices in tourism sector that are sustainable
SDG13: Climate action (*environmental*)	6. Carbon-neutral production 7. Resilience and adaptive responses to natural hazards	Reduced global warming Closed loop production (circular economy) and supply chain systems Mitigation of climate change impacts on tourism infrastructure, operations and tourists	• Reduction in tourism sector GHG emissions contributions (to zero) • Percentage of tourism businesses industries using renewable energy • Percentage of tourism businesses with a climate action plan

This includes their supply chains. Refer to Chapter 4 for an explanation of SDG goals, targets and indicators (see also Box 4.5). Triple-bottom line implications for each SDG are shown in italics.

SUMMARY

- The Australian economy is dominated by its highly diversified services sector, which generates about 63% of its GDP and employs nearly 80% of its labour force.
- Climate change poses a significant threat to Australia's economy, with some economic sectors considered more exposed than others. A respected study estimated that by 2070, Australia's economy could lose AU$3 trillion from unchecked climate change. However, by adopting a new growth (or climate-resilient) development pathway, an extra AU$680 billion and 250,000 jobs added to its economy by that date.
- The energy, transportation, agriculture, mining, manufacturing, construction and tourism sectors are considered most vulnerable to changes in climate and the increasing frequency, intensity and duration of severe weather events. All these sectors will require significant structural adjustment policies to respond to climate change, ensuring that Australia meets its Paris Agreement target of net-zero carbon emissions by 2050 and that they survive and thrive in various forms.
- GHG emissions vary across Australia's economic sectors. Energy and transportation produced 61% of the country's net national emissions in 2019, agriculture accounted for 15%, mining was 10%, manufacturing and construction was 25%, and tourism was about 5%. In the case of energy and transportation, most of the GHG emissions were from the combustion of fossil fuels from a range of sources, notably coal, natural gas, diesel and petroleum products.
- Climate mitigation pathways for Australia's economic sectors will require incremental, structural and transformative changes. Achieving an electricity supply with zero emissions will be fundamental for the country by 2050.

This target will require investments in renewable energy sources (solar and wind) and storage and a rapid transition away from fossil fuel energy generation. Improvements in energy storage are considered a long-term solution to the intermittency in energy supply, with gas likely to play a decreasing role as new renewables technology evolves. Similarly, the transportation sector must seek opportunities to reduce dependence on fossil fuels, e.g. electrification of vehicles and rail systems and green hydrogen-powered heavier vehicles. The aviation sector will rely on offsetting while new aircraft fuel and engine alternatives are developed. Reducing GHG emissions from crucial agricultural industries is challenging, particularly for cattle and sheep production, and the red meat sector will have to rely on offsetting in the medium term. However, the agricultural and land sector can significantly contribute to Australia's pathway to net-zero emissions, including manure and fertiliser management, carbon farming initiatives and shifting to EVs or hydrogen-powered tractors and harvesting machinery. Likewise, the mining, manufacturing and construction industries will be largely reliant on offsetting to mitigate their emissions. However, there are also opportunities to employ renewable energy in their operations, including EVs and hydrogen-powered heavy vehicles and machinery. The biggest mitigation challenge to Australia's tourism industry is its high dependence on air travel, so the industry will continue to rely on offsetting and adopting renewable energy in its infrastructure and operations.

- Planned climate adaptation is vital across the Australian economy, and all sectors must play their role in responding to the challenges. The energy sector is highly exposed to climate change and pressures to meet the net-zero target by 2050. This high exposure has been created by a decade of weak federal energy policy, making it difficult for businesses and industries to invest in renewables and other energy technologies. The most urgent requirement is the rapid construction of a modern, efficient electricity system for the country's two main grids.

- The inclusion of climate change risks in the design, location and rating of future energy, agricultural, mining, manufacturing, construction and tourism infrastructure will be paramount to building resilience to extreme weather events. Agriculture is highly exposed to climate change. As time goes by, incremental adaptation to changing conditions may not be enough, and transformational responses will be required, e.g. relocating to more suitable climates for food and fibre production. In the meantime, "no regrets" practices (e.g. best practice natural resource management) will

remain relevant across the agricultural and land sectors. The mining, manufacturing, construction and tourism industries will also have to invest in climate change adaptation to increase their resilience to extreme weather events. Transformative adaptation strategies will be essential for northern and central Australian industries due to increasing exposure to heat risk and water shortages.

- Addressing the SDGs for vulnerable industries inform incremental, structural and transformative strategies to avoid mal-mitigation and maladaptation pathways in the future. A whole-of-economy perspective will be critical in driving innovation and change towards a net-zero world. There will also need to be incremental and transformational changes in various social systems that depend on Australia's main industries (see the next chapter).

7

Social Systems

Close to 90% of the Australian population lives in urban areas. The impacts of climate change will be felt by most people within urban environments, and it is from those environments that solutions are likely to emerge.

Australian Academy of Science (2021)

INTRODUCTION

Australia often presents itself to a global audience as a place where people are disconnected from urban spaces. This portrayal includes images of kangaroos jumping through vast grassy plains, people with big hats mustering cattle, vistas of deserted sandy beaches and splendid aerial views of the iconic Great Barrier Reef and Red Centre. This worldview, characterised by several Tourism Australia media campaigns, does not represent how Australians live and work. About 90% of Australians live in cities and towns, making the country among the most urbanised in the world (outside several microstates). Of those, 85% live within 75 km of the coastline. Unlike the contiguous United States of America, no inland capital cities occur in Australia other than Canberra, the deliberately planned national capital. Almost 80% of Australians live and work in the eastern mainland states of Queensland, New South Wales (NSW), Australian Capital Territory (ACT) and Victoria (see map Chapter 3, Figure 3.1).

Australia's population reached 25,739,256 persons at 30 June 2021 (Australian Bureau of Statistics, 2021b). The latest demographic data shows the continued growth of capital and regional cities in the mainland eastern states (Australian Bureau of Statistics, 2021b). There is now very little difference in population between Melbourne and Sydney, Australia's most prominent capital cities (both near 5 million). However, the population growth rate for Melbourne is higher (24% from 2010 to 2020) and is projected to approach 6 million by 2030, compared with a 5.9 million estimate for Sydney (Australian Bureau of Statistics, 2021b). In June 2019, both cities combined had 41% of the national population (Table 7.1). However, in terms of conurbations, greater Sydney (Sydney and Central Coast) had about 5.4 million residents in June 2020, compared with greater Melbourne (Melbourne, Bacchus Marsh, Gisborne-Macedon and Melton) with about 5.2 million residents (Australian Bureau of Statistics, 2021b). Persons living in the capitals (Table 7.1) increased by 1.4% in 2019–2020, while regional areas grew by 1.1% (Australian Bureau of Statistics, 2021b).

Other state and territory capitals followed a similar growth trend to Melbourne over the past decade (2011-2020), but rates varied. Brisbane, Sydney and Perth each grew about 19%, compared with Darwin (23.2%), Canberra (16.8%), Hobart (9.9%) and Adelaide (6.6%). Many regional cities on the east coast also grew rapidly over 2010–2020. By rank, the top ten fastest growing regional cities over 50,000 population were Melton (58.2%), Geelong (54.7%), Toowoomba (29.1%), Sunshine Coast (23.1%), Newcastle-Maitland (22.1%), Gold Coast-Tweed Heads (21.8%), Bendigo (15.2%), Ballarat (14.9%), Cairns (14.1%) and

Table 7.1 Population statistics for Australia's capital cities as of June 2020

Rank	Capital	Jurisdiction	Population (June 2020)	2011 Census	Growth (2010–2020) (%)	Percentage of National Population (June 2019)
1	Melbourne	Victoria	4,969,305	3,999,982	+24.08	19.86
2	Sydney	New South Wales (NSW)	4,966,806	4,231,954	+19.10	20.93
3	Brisbane	Queensland	2,475,680	2,065,996	+19.20	9.85
4	Perth	Western Australia	2,083,645	1,728,867	+19.12	8.24
5	Adelaide	South Australia	1,357,504	1,262,940	+6.56	5.38
8	Canberra– Queanbeyan[a]	Australian Capital Territory	464,995	391,645	+16.83	1.83
13	Hobart	Tasmania	219,071	211,656	+9.90	0.93
17	Darwin	Northern Territory	133,268	120,586	+23.20	0.59

Population sizes are given for 2011 and 2020, growth rates over 2010–2020, and percentage of the national population for each city as of June 2019. For geographical locations, refer to Chapter 3, Figure 3.1. Source of data (Australian Bureau of Statistics 2021b).

[a] Queanbeyan is adjacent to Canberra but is part of New South Wales.

Albury-Wodonga (14.0%) (Australian Bureau of Statistics, 2021b). Outside mainland eastern states, the fastest growing regional cities with over 50,000 population were Bunbury in Western Australia (13.3%) and Launceston in Tasmania (6.2%).

Australia's population almost doubled in the past 50 years, from 1971 to 2021. The average annual growth rate during this time (1.4%) was higher than in many developed countries (Australian Bureau of Statistics, 2021b). This strong growth was driven mainly by natural increases (births minus deaths, 60% of growth) before 2001. Since then, overseas migration was the major contributor (57%) until international borders closed due to the COVID-19 pandemic in March 2020 (Australian Bureau of Statistics, 2021b). Pre-pandemic overseas migration levels are poised to resume as international travel returns in 2022.

Like all highly developed countries, there has been a steady increase in the number and proportion of older people in Australia (Commonwealth of Australia, 2015b). The median age was 27.5 years in 1971 and has increased to 38.2 years in 2021 (Australian Bureau of Statistics, 2021b). Life expectancy has also increased over the past 50 years and was 81.2 years for males and

85.3 years for females in 2018–2020 (Australian Bureau of Statistics, 2021b). However, the arrival of younger migrants has only partly offset the ageing population. Regional cities such as Sunshine Coast, Wollongong, Geelong, Launceston, Newcastle-Maitland, Bendigo and Gold Coast-Tweed Heads have the highest proportion of their population over 65 years of age (Commonwealth of Australia, 2015b). Declining birth rates in recent decades have accompanied declining mortality rates. In 2019–2020, the total fertility rate was 1.58 births per woman for all Australian women. For Aboriginal and Torres Strait Islander women (see Indigenous Peoples below), the rate was 2.25 births per woman (Australian Bureau of Statistics, 2021b). An ageing population also brings many challenges for transport and infrastructure across the country and health and aged care spending (Commonwealth of Australia, 2015b).

Significant demographic differences exist between the capital cities (Table 7.1) and the regions. The median age for persons in capital cities is 36.5 years, compared with regions at 41.4 years (Australian Bureau of Statistics, 2021b). The youngest capital was Darwin (median age is

34.7 years), and Hobart was the oldest (39.9 years). In 2019–2020, the capital cities and regions had similar sex ratios for all age groups (sex ratio was 98.2 males per 100 females), except for those over 79 years, where more women are represented (Australian Bureau of Statistics, 2021b). Adelaide had the lowest sex ratio of all capital cities (96.5 males per 100 females), while Darwin had the highest (107.9 per 100).

The latest ABS population projections for Australia show continued growth. Based on 2017, projections make assumptions of fertility, mortality and migration through to 2066 (Australian Bureau of Statistics, 2018). However, the projections pre-date the pandemic, so they may need to be revised in the future. The country's population will likely reach between 37.4 and 49.2 million persons by 2066 (Australian Bureau of Statistics, 2018). Over the same period, the average annual growth rate (1.7% in 2017) is projected to decline between 0.9 and 1.4%. Notably, the median age (37.2 years in 2017) is likely to reach between 39.5 and 43.0 years by 2066 (Australian Bureau of Statistics, 2018).

Population growth will vary among the states and territories (Australian Bureau of Statistics, 2018). Mid-range population projections for 2066 are NSW (13.1 million), Victoria (12.0 million), Queensland (8.7 million), Western Australia (4.8 million), South Australia (2.2 million), ACT (0.76 million), Tasmania (0.58 million) and Northern Territory (0.44 million). The population distribution among capital cities and regions is also projected to change from a relatively high base level. In 2017, 67% of Australians lived in capital cities (Australian Bureau of Statistics, 2018). This proportion of people is likely to increase between 69% and 70% by 2027. All capital cities will grow more than their respective regions by 2027. Notably, Brisbane is on track to increase from 49% of Queensland's population to 51% in 2027, accommodating most of Queensland's people for the first time (Australian Bureau of Statistics, 2018). Darwin will also increase its share of the Northern Territory's population by more than any other capital city. Its share of the population is likely to increase from 60% in 2017 to between 63% and 64% in 2027. The remaining capital cities will likely increase their share of their state and territory populations by about 2% by 2027 (Australian Bureau of Statistics, 2018).

From a population size, distribution and average age perspective, most Australians exposed to climate change, and extreme weather events (see Chapter 3, Table 3.2, Box 3.1) are 'ageing urban dwellers' living in the continental southeast of the country (see map Chapter 3, Figure 3.1). This concentrated (and mostly coastal) region occurs south of the Bundaberg/Wide Bay area in southeast Queensland to Geelong in southern Victoria (including Brisbane, Sydney, Canberra and Melbourne). Clusters of other higher density populations also occur in coastal settlements, including:

1. The southwest of Western Australia (WA), from Geraldton to Esperance (including Perth);
2. Southern parts of South Australia (including Adelaide);
3. Coastal northeast Queensland, notably Cairns, Townsville, Mackay, Rockhampton–Yeppoon and Gladstone.
4. Launceston and Hobart in Tasmania; and
5. Darwin in the Northern Territory.

Many remote regional settlements and Aboriginal and Torres Strait Islander (Indigenous) communities in the north are also highly vulnerable to climate change (see Chapter 6, Box 6.3). Northern Australia refers to the large region north of the Tropic of Capricorn and all the Northern Territory (see map Chapter 3, Figure 3.1). This region comprises 45% of the total area of Australia but contains only 1.3 million or 5% of its residents (Australian Bureau of Statistics, 2021b). Most persons (about 70%) in this large region live in cities and towns in northeast Queensland and in and around Darwin in the Northern Territory (see above). Smaller urban areas (Australian Bureau of Statistics, 2021b) include Alice Springs (26,448 persons), Karratha (17,482 persons), Mt Isa (18,334 persons) and Broome (14,403 persons). Most Australian Indigenous peoples live in urban areas in southern and eastern Australia but are the predominant population in remote regions of Northern Australia (Lawrence et al., 2022). For example, about 30% of the Northern Territory's population are Indigenous, while Indigenous persons only make up 3% of the country's total population or about 772,000 persons (Australian Bureau of Statistics, 2021b). The population of Indigenous Australians is projected to increase to between 907,800 and 945,600 people in 2026, at an average growth rate of between 2.0% and 2.3% per year (Australian Bureau of Statistics, 2021b).

Appreciating the human geography of Australia is essential to gaining a better understanding of observed and the likely future impacts of climate change on the country's social systems (see Chapter 4, Figure 4.1). Economic, demographic and socio-cultural trends affect the exposure, vulnerability and adaptive capacity of individuals and communities across the country (Cresswell et al., 2021, Lawrence et al., 2022, Naughtin et al., 2022). Notably, remote Indigenous communities face severe housing, health, education, employment and services issues (Kotey, 2015, Matthews et al., 2021). Their significant social disadvantage adds to their high exposure to climate change.

This chapter discusses four socially orientated themes that vary in their vulnerability to climate change and increasingly extreme climate-driven events (Steffen et al., 2019, Australian Academy of Science, 2021, Matthews et al., 2021):

1. Cities, settlements and built environments;
2. Human health and well-being;
3. Indigenous peoples; and
4. Cross-cutting social issues.

Evaluation of each theme will follow the conceptual framework for climate change impacts, vulnerability and climate risks (see Chapter 4, Figure 4.1). The emphasis will be on the local government area (LGA) scale. This local to regional scale is more appropriate when examining social systems' responses to climate change, representing the lowest governance level. LGAs are also at the forefront of climate change and face many institutional, financial and logistical challenges when responding to extreme weather events. By themes, the discussion will first examine:

1. Any observed effects of climate change, along with any significant issues; and
2. Potential climate impacts, threats and risks in the context of their vulnerability (i.e. exposure and sensitivity) to climate change and adaptive capacity.

The second section evaluates climate-resilient development pathways for social systems at the LGA scale. This discussion includes identifying climate mitigation and adaptation pathways for vulnerable infrastructure, properties, communities and populations. The last section will integrate the four themes with the relevant, sustainable development goals (see Chapter 4, Sustainable Development Goals, SDGs). Addressing the SDGs at the LGA scale will inform incremental, structural and transformative strategies to avoid mal-mitigation and maladaptation pathways in the future (see Chapter 4, Figures 4.1–4.3, Box 4.1).

CITIES, SETTLEMENTS AND BUILT ENVIRONMENTS

Climate impacts and risks

The world's cities are powerhouses for production, trade, employment and investment, generating about 80% of the world's GDP (Stock et al., 2017). Large urban areas also produce large amounts of GHGs from building energy use, transportation and manufacturing industries (see Chapter 6, Energy, Transportation and Manufacturing sectors). Australia's cities, towns and shires (LGAs) are at high risk from worsening climate impacts, such as rising sea levels, floods, bushfires and extreme heatwaves (Stock et al., 2017, Australian Academy of Science, 2021, Steffen et al., 2019, Cresswell et al., 2021, Lawrence et al., 2022, Rice et al., 2022). This risk is exacerbated by the high concentration of cities, towns and settlements near the coast (Stock et al., 2017, Steffen et al., 2019, Rice et al., 2022). Climate change and extreme weather events are also a significant threat to Australia's finance, insurance and property sectors (Box 7.1). These climate threats compromise the long-term sustainability and viability of many exposed residential and commercial assets and critical public infrastructure (Stock et al., 2017, Steffen et al., 2019, Australian Academy of Science, 2021). A new report (CoreLogic, 2022) found that AU$25 billion worth of residential property in Australia faces coastal climate risk over the next 30 years due to sea-level rise, increasing storm surges and coastal erosion. There is also growing inequality in the availability of affordable property insurance (Box 7.1). On average, about 20% of households north of the Tropic of Capricorn (see map Chapter 3, Figure 3.1) are foregoing home insurance (Naughtin et al. 2022). This is about twice the percentage of households compared to those in southern Australia.

BOX 7.1: Economic impacts of climate change on the Australian finance, insurance and property sectors

Finance and insurance risks and costs

The finance sector is highly exposed to climate change: Aggregated insured losses from weather-related hazard events from 2013 to 2020 were almost AU$15 billion for Australia (1.2% of GDP) [1].

Banks and insurers are starting to undertake climate risk analyses and disclosing their risks [1].

Risks for the finance sector are projected to increase [1].

Climate adaptation finance is not evident [1].

Dealing with legacy developments in vulnerable areas [1].

Insurance costs and availability will increase: Reinsurance companies assert that insurers cannot keep using the past as a 'yardstick' for the future.

In the absence of climate adaptation, insurance premiums will have to increase [2,3].

Insurance may eventually be unavailable in some exposed places and regions [3].

Costs to governments as insurers of last resort [1].

Not all hazards are covered by commercial insurance: Hazards such as bushfires, riverine flooding and storm damage are generally *covered*. Events such as coastal inundation, erosion, landslip and subsidence are all generally excluded. These arrangements will have to change in the future [2].

An overall focus on flood insurance leaves unanswered questions about the different insurance challenges posed by storms and wildfire [4].

Insurance damage bills will increase: Modelling estimates that the annual average cost of damage from extreme weather and climate hazards to properties will rise to AU$85 billion in 2030, AU$91 billion in 2050 and AU$117 billion in 2100 [2].

Property risks and costs

Relative property values will decrease: Banks in principle will have to lend less for houses with higher insurance costs. Thus, for two equivalent homes, the one that is more exposed and vulnerable to extreme weather impacts, driven by climate change, is expected to decrease in value compared to less affected properties [2].

The total estimated damage-related loss of property value (excluding any disruptions to productivity or physical damage or replacement costs of buildings) is expected to rise to AU$611 billion by 2050 and AU$770 billion by 2100 [2].

Loss in value of property market: The property market is expected to lose AU$571 billion in value by 2030 due to climate change and extreme weather events. It will continue to lose value in the coming decades under high GHG emissions scenarios [2].

Major public and private assets are at risk:

An estimated 160,000–250,000 Australian properties are at risk of coastal flooding with a sea-level rise of 1 m by 2100 [5].

More than AU$226 billion in commercial, industrial, road, rail and residential assets will be at risk from sea-level rise alone by 2100, under high GHG emissions scenarios (equates to a 1.1 m sea-level rise) [2].

There is likely to be a 111% rise in inundation cost from 2010 to 2100 [1].

A 2022 study showed that over the next 30 years, AU$25 billion of residential property faces coastal risk due to increasing storm surges and coastal erosion. Suburbs at most risk include Paradise Point (Queensland), Cronulla (NSW) and Port Melbourne (Victoria) [6].

Heat risk is an issue for many properties: The design of buildings has a strong impact on the cooling needs of occupants, and most existing dwellings are not able to adequately protect people from severe heatwaves (see also Chapter 6, Construction sector) [2].

Finance and insurance risks and costs	Property risks and costs
Insurance costs will be uneven: Increased costs will be concentrated on about 5%–6% of properties, which will experience average annual risk costs that are effectively unaffordable (costing the equivalent to 1% or more of the property value per annum). By 2030, about 1 in every 19 properties is projected to fall into this category, by 2050 about 1 in every 18 properties, and by 2100 about 1 in every 15 properties [2,3]. These properties are concentrated in areas that are highly exposed to current and future hazards – like riverine flooding, tropical cyclones, bushfires and coastal inundation, and that were not built to withstand such events [2]. Insurance premiums in Northern Australia are almost double those in the rest of Australia, and rising, mainly due to cyclone damage [1].	

Sources: [1] Lawrence et al. (2022), [2] Steffen et al. (2019), [3] Rice et al. (2022), [4] Lucas et al. (2021), [5] Australian Academy of Science (2021), [6] CoreLogic (2022).

RISING AVERAGE TEMPERATURES AND EXTREME HEATWAVES

Australian land areas have warmed by 1.4°C between 1910 and 2019, which is higher than the global average of 1.1°C over that period (see Chapter 2, Figure 2.6c). The country's urban populations are highly exposed to a range of future climate risks (see Chapter 3, Table 3.2, Box 3.1). As baseline temperatures increase, so does the threat of more extreme weather events (Intergovernmental Panel on Climate Change, 2021). Heatwaves in Australia are becoming more frequent, more severe and longer-lasting (Stock et al. 2017, Australian Academy of Science, 2021). This increasing trend is occurring at locations across the country. Notably, 2019 experienced 43 very hot days (>40°C) nationwide, more than triple the number in any of the years before 2000 (Bureau of Meteorology and Commonwealth Scientific and Industrial Research Organisation, 2020).

One study (Perkins and Alexander, 2013) compared heatwave days (days >35°C) between 1950–1980 and 1981–2011 for the country's capital cities. Heatwave patterns varied across the capital cities. For example, the number of heatwave days has more than doubled in Darwin and Canberra; it has nearly doubled in Adelaide and increased by 50% in Perth. Other capitals show changes in the onset of heatwaves in summer. In Sydney, heatwaves now start 19 days earlier, compared with 17 days for Melbourne and 12 days for Hobart. One significant finding is that the hottest heatwaves in Adelaide are 4.3°C hotter than in other capital cities (Perkins and Alexander, 2013).

All capital cities can expect an increase in their annual number of hot days (>35°C) due to global warming. This increase will depend on future GHG emissions (see SSPs in Chapter 2, Table 2.6). Relative to the 1981–2010 average, and under the highest emissions scenario (SSP5–8.5), Sydney may expect 11 hot days per year by the 2090s (1981–2010 average: 3.1), Melbourne 24 hot days (average: 11), Perth 63 hot days (Mean: 28), Brisbane 55 (average: 12), and Darwin 265 hot days (average: 11) (Australian Academy of Science, 2021). Even if the global average temperature is held at 1.5°C above pre-industrial by 2100, the city of Darwin is locked into over 110 hot days per year (Watterson,

2015). A combination of high temperature and high relative humidity will make Darwin, other towns and more remote communities across Northern Australia highly unsuitable for human habitation for a large part of the year (see Chapter 6, Box 6.3).

Large urban areas also produce an urban 'heat island' effect due to concrete and asphalt that retain heat (Oke, 1987). Urban heat islands are usually metropolitan areas that are significantly warmer than their surrounding rural areas due to a range of human activities (Cresswell et al., 2021). This effect means many people in Australia's largest cities may experience ambient temperatures 1°C–3°C higher than their surrounding rural areas. The rapidly growing area of Western Sydney is particularly exposed to the heat island effect due to its poor urban design and distance from coastal sea breezes that temper heatwave events in the more prosperous eastern and northern suburbs (Box 7.1). For example, on 4 January 2020, Penrith in Western Sydney reached 48.9°C, and in 2019 Parramatta (Sydney's second CBD) sweltered for 47 days with temperatures over 35°C (Climate Council, 2021b). Many of the residents in these western centres are from lower socio-economic and disadvantaged groups.

Heatwaves often cause costly disruption of critical infrastructure, notably electricity and transportation (see Chapter 6, the 'Energy and Transportation' sections). Losses of electricity are typical during severe heatwaves mainly because of the country's dependence on coal-fired generators (that can fail during extreme heat) and high demand for air conditioning. For example, the January 2009 heatwave in Melbourne caused financial losses estimated to be AU$800 million, mainly driven by power outages and disruptions to the transport network (Stock et al., 2017). Specifically, half a million residents in western and central Melbourne lost electricity supply from 1 hour to 4 days. Electrical faults caused technical failures in rail signalling; some older train air-conditioners failed, and there were 29 cases of buckled rail tracks. Due to the inability to meet power demand for air conditioning, electricity loss during extreme heatwaves exposes vulnerable communities and people (see the 'Human Health and Well-Being' section below).

CHANGES IN RAINFALL

Heavy rainfall and river floods are very likely to increase across Australia (see Chapter 3,

Table 3.2, Box 3.1). Extreme rainfall events will also have severe implications for the country's cities (Cresswell et al., 2021). Cities are characterised by hard, impermeable surfaces, like concrete, asphalt and compacted soil (Australian Academy of Science, 2021). The lack of permeability exacerbates any effects of heavy rain across urban areas. These changes will affect rainfall-runoff following extreme rainfall events, contributing to flash flooding and riverine flood events (Figure 7.1). In the future, floods will be more common and severe in terms of the area affected, speed of onset and depth of water (Australian Academy of Science, 2021, Rice et al., 2022). Increased flooding risk has significant implications for insurance, as many locations will be deemed uninsurable, and some may become prohibitively expensive for residents and businesses (Box 7.1). Variable insurance costs will disfavour some regions (e.g. Northern Australia), and disadvantaged communities that can ill-afford rising costs of premiums (see the 'Human Health and Well-Being' and the 'Indigenous Peoples' sections below). Urban areas are also vulnerable to droughts, which are expected to become more frequent and severe with climate change (see Chapter 3, Table 3.2). Prolonged dry periods can also result in structural impacts, including damage to building foundations when clay soils dry out and crack (Australian Academy of Science, 2021).

WATER SCARCITY

The most populated regions of Australia, including its largest cities (Table 7.1), are all in southern Australia. There has already been a significant drying and warming trend across the south (see Chapter 3, Historical Changes in Climate). This trend is expected to continue in the future, with severe implications for water security for urban areas in the southwest, south and continental southeast (see Chapter 3, Table 3.2. Box 3.1). Hence, water security is a significant climate risk for all capital cities, except for Darwin (see Chapter 5, Box 5.4). Lower rainfall lessens runoff and streamflow disproportionately more than the reduction in rainfall. For example, rainfall decline in southwest Western Australia of 19% since the mid-1970s has dramatically reduced the annual average streamflow into Perth's dams by nearly 80% (Stock et al., 2017). As the country's cities grow, water scarcity will grow as urban water demand threatens to outstrip catchment water supply during droughts

Figure 7.1 Brisbane River, Queensland. Brisbane, Australia's third largest city (see Table 7.1) has experienced several severe floods in its 200-year colonial history, including 1893, 1974, 2011 and 2022. The latter two events occurred despite flood mitigation measures upstream of the city's flood-prone areas. (Credit: http://teamblm.com.au/Unsplash/Public Domain.)

(Lawrence et al., 2022, Naughtin et al., 2022). Many locations in Australia at global warming of 3°C (see Chapter 3, Box 3.1) would be challenging to live in due to extreme projected water shortages (Australian Academy of Science, 2021). With increasing water scarcity in most years, there are likely to be competing demands among urban, agricultural and ecosystem needs for freshwater resources (see Chapter 5, Box 5.4).

BUSHFIRE RISK

The intensity, frequency and duration of fire weather events are likely to increase throughout Australia in the future (see Chapter 3, Table 3.2). The southern half of the country is particularly vulnerable, where there has already been an increase in the frequency of extreme fire days and lengthening of the fire season since the 1950s (see Chapter 5, Box 5.3). The outer suburbs of cities and large towns will be at increasing risk as temperatures rise (Australian Academy of Science, 2021). Since 2000, large and uncontrollable fires have destroyed property in cities, towns and shires across Australia (Stock et al., 2017, Lawrence et al.,

2022). These bushfires have occurred all over the country, particularly in the southern half.

The 2019–2020 bushfire season (Black Summer Fires) is a historical standout (Canadell et al., 2021). This assertion was due to its unusual intensity, size, duration and uncontrollable nature (Filkov et al., 2020). The unique combination of fire characteristics meant it was a mega-fire on a global scale (Canadell et al., 2021). Throughout the Black Summer, hundreds of fires burnt, mainly in the southeast of the country, with devastating impacts on settlements and infrastructure (Lawrence et al., 2022). The major fires peaked from December to January. By March 2020, the bushfires had affected almost 19 million hectares and destroyed over 3,500 houses (Filkov et al., 2020). The total cost of the bushfires was over AU\$8 billion (Lawrence et al., 2022). Because of the 2019–2020 mega-fires, bushfire-prone areas of Australia are now a significant risk for the finance and insurance sector (Box 7.1). Notably, some fire-prone areas may become uninsurable in the future, and this will eventually cause property values to decline in bushfire exposed places.

RISING SEA LEVELS

Sea-level change around Australia is comparable to global rates, with some regional differences (see Chapter 3, Historical Changes in Climate). The country is not only highly urbanised, but it is also largely coastal-centric. Hence, many cities, towns and shires are highly vulnerable to sea-level rise and increased risk of inundation and coastal erosion (Australian Academy of Science, 2021, Cresswell et al., 2021). Future sea-level rise is unavoidable and will be dependent on future GHG emissions (see Chapter 3, Table 3.2). From a risk perspective, there are two interacting stressors to be considered when considering the effects of rising sea levels on cities, settlements and built infrastructure (Figure 7.2). First is the gradual inundation of coastal areas and inlets and its impact on property and public infrastructure (a chronic stressor). Second, there are the impacts of individual high-energy coastal storm events on coastal communities and assets (acute stressors). As sea levels steadily increase, high-energy coastal storm events will occur on a higher baseline sea level (Turton, 2019, Australian Academy of Science, 2021). Over time, these storm events will become more damaging as they penetrate further inland affecting more critical infrastructure, properties and people (Stock et al., 2017, Australian Academy of Science, 2021).

Research shows that a 0.5 m sea-level rise around Australia's largest cities (Table 7.1) would result in substantial increases in the frequency of extreme coastal flooding events (Stock et al., 2017). This frequency is by a factor of several hundred for most places and up to one thousand for

Figure 7.2 An aerial oblique view (looking north) of Surfers Paradise on Queensland's iconic Gold Coast. Rising sea levels, storm surges and erosion are a major threat to highly developed coastal communities in Australia. (Credit: Caleb Russell/Unsplash/Public Domain.)

some vulnerable locations. To provide context, a multiplying factor of 100 means that an extreme event (e.g. a storm surge) with a current probability of 1-in-100 years would occur on average every year. Under a high emissions scenario (see SSPs in Chapter 2, Table 2.6: SSP5–8.5), sea levels are likely to be about 1 m higher by 2100. This sea-level rise will result in significant property and infrastructure damage in exposed cities, towns and shires (Stock et al., 2017, Australian Academy of Science, 2021). The large number of buildings exposed to sea-level rise in Australia was quantified in 2011 (see Lawrence et al., 2022). There were 187,000–274,000 residential buildings, 5,800–8,600 commercial buildings and 3,700–6,200 light industrial buildings considered exposed. In terms of local government assets, there were 27,000–35,000 km of roads and 1,200 –1,500 km of rail lines and tramways. These figures are likely to be much higher today.

It is highly likely that rising sea levels and more frequent and severe high-energy coastal flooding events will adversely impact the availability of finance and insurance for homes and businesses in exposed places, with knock-on effects on their property values (Box 7.1). As the sea level rises, the risk of coastal flooding during high tides and storm surges increases (Australian Academy of Science, 2021, Rice et al., 2022). These risks are usually associated with high-energy events, such as East Coast Lows and tropical cyclones (see Chapter 3). Impacts of extreme coastal flooding events can include loss of life and disruption of health and social services (see the 'Human Health and Well-Being' section below). There is also a significant risk of inundation of property and coastal infrastructure, such as houses, businesses, ports, airports, railways and roads (Stock et al., 2017, Steffen et al., 2019, Lawrence et al., 2022). However, the effects of sea-level rise are not confined to the exposed coast. Given a 1 m sea-level rise, the number of properties along New South Wales estuaries impacted by a 1-in-100-year storm surge would triple to 74,700, compared to 24,300 for 2020 sea levels (Australian Academy of Science, 2021). As the frequency and severity of coastal flooding events increase over time, there will be negative implications for the finance, insurance and property sectors (Box 7.1).

Adverse impacts of rising sea level are already occurring on coastal, estuarine and freshwater ecosystems (see Chapter 5, Natural Systems). These coastal ecosystems play an essential role in 'buffering' cities and settlements from storm surges and riverine flooding. If these natural buffers become degraded or disappear, there will be knock-on effects for adjacent urban areas, causing an increased exposure to storm surge events. A related phenomenon known as 'coastal squeeze' can force coastal ecosystems such as mangroves, salt marshes and seagrass beds to move inland as sea level rises (Turton, 2012, Australian Academy of Science, 2021). Over time, these coastal ecosystems push up against human infrastructures, such as roads and buildings (see Chapter 5, Box 5.6). With nowhere to go, these coastal ecosystems become highly degraded or disappear.

Some of the most catastrophic coastal flooding events are driven by a 'double whammy' of synchronised high-energy coastal flooding events and heavy rainfall events in the catchments inland of coastal settlements (Stock et al., 2017, Rice et al., 2022). Under these combined conditions, coastal settlements can be inundated by water from a storm surge on a high tide and riverine and flash flooding from the catchments. Similarly, flood inundation levels can go up and down after a storm event as high tides push freshwater runoff from the catchments upstream. This phenomenon can also occur without a high-energy coastal flooding event, as normal high tides interact with high-volume riverine flooding from heavy rainfall in the upper catchment.

HUMAN HEALTH AND WELL-BEING

Climate impacts and risks

The increasing frequency and intensity of extreme weather events affect human health and well-being and the livelihoods of exposed communities and their people (Australian Academy of Science, 2021, Cresswell et al., 2021, Lawrence et al., 2022). These adverse effects on public health are compounded by an ageing population (Australian Bureau of Statistics, 2021b), as generally, older people cannot tolerate extreme weather events like younger groups (Australian Academy of Science, 2021). Other socially disadvantaged groups are particularly exposed to climate change, including low-income earners living in high climate risk suburbs, people with disabilities, remote settlements and Indigenous communities (see the 'Indigenous Peoples' section below). For many of these disadvantaged groups, building and contents insurance is often beyond their reach (Box 7.1), and they are often neglected in the aftermath of natural disasters (Caruana, 2010). Climate risks and threats to human health and well-being include heatwaves, bushfires, water scarcity, floods and storms (Stock et al., 2017, Australian Academy of Science, 2021, Matthews et al., 2021, Rice et al., 2022).

HEATWAVES

Australia's most significant threat to human health is projected increases in more frequent, prolonged and intense heatwaves (see Chapter 3, Table 3.2, Box 3.1). Heatwaves kill more people than other natural disasters combined (Coates et al., 2022). Vulnerable groups include the socio-economically disadvantaged (e.g. people in rental housing), outdoor workers, individuals with pre-existing medical conditions, infants and the elderly (Australian Academy of Science, 2021, Cresswell et al., 2021). Climate change has increased the frequency and intensity of heatwaves in Australia and associated deaths. Vicedo-Cabrera et al. (2021) found that between 1991 and 2011, about 36% of heat-related mortality in Brisbane, Sydney and Melbourne was blamed on climate change. This attribution equated to nearly 106 deaths a year on average over the 20-year survey. More recent research (Coates et al., 2022) found that from 2000 to 2018, at least 354 people have died in heatwaves across the country. Increases in extreme heatwave events will likely increase human morbidity and mortality, particularly in vulnerable populations (Varghese et al., 2020, Matthews et al., 2021).

Health effects of extreme heat include acute heatstroke and the worsening of chronic medical conditions such as heart and kidney disease (Stock et al., 2017, Australian Academy of Science, 2021). Other risk factors during heatwaves include social isolation, geographical remoteness (e.g. Indigenous communities), the presence of disabilities (physical or mental), adverse interactions with some prescribed medications and the absence or non-use of air conditioning or other building heat protection (Matthews et al., 2021, Coates et al., 2022). Workplace health and safety consequences during heatwaves may include more accidents because of concentration lapses and poor decision-making ability due to higher levels of fatigue (Steffen et al., 2019, Lawrence et al., 2022). The productivity of outdoor workers and others working in non-air-conditioned environments is expected to decline in response to increases in the frequency and severity of heatwaves (Australian Academy of Science, 2021). Notably, declining productivity in outdoor workforces is highly likely across northern Australia (see Chapter 6, Box 6.3).

During heatwave events, there are knock-on effects that affect social and public health services – for example, emergency departments dealing with large numbers of affected people (Australasian College for Emergency Medicine, 2020). At the community level, socially isolated groups are exposed to heatwaves due to economic challenges and limited access to culturally and linguistically diverse information (Matthews et al., 2021, Coates et al., 2022). Other compounding health effects in communities exacerbated by heatwaves include acute respiratory illnesses due to poor air quality from bushfires in urban and peri-urban areas of the country (see below). Notably, high temperatures magnify the health risks of local air pollution (Lawrence et al., 2022). One study found that without adaptation, ozone-related deaths in Sydney may increase by up to 60 per year by 2070 (Physick et al., 2014). Significantly, an ageing population will worsen many of these extreme compound events (Australian Academy of Science, 2021). For example, by 2056, about 25% of Australians will be older than 65 (Australian Bureau of Statistics, 2021b).

BUSHFIRES

Increases in the frequency, intensity and duration of bushfires in Australia are causing rises in reported adverse impacts on human health and well-being (Stock et al., 2017, Australian Academy of Science, 2021, see Chapter 5, Box 5.3). These effects include direct loss of life and fires exacerbating pre-existing conditions such as lung and heart disease due to air pollution from prolonged and more severe fire seasons. For example, the unprecedented Black Summer bushfires in 2019–2020 directly killed 33 people (Davey and Sarre, 2020). The same series of events exposed millions of people to heavy particulate pollution (Vardoulakis et al., 2020). During the same bushfires, atmospheric air quality was worsened by dust storms from inland areas mixing with bushfire smoke over highly populated areas (Australian

Academy of Science, 2021). The bushfires and dust storms indirectly caused a further 417 deaths, 3,151 hospitalisations resulting from respiratory or cardiovascular conditions and over 3,100 admissions for asthma due to smoke exposure (Borchers Arriagada et al., 2020, Vardoulakis et al., 2020). Johnston et al. (2020) estimated the smoke-caused health costs for the 2019–2020 bushfires to be AU$1.95 billion.

Bushfires can also impact urban and rural drinking water supplies (see the 'Water Scarcity' section below). When widespread fires occur in dam or reservoir catchments, they usually disrupt potable water supplies. Heavy rains soon after bushfires often cause water supply problems due to enhanced erosion of soils and nutrients and then runoff into dams from burnt-out catchments. Canberra's main reservoirs were affected by bushfires and sedimentation after the devastating 2003 bushfires (White et al., 2006). On a grander scale, the Black Saturday bushfires in 2009 directly affected 30% of the catchments that provide drinking water for Melbourne (Stock et al., 2017).

WATER SCARCITY

Climate change is already leading to declines in rainfall in the southwest and continental southeast of Australia (see Chapter 3, the 'Historical Changes in Climate' section). These trends are expected to continue in the future (see Chapter 5, Box 5.4). Threats to agriculture and water security (see Chapter 6, see 'Agriculture and Land' section) have many adverse health effects, including those precipitated by declining economic livelihoods and social cohesion (Australian Academy of Science, 2021). In recent years, there have already been cases of small communities running out of water and having to import water from elsewhere for drinking supplies, such as Stanthorpe in Queensland's Southern Downs in early 2020 (Cresswell et al., 2021). Future extreme droughts are highly likely, particularly when global warming exceeds 1.5° C above pre-industrial, in the 2030s. Water scarcity is expected to affect local rural communities adversely and may be the trigger to force people to move to places with climate-proof water supplies (Australian Academy of Science, 2021). Loss of social capital and cohesion will adversely affect many rural and remote economies (Steffen et al., 2019).

Climate change is increasing the frequency, severity and duration of droughts in many parts of southern Australia. Depending on future GHG emissions, extreme droughts are highly likely to occur in the future (see Chapter 3, Table 3.2, Box 3.1). Prolonged droughts may pose risks to physical and mental health in rural and remote communities (Stanke et al., 2013, Australian Academy of Science, 2021, Matthews et al., 2021). During droughts, mental health among farmers is affected by a range of socio-demographic and community factors that exacerbate stress related to their occupation and rural context (Austin et al., 2018). Notably, younger farmers who live and work on a farm and experience financial hardship are more exposed to drought-related stress.

FLOODS AND STORMS

Extreme riverine flash floods from high rainfall events and coastal inundation from storm surges are increasing in frequency and intensity with climate change (Australian Academy of Science, 2021, Rice et al., 2022). These extreme events are expected to increase in the future (see Chapter 3, Table 3.2, Box 3.1). Australian cities, towns and settlements are highly exposed to floods and storms due to the location of most of the population and critical infrastructure near the coast and on flood plains. Adverse human health and well-being impacts from floods and storms in vulnerable areas include injuries, depression, anxiety and poor physical health (Stock et al., 2017). These health effects mainly occur during extreme events, e.g. direct exposure to a storm or flood. However, human suffering also occurs after such events, e.g. exposure to contaminated drinking water and mosquito-borne diseases and health and safety risks from clean up and recovery activities. Long-term negative issues for communities include mental health issues following traumatic and stressful experiences (Stock et al., 2017, Rice et al., 2022).

Riverine and flash flood events are Australia's most expensive extreme weather events, accounting for nearly 30% of economic damages over the past 10 years (Rice et al., 2022). Totals flood costs from 2010 to 2019 exceeded AU$10 billion in damages. The recent 2022 floods in Queensland and New South Wales are expected to reach AU$3 billion, making the floods one of the country's most expensive natural disasters (Australian Financial Review, 2022). However, due to a lack of insurance and underinsurance (see Box 7.1), the final cost will be many times higher (Rice et al., 2022). In Queensland alone, the state government estimated rebuilding costs of about AU$2.5 billion. Previous flooding events in Queensland, notably the January/February 2019 flooding event in north Queensland (see Turton, 2019b), had a total social and economic cost of about AU$5.7 billion, representing 14% of the region's annual economic output (Deloitte Access Economics, 2019). Included in this figure were damage to homes and critical infrastructure, effects on human health and well-being and the loss of half a million cattle (Turton, 2019b, Rice et al., 2022). Ongoing effects of floods include business closures, loss of work productivity through absenteeism and reduced agricultural productivity, increased emotional stress and physical illnesses (Rice et al., 2022).

Future extreme flood events are inevitable as the oceans and atmosphere continue to warm (see Chapter 2). Floods will likely be Australia's most costly natural hazards, and rising sea levels will exacerbate coastal flooding events from oceanic storms. Kompas (2020) has estimated annual damages from extreme weather events in Australia, including sea-level rise and other adverse impacts driven by climate change (e.g., bushfires, droughts, floods), could exceed AU$100 billion by 2038. Floods alone could cost the national economy as much as AU$40 billion per year by 2060 (Deloitte Access Economics, 2021).

There is already evidence that extreme floods and other natural disasters worsen inequality in Australia (Rice et al., 2022). Disadvantaged groups include women and children, Indigenous Australians (see below), some ethnicities (e.g. those with limited English) and low-income families (Matthews et al., 2021, Rice et al., 2022). Households on low incomes are more likely to live in areas exposed to flooding and other climate hazards and are much less likely to have home and contents insurance (see Box 7.1). These vulnerable groups have a low adaptive capacity (see Chapter 4, Figure 4.1) to deal with extreme weather events and inadequate financial resources to recover fully.

INDIGENOUS PEOPLES

Background context

Aboriginal and Torres Strait Islander Peoples belong to the world's oldest living cultures, surviving continually on their ancestral lands or 'Country' for over 65,000 years (Lawrence et al., 2022). This multi-millennial year period of human occupancy corresponds with about half of the last glaciation (100–18 ka BP) during the late Pleistocene epoch. This extended occupancy coincides with the Last Glacial Maximum and has persisted into the interglacial Holocene epoch (see Chapter 2, Box 2.1). Indigenous Australians have therefore survived and adapted to past climate changes, notably the significant sea-level rise and extreme rainfall variability during the late Pleistocene epoch (Australian Academy of Science, 2021, Lawrence et al., 2022, see also Chapter 3, Past Climates). Over many millennia, this ability to adapt to climate change relied on intimate place-based Indigenous knowledge in practice and gained while losing traditional land and sea 'Country' during the Last Deglacial Transition (18–11 ka BP) (Liedloff et al., 2013, Golding and Campbell, 2009, Nunn and Reid, 2016). During the 7,000-year deglacial transition, the global mean sea level rose from about −120 m to be near −50 m, relative to that for 1850–1900. Therefore, the first Australians living in coastal areas contended with relatively rapid sea-level change during the early Holocene. This rate equates to roughly 1 m of sea-level rise per century. This rapid upward change corresponds with the high emissions projections for sea-level rise at the end of this century of 0.6–1.1 m (Intergovernmental Panel on Climate Change, 2021).

On behalf of Australian Indigenous peoples and written by Indigenous scholars, a recent discussion paper (Matthews et al., 2021) describes climate change in Australia and its impact on the health and well-being of Indigenous Australians. The paper highlights Indigenous-led initiatives in climate change adaptation and mitigation that aims to strengthen well-being and benefit the global community. Despite decades of advocacy, Australia does not have a formal treaty (or legal framework) that preserves Aboriginal and Torres Strait Islander Peoples' rights to land and sea (Matthews et al., 2021). The Uluṟu Statement from the Heart (Uluṟu Statement, 2017) continues the call for truth, a treaty and a voice for Indigenous Australians. As of April 2022, the Uluṟu Statement has been rejected by two successive Australian prime ministers.

Inequitable power structures are a significant form of uncertainty undermining Indigenous communities' ability to build adaptive capacity and respond to climate challenges (Lyons et al., 2019, Australian Academy of Science, 2021, Cresswell et al., 2021). In this regard, the United Nations Declaration on the Rights of Indigenous Peoples (2007) supports Aboriginal and Torres Strait Islander participation in planning and implementing climate change responses (Matthews et al., 2021). Other international human rights instruments relevant to Indigenous people and climate change include the International Covenant on Civil and Political Rights (United Nations Human Rights, Office of the High Commissioner, 1966a) and the International Covenant on Economic, Social and Cultural Rights (United Nations Human Rights, Office of the High Commissioner, 1966b). The latter instrument outlines rights to self-determination and adequate housing, health, food and water. (Matthews et al. 2021). Despite these international human rights proclamations, the role of Indigenous Australians in identifying solutions to climate change threats and risks is only progressing slowly (Australian Academy of Science, 2021, Cresswell et al., 2021, Moggridge et al., 2022). As a signatory to the Declaration of Rights of Indigenous Peoples, the Australian Government *de facto* recognises the self-determination of its Indigenous peoples on issues that are important to them. Hence, Indigenous Australians should be crucial stakeholders in climate change decision-making, including the participation of Indigenous culture and science experts as lead authors in future IPCC Assessment Reports (Moggridge et al., 2022).

Australian Indigenous peoples have distinctive sources of resilience and vulnerability to climate change (Hill and Lyons, 2014). Their resilience to climate change is based on their unique knowledge and cultural practices honed over millennia. 'Healthy Country' is all-encompassing to Indigenous peoples' livelihoods, stewardship responsibilities and health and well-being (Lawrence et al., 2022). However, social inequalities and injustices are compounded by climate risk and exposure for Indigenous communities and their people (Matthews et al., 2021). For example, Indigenous communities are prone to chronic health issues. These include cardiovascular diseases, substance use disorders, mental health conditions, cancer, chronic kidney disease, diabetes, vision and hearing loss and selected musculoskeletal, respiratory, neurological and congenital disorders (Australia Institute for Health and Welfare, 2022). Indigenous Australians are also at greater risk of disability due to a higher prevalence of low birth weight, chronic disease, infectious diseases, injury and substance use (Australia Institute for Health and Welfare, 2022).

Climate impacts and risks

Climate-related impacts on Aboriginal and Torres Strait Islander Peoples, traditional lands (Country) and cultures observed across Australia are pervasive, complex and compounding (Matthews et al., 2021, Lawrence et al., 2022). Examples include (see Table 7.2):

- Loss of bio-cultural diversity (see Chapter 5, Box 5.1 and Chapter 6, Box 6.7);
- Nutritional changes through the availability of traditional foods and forced diet change;
- Water security (see Chapter 5, Box 5.4); and
- Loss of land and cultural resources through coastal erosion and sea-level rise.

Many of the compounding effects of climate change on the health and well-being of Indigenous communities are indirect. These adverse effects include (Matthews et al., 2021):

- *Housing and community infrastructure*: Indigenous housing is substandard, and there are overcrowding problems across remote, rural and urban areas. Health impacts on communities will likely increase due to the inability to safely shelter from extreme weather such as heatwaves, floods and tropical cyclones. Another issue is unsuitable housing design and energy poverty that affect people's capacity to adapt to heatwaves.
- *Intangible loss and social and emotional well-being*: Extreme weather events often compromise community ability to carry out cultural responsibilities in 'Caring for Country'. This compromise will further compound ecosystem degradation, reduce biodiversity and adversely affect overall social and emotional well-being (see Chapter 5, Boxes 5.1 and 5.2).
- *Impacts on health service provision*: Climate change threatens the integrity of health system infrastructure and the effectiveness of service operations through environmental interruptions, increased demand and reduced workforce capacity. Health care services in remote areas

Table 7.2 Impacts of climate change on the health and well-being of Australian Indigenous peoples

Impacts	Implications for Indigenous peoples' country and cultures
Loss of bio-cultural diversity (land, water and sky)	Damage to bio-cultural diversity can magnify the loss of spiritual connection to country and cause disruption of cultural structures. Climate change impacts can exacerbate or accelerate existing threats of habitat degradation and biodiversity loss, and create challenges for traditional stewardship of landscapes (see Chapter 5, Box 5.1, Chapter 6, Box 6.7).
Climate-driven loss of native title and other customary lands (sea-level rise)	Traditional coastal lands lost through erosion and rising sea level, with associated mental health implications from loss of cultural and traditional artefacts and landscapes, including the destruction and exhumation of ancestral graves and burial grounds, e.g. already occurring and likely to intensify in the low-lying islands of the Torres Strait, Kakadu National Park, and was also noted during the Black Summer (2019–2020) bushfires in Eastern Australia.
Changing availability of traditional foods and forced diet change	Adverse human health effects can be worsened by climate change through changing availability of traditional foods and medicines, while outages and high costs of electricity can limit the storage of fresh food and medication. Communities becoming more reliant on stored foods that can be of poor nutritional quality, as well as unreliable and expensive in rural and remote areas due to transportation and storage costs.
Changing climatic conditions for subsistence food	Climate change-induced sea-level rise and saltwater intrusion can limit the capacity for traditional floodplain pastoralism and affect food security, access and affordability to healthy, nutritional foods.
Extreme weather events triggering disasters	Increasing frequency or intensity of extreme weather events (floods, droughts, cyclones, heatwaves) can cause disaster responses in remote communities, including critical infrastructure, such as water and energy systems and health facilities.
Heatwave and bushfire impacts on human health	Heatwaves are a national risk. Tropical regions can experience prolonged seasons of high temperatures and humidity levels, resulting in extreme heat stress risks. Heatwaves and poor air quality have increased emergency department admission rates for Aboriginal people in Perth, and Aboriginal children in Brisbane due to respiratory conditions. Aboriginal people were impacted by the Black Summer Fires in NSW and Victoria, comprising 5.4% of the population in the fire-affected areas.
Health impacts from changing conditions for vector-borne diseases	A likely increase in exposure and increase risk for remote Indigenous peoples to infection from waterborne and insect-borne diseases, especially if medical services are limited or damaged by extreme weather events, e.g. in the Torres Strait Islands the changing climate is affecting the range and extension of the *Aedes* spp. mosquitoes that can carry and transmit dengue and other viruses.
Unadaptable infrastructure for changing environmental conditions	Poorly designed, inferior quality and unmaintained housing can create health challenges for tenants in extreme heat. Essential community-scale water and energy service infrastructure, unpaved roads, sea walls and storm water drains can fail in extreme weather events.

(Continued)

Table 7.2 (*Continued*) Impacts of climate change on the health and well-being of Australian Indigenous peoples

Impacts	Implications for Indigenous peoples' country and cultures
Drinking water security	Communities are already experiencing shortages of safe drinking water. There are concerns about the drying up of ancient waterholes and the safety of groundwater-dependent sacred sites, further risking health and well-being. This groundwater can also have microbial contamination from sewage and chemicals supporting bacterial growth that causes melioidosis in humans. In the Torres Strait, increasing reliance on desalination for drinking water raises costs for fuel and its associated transport.

Sources: Matthews et al. (2021), Cresswell et al. (2021), Lawrence et al. (2022).

(particularly in northern and central Australia) are at risk due to climate extremes and isolation levels. In addition, the healthcare workforce is inequitably distributed within Australia, with chronic staff shortages in rural and remote regions, including Indigenous communities.

Climate mitigation and adaptation

Aboriginal and Torres Strait Islanders have considerable bio-cultural knowledge and expertise to contribute to climate adaptation in Australia (Uluru Statement, 2017). Indigenous knowledge has been gained over 60,000 years of 'Caring for Country' that includes traditional practices for holistic land and cultural protection to manage past changes in climate. Lawrence et al. (2022) summarise at least six Indigenous stewardship practices that are demonstrating effective climate-resilient development pathways, both from a mitigation and adaptation perspective:

1. Indigenous Protected Area[1] (IPA) management plans enable culturally and ecologically appropriate development pathways that contribute to local Indigenous economies;
2. IPAs can avoid the potential for 'nature-culture dualism' that locks out Indigenous access in

[1] Indigenous Protected Areas (IPAs) are areas of land and sea managed by Indigenous groups as protected areas for biodiversity conservation through voluntary agreements with the Australian Government. There are currently 78 dedicated IPAs over 74 million hectares that account for more than 46% of the National Reserve System (Department of Agriculture, Water and the Environment, 2022).

some state government protected area legislation, as IPA's are informed by local Indigenous knowledge of land and sea country;

3. Fire management using cultural practices can achieve greenhouse gas emission targets while also maintaining Indigenous cultural heritage and contributing to managing increasing fire risk due to climate change (see Chapter 5, Box 5.3);
4. Indigenous Ranger programmes provide a means for Indigenous-guided land management, including for fire management and carbon sequestration, fauna studies, medicinal plant products, weed management and recovery of threatened species (see the 'Mitigation and Adaptation pathways' sections below);
5. Faunal field surveys can engage local, bounded and fine-scale intuitive species location by Indigenous knowledge holders and their knowledge applied to conservation planning across landscapes; and
6. 'Cultural flows' in waterways are a profound demonstration of cultural knowledge, values and practice in action as they are informed by Indigenous knowledge and bound by water-dependent values. They define when and where environmental water must be delivered and have enormous benefits under a changing climate (see Chapter 5, Box 5.4).

CROSS-CUTTING SOCIAL ISSUES

Compound extreme events

There are likely to be cascading, compounding and aggregate climate impacts on Australian cities, settlements, built environments and the health and well-being of their inhabitants (Australian Academy of Science, 2021, Lawrence et al., 2022, Rice et al., 2022, Naughtin et al., 2022). Effects of compound extreme events will also extend to supply chains and services in urban centres and between urban and rural communities (see Chapter 6, the 'Agriculture and Land' and the 'Energy and Transportation' sections). Such events will significantly increase in the future (see Chapter 3, Table 3.2, Box 3.1), leading to widespread and pervasive damage and disruption to human activities (Lawrence et al., 2022). The consequences of extreme compound events will be exacerbated by the highly interdependent and interconnected characteristics of natural, economic and social systems (see Chapters 5 and 6). Examples of adverse compound impacts include (Lawrence et al., 2022):

- Failure of transport, energy and communication infrastructure and services, heat stress, injuries and deaths, air pollution, stress on hospital services, damage to agriculture and tourism and insurance loss from heatwaves and fires (see Box 7.1);
- Failure of transport, stormwater and flood-control infrastructure and services from floods and storms;
- Water restrictions, reduced agricultural production, stress for rural communities, mental health issues and lack of potable water from droughts;
- Damage to buildings, roads, railways, electricity and water infrastructure, loss of assets and lives, displacement of people, reduced social cohesion and degraded ecosystems from extreme sea-level rise; and
- Large aggregate costs due to lost productivity and significant disaster relief expenditure, creating unfunded liabilities and supply chain disruption, e.g. the 2019–2020 'Black Summer' Australian bushfires and the February–March 2022 eastern Australian floods.

Populated areas, rural and remote settlements, Indigenous lands and sacred sites are all exposed to compound climate extremes (Lawrence et al., 2022, see Table 7.2). Greater urban density and population growth increase exposure in high-risk areas (e.g. near the coast and rivers). There are also different levels of exposure to various climate hazards

(Australian Academy of Science, 2021, Lawrence et al., 2022). For example, heatwaves are mainly a risk for inhabitants in urban and peri-urban areas. At the same time, bushfires are a significant threat to people, property and essential infrastructure in peri-urban regions and settlements near forests (see Chapter 5, Box 5.3). Riverine and flash floods are a substantial risk for people, property and infrastructure in cities and settlements on floodplains (e.g. Brisbane, Figure 7.1). Buildings and essential infrastructure in cities, towns and settlements remain exposed to severe winds associated with supercell thunderstorms, East Coast Lows and tropical cyclones (see Chapter 3, Table 3.2, Box 3.1). Importantly, the frequency and intensity of climate-driven natural disasters are increasing, and the recovery time between extreme events is declining (Rice et al., 2022).

The economy-wide projected costs of climate change in Australia are daunting, but long-term estimates depend on both impacts and adaptation (Lawrence et al., 2022). Residential, light industrial and commercial buildings in the coastal zone are highly vulnerable to climate change (see Cities, Settlements and Build Environments above). Damage-related loss of property value in Australia is projected to be AU$571 billion in 2030, AU$611 billion in 2050 and AU$770 billion in 2090 (Steffen et al., 2019, Box 7.1). This cost equates to AU$91 billion per year by 2050 and AU$117 billion per year by 2090. Australia's road infrastructure is exposed to sea-level rise. In the case of a 1.1 m sea-level rise by 2100, the loss of asset value for road infrastructure (i.e. freeways, main roads and unsealed roads) could be between AU$46 to 60 billion (Department of Climate Change and Energy Efficiency, 2011). For the same sea-level rise, loss of asset value of rail and tramway infrastructure could range between AU$4.9 and 60 billion (Department of Climate Change and Energy Efficiency, 2011).

Breakdown of institutions and governance systems

One of the most significant challenges of climate change is the inability of institutions and governance systems to manage climate risks (see Chapter 4, Box 4.2: barriers). The social and economic consequences of the breakdown of institutions and governance systems include (Lawrence et al., 2022):

- Overwhelmed capacity to provide necessary policies, services, resources, coordination and leadership. Examples include fragmented institutional and legal arrangements, under-resourcing of services, lack of dedicated adaptation funding instruments and resources to support communities and local government (see Table 7.3, Box 7.3).
- Vulnerability to climate hazards is exacerbated by an uneven capacity to manage uncertainty, conflicting values and competing policy and political interests.
- In cases of failed adaptation at the institutional and governance level, there are risks for widespread and pervasive impacts for all of society.
- There is often a reliance on reactive, short-term decision-making that locks in existing climate risk exposures, leaves perverse incentives and interconnected and systemic impacts unaddressed.
- If systemic impacts are ignored, this worsens vulnerability to climate change and leads to maladaptation (see Chapter 4, Figure 4.1), inequities and injustices within and across generations. These injustices include the rights, interests, values and practices of Indigenous peoples (see Table 7.2).
- Failure to take climate adaptation action generates litigation risk (see Box 7.1).

CLIMATE-RESILIENT DEVELOPMENT PATHWAYS AND OPPORTUNITIES

Mitigation pathways

Chapter 6 dealt with GHG emissions and climate mitigation pathways for Australian economic sectors most exposed and vulnerable to climate change. Much of that discussion is relevant to cities, settlements and built environments, e.g. energy, transportation, agriculture, mining, manufacturing and construction (see also Chapter 4, Box 4.4 for global context). Due to their population size and various economic activities, Australia's urban areas are significant sources of GHGs (Stock et al. 2017). Many industries and value chains that supply the country's major cities and towns with raw goods and materials (e.g. agriculture and mining) also produce GHGs outside the urban footprint.

While Australia's urban areas are major emitters of GHGs, its cities and towns can deliver about 70% of required GHG reductions, thereby contributing significantly to the country-level pledge to achieve net-zero emissions by 2050 (Stock et al., 2017). Australia's LGAs are particularly well placed to deliver on GHG reductions through embracing renewable energy, energy efficiency and sustainable transport systems (Stock et al., 2017, see also Box 7.2). There are currently 566 LGAs covering the whole of Australia, including unincorporated areas. These LGAs fall under the six Australian states and Northern Territory jurisdictions by law. Due to their focus on local-scale issues, LGAs are at the forefront of climate change, dealing with many adverse climate risks and threats. However, they are often the least equipped to deal with climate risk, particularly those with small rate bases that dominate many climate-exposed regional and rural areas. These LGAs include many remote councils, notably Aboriginal and Torres Strait Islander Councils (see Indigenous Peoples above). For example, deep bio-cultural knowledge of 'Country' held by Indigenous peoples is beginning to influence national policies for fire management according to the understanding of seasonal calendars, implementation of carbon reduction and capture ecosystem services (see Chapter 5, Box 5.2) and culturally based fire economies (see Chapter 5, Box 5.3) embracing Caring for Country (see Dr Emma Lee *Tebrakunna Country*, in Australian Academy of Science, 2021).

Australian LGAs have been particularly active in climate mitigation activities (Box 7.2), despite periods of instability and policy changes at state and federal levels (Stock et al., 2017). In line with LGAs in many other developed countries, Australian local governments have led the way in climate mitigation. Examples include The Cities for Climate Protection and Cities Power Partnership (Box 7.2). LGAs have provided leadership in their communities by delivering renewable energy targets to improve the energy efficiency of their council buildings and vehicle fleets. The larger ones have provided new public transport infrastructure to encourage their residents to reduce individual carbon footprints. Importantly, cities and towns can also be part of the climate mitigation solution (see Chapter 4, Box 4.4). High-density urban living translates to a lower per capita greenhouse gas emissions 'footprint', and innovative solutions are much easier to implement in urban environments (Australian Academy of Science, 2021). Examples include passive cooling

BOX 7.2: Australian LGAs initiatives in Climate Change Mitigation

Climate mitigation programmes

Cities for climate protection:

An initiative of the International Council for Local Government. Participating LGAs plan and implement actions to reduce GHGs across council operations, households and businesses.

Most capital city LGAs have now committed to policies and targets to increase renewable energy, energy efficiency and sustainable transport uptake and to reduce emissions [1].

Emissions reductions initiatives implemented by LGAs include: solar PV for council buildings; behaviour change programmes (e.g. encouraging residents to switch to efficient electrical appliances); energy-efficient street lighting; and sustainable transport solutions such as councils choosing more efficient fleet vehicles and encouraging people to walk, cycle or take public transport [1].

In 2016, a survey by Beyond Zero Emissions found one in five councils had adopted either net-zero emissions or 100% renewable energy targets [1].

Cities power partnership:

An initiative of the Climate Council is the largest network of LGAs leading the way to a thriving, zero emissions future. It is made up of over 165 LGAs from across the country, representing almost 65% of the population [2].

LGAs that join the partnership make action pledges to tackle climate change.

Each LGA must select five key actions from a list grouped under renewable energy, energy efficiency, transport and working together and influence. [1,2]

Cities power partnership in action [1]

Renewable energy actions:

- Promote renewable energy – both at the residential, commercial and larger scale, e.g. use land use planning measure to encourage uptake, such as streamlining approvals processes and removing barriers (see Chapter 6, Energy sector).
- Power council operations by renewable energy, e.g. set targets to increase the level of renewable power for council operations and the broader community over time (see Chapter 6, Energy sector).
- Collective tendering, i.e. facilitate large energy users collectively tendering and purchasing renewable energy at a low cost.
- Powering electric vehicles with renewable energy, e.g. electrify transport systems such as council buses and fleet vehicles and power these by 100% renewable energy (see Chapter 6, Transportation sector).
- Lobby electricity providers and state government to address barriers to renewable energy take up at the local level (see Chapter 6, Energy sector).

Energy efficiency actions:

- Promote energy efficiency – both at the residential, commercial and larger scale, e.g. encourage local businesses, community facilities and residents to take up energy efficiency measures by providing incentives (such as grants, solar bulk buy schemes or flexible payment options) (see Chapter 6, Energy sector).
- Energy efficiency in council operations, e.g. adopt best practice energy efficiency in council buildings (see Chapter 6, Energy and Construction sectors).

Climate mitigation programmes	Cities power partnership in action [1]
	Sustainable transport actions:
	• Promote sustainable transport options in the community, e.g. provide fast-charging infrastructure (powered by 100% renewable energy) throughout the city at key locations for electric vehicles (see Chapter 6, Transportation sector).
	• Promote sustainable transport options within council, e.g. ensure council fleet purchases meet strict GHG emissions requirements and support the uptake of electric vehicles (powered by renewable energy) (see Chapter 6, Transportation sector).

Sources: [1] Stock et al. (2017), Cities Power Partnership (2022). Refer to Chapter 4 (Box 4.4) for international context.

applications in cities, such as street trees, green spaces and green walls and roofs, to reduce city temperatures (see Chapter 6, Construction sector).

Adaptation pathways

A LEADERSHIP ROLE FOR LOCAL GOVERNMENT

While climate mitigation initiatives by Australian LGAs are to be commended (Box 7.2), when it comes to climate adaptation, the country's LGAs are lagging behind those in many developed countries. Norman (2022) asserts the lack of climate adaptation being witnessed at the local or community level is because of non-existent policies and plans at the national level, including:

• No national coastal plan for coastal erosion and inundation;
• No national urban policy for climate-resilient development;
• No national requirement for climate change to be considered in urban and regional land use plans; and
• No funded national support programme for urban and regional communities to adapt to current and future climate risk.

The lack of national policies and plans to guide state, territory and local governments with

developing adaptation strategies to manage climate risks is a significant concern (see Chapter 4, Box 4.2: barriers). Climate change is a pervasive national threat, and it covers all jurisdictions. Therefore, having national policies for tackling climate risk and impending threats is needed to ensure uniformity of climate adaptation around the country. Norman (2022) recommends a five-point strategy that should underpin a national plan to build climate-resilient development in Australia at the LGA scale (Box 7.3). This strategy includes an integrated national climate action plan, a national coastal strategy, a review of urban planning legislation and city/regional plans, enhancement of links between organisations and funding for applied research and community plans.

The Planning Institute of Australia (PIA) accepts the IPCC assessments (Intergovernmental Panel on Climate Change, 2021, 2022a, 2022b) that human activities are driving climate change and the increase in more extreme weather events (Planning Institute of Australia, 2021). PIA also acknowledges that appropriate town and regional planning and coordinated leadership are essential strategies to mitigate and adapt to adverse climate change impacts. Planning systems in the country can potentially reduce many of the drivers of carbon emissions, including urban sprawl, car dependency and loss of vegetation (Planning Institute of Australia, 2021). On the adaptation side, sound

BOX 7.3: A national plan for Australia to build climate-resilient development at the local government (community) level

1. *An integrated national climate action plan:*
   - This would involve funding and programmes for state and territory governments, local councils and industry, enabling them to work with communities to prepare for climate change.
2. *A national coastal strategy:*
   - Coastal communities are especially vulnerable to sea-level rise, storms, floods and bushfires, which will worsen under climate change (see Chapter 3, Table 3.2, Box 3.1). Leading experts last year outlined the need for a climate change plan tailored to these communities. It would include a national agency to coordinate ocean and coastal governance across tiers of government.
3. *Review urban planning legislation and city plans:*
   - Planning experts and others have called for climate change to be considered when making everyday decisions about the built environment. This would lead to more sustainable, pleasant and healthy urban and regional communities, as well as minimising disaster risks.
   - These decisions include where to locate new housing developments, as well as investing in green buildings and water-sensitive urban design (see Chapter 6, Construction sector).
   - Bureaucrats need to start conversations with communities at risk, such as those on floodplains or in bushfire-prone areas, or vulnerable and socially disadvantaged groups (e.g. Indigenous communities, see Table 7.2), to prepare city and regional plans that incorporate future risks.
4. *Stronger links between organisations:*
   - Greater cooperation is needed between emergency management, climate scientists and land use and urban planners, so they can effectively work together to prepare climate-resilient community plans. Better communication is also needed to ensure knowledge is shared, and best practice is celebrated and maintained.
5. *More funds for research and community plans:*
   - Governments must fund the development of cutting-edge applied research to better understand and map climate risks (e.g., NCCARF Mark 2). In addition, funding is needed for climate-resilient urban development and to support vulnerable communities through long-term adaptation plans.

Source: Adapted from Norman (2022).
Refer to Chapter 4 (Box 4.2) for examples of enablers.

planning can provide people and communities with healthier, aesthetically pleasing, walkable, connected and resilient communities. Lastly, PIA maintains that planning systems should integrate local knowledge and understanding with policy to deliver regionally specific outcomes that increase climate resilience and empower communities (Planning Institute of Australia, 2021). This extension must also include Indigenous understanding of 'Land and Sea Country'.

Australia's local government planning legislation is governed at the state and territory level. This constitutional arrangement means statutory planning legislation, and requirements can differ among jurisdictions (see Chapter 4, Box 4.2: barriers). The lack of national uniformity of local government statutory planning responses to climate change is an issue of significant concern (Box 7.3). State and territory governments frequently compete for business investments in their jurisdictions.

While healthy competition is the cornerstone of a vibrant market-based economy, there is always a risk of perverse outcomes. For instance, it may be more convenient for property developers to gain approval for high climate risk residential or commercial developments in some states or territories than others, resulting in maladaptive adaptation pathways (see Chapter 4, Figures 4.1–4.3).

Despite the lack of national leadership on climate change and local government (Box 7.3), urban planning professionals are spearheading action on climate change. In 2020, PIA declared a climate emergency. A year later, PIA called for every Australian state and territory planning system to undertake ten key reforms to become climate-conscious. In late 2021, PIA (Queensland Division) was the first division to follow the national recommendation. It produced a 10-point plan for climate action at the local government level in Queensland (Table 7.3). This innovation plan identifies existing climate-responsive design and planning courses offered in the planning degrees in Australia. It also gauges how planners see their role in responding to the future climate challenges, how much they know about the topic and how prepared they feel to face the urban climate challenges. There is much work to be done to operationalise the 10-point plan. Nonetheless, it should be considered a decisive step in the right direction to better prepare Australia's cities, towns and shires for the inevitable changes in climate over the coming decades.

In addition to local government, it is also crucial that the Australian property sector engages in climate adaptation leadership, given its very high exposure to climate risk (see Box 7.1). Rice et al. (2022) propose five climate adaptation strategies for the property sector that will complement activities being driven by governments:

1. Mapping physical risks for current portfolios and potential acquisitions.
2. Incorporating physical adaptation and resilience measures for assets at risk.
3. Including climate risk in their due diligence processes.
4. Investing directly in adaptation measures for specific assets.
5. Getting involved in local resilience strategies with policymakers.

Coastal councils and shires currently take sea-level rise into account in their decision-making about land use in the near-coastal zone. However, for many LGAs, the decision as to which mix of strategies to apply is difficult (Australian Academy of Science, 2021). Decision-makers should consider costs and benefits, conflicting interests, threatened ecosystems and the long-term reality of sea-level rise (Australian Academy of Science, 2021). In the case of conflicting interests, Cradduck et al. (2020) note that disputes between local government, residents and developers are common in Australia. Given future climate change projections, conflict among these core groups may be expected to continue in the future and include litigation cases (see Box 7.1).

ADAPTATION STRATEGIES FOR LOCAL GOVERNMENT AREAS

Australia's cities, settlements and built environments and Indigenous communities are highly vulnerable to climate change and associated risks to people and property and health and well-being (Table 7.2). Climate projections (see Chapter 3, Table 3.2, Box 3.1) indicate that future climate conditions and the rise in extreme weather events will be challenging for many places in the country (Lawrence et al., 2022). Climate adaptation will be needed to reduce the exposure of millions of Australians who are living in cities, towns and shires that are at risk of significant climate change. Incremental adaptation is already happening in many places, but as time progresses and climate threats increases, transformative adaptation will be required to reduce risks (see Chapter 4, Figure 4.1, Box 4.1).

Ineffective planning regulations have increased the vulnerability of human communities (see Box 7.3 and Table 7.2), including reduced availability and increased cost of insurance (see Box 7.1), and inadequate investment in avoidance and preparedness worsening underlying social disadvantages (Lawrence et al., 2022). Climate adaptation will need to focus on coastal urban areas at risk of sea-level rise, increasing risk for riverine and flash flooding, heat stress to human communities, bushfire risk and water scarcity during droughts. Climate adaptation strategies across the range of threats will need to be incremental and transformative (Table 7.4).

For many Australians, socio-economic inequality, low incomes and high household debt, poor health and disabilities increase vulnerability and

Table 7.3 Planning to tackle climate change, including ten actions for a climate-conscious planning system in the State of Queensland, Australia

Planning actions	What needs to happen
1. Adopt a common climate change goal across all planning legislation	Adopt consistent climate change purpose statements across the *Economic Development Act 2012* and the *State Development and Public Works Organisation Act 1971* in line with the focus of the *Planning Act 2016* on the achievement of ecological sustainability.
2. Provide new strategic planning guidelines for mitigating and adapting to climate change	Strengthen the strategic planning framework to protect Queenslanders from the adverse impacts of climate change, including: • Develop a Queensland Settlement Strategy. • Prepare new guidance material for local government about the incorporation of climate change in strategic planning. • Review the practical implementation of s30 of the *Planning Act 2016*. • Expand the information in Planning and Development Certificates to include resilience and climate change.
3. Assess new infrastructure for climate impacts	Include clear and measurable climate assessment requirements for new state and local government infrastructure in the proposed State Infrastructure Strategy and ensure these requirements are reflected in the Project Assessment Framework, infrastructure project procurement and infrastructure programme funding processes.
4. Introduce stronger sustainability standards for new buildings	Review the Queensland Development Code to identify opportunities to lift sustainability and resilience standards for new buildings (see Chapter 6, Construction sector).
5. 5) Plan for the impacts of heat on Queensland	Update the State Planning Policy to incorporate heat (including heatwave and heat island) as a state interest for natural hazards, including guidance mapping, policies and assessment benchmarks.
6. Invest more to address climate change risks and build resilience at the local level	Deliver "Planning for Climate Change Grants" to support training, capacity-building and local planning projects that address climate change risks and improve resilience to future extreme climatic events and hazards.
7. Streamline development assessment for low and zero-carbon development proposals	Introduce streamlined development assessment pathways for renewable energy projects and other low or zero-carbon development to incentivise high performance and fast-track jobs. (see Chapter 6, Energy sector).
8. Deliver a low or zero-carbon precinct as a demonstration project to showcase best practice	Partner with the private sector to pilot a low-carbon or zero-carbon precinct to demonstrate the potential to achieve precinct-scale carbon reduction (see Chapter 6, Energy and Construction sectors).

(Continued)

Table 7.3 (*Continued*) Planning to tackle climate change, including ten actions for a climate-conscious planning system in the State of Queensland, Australia

Planning actions	What needs to happen
9. Introduce better planning for protecting and expanding green and open space	Deliver new planning initiatives that expand and protect green and open space, including: • Progress and finalise the proposed Strategic Assessment for South East Queensland. • Develop Green Grid strategies and tools to plan for interconnected networks of green and open space. • Encourage local governments to include mature street tree planting (or trees able to grow to provide canopy and shade) in the Desired Standards of Service under Local Government Infrastructure Plans and/or planning schemes.
10. Retrofit existing neighbourhoods with climate-responsive urban design	Establish a 'Walkable Communities Fund' to invest in projects that retrofit climate-responsive urban design via more walkable and accessible neighbourhoods, including footpaths and tree planting (see Chapter 6, Transportation sector).

Source: Reproduced by kind permission of the Planning Institute of Australia (2021), Queensland Division, © 2021 Planning Institute of Australia, Queensland Division.

limit adaptation to climate change (Cresswell et al., 2021, Lawrence et al., 2022). The lack of essential services (e.g. education and primary health care) between metropolitan and rural areas worsens inequalities among socio-economic groups and geographical location. In the absence of proactive adaptation, adverse impacts of climate change are highly likely to exacerbate inequalities between Indigenous and non-Indigenous people and other vulnerable groups (Matthews et al, 2021, Lawrence et al., 2022).

Lawrence et al. (2022) suggest a range of adaptation options to enhance the ability of institutions and governments to manage climate risks (see also Chapter 4, Box 4.2: enablers). The main adaptation options include:

- Pre-emptive options that avoid and reduce risks;
- A redesign of policy and statutory frameworks, and funding instruments for addressing changing risks and uncertainties that enable fair and collaborative governance across scales and domains (see Table 7.3, Box 7.3);
- Addressing existing vulnerabilities, and capacity, capability and leadership deficits within and across all levels of government (see Box 7.3), all economic sectors, Indigenous peoples and communities (see Table 7.2);

- Development and implementation of risk and vulnerability assessment methodologies and decision-making tools that build resilience and address changing risks and vulnerabilities; and
- Co-designed climate adaptation approaches implemented with local communities, including Australian Aboriginal and Torres Strait Island peoples (see Dr Emma Lee *Tebrakunna Country*, in Australian Academy of Science, 2021).

Integration with the sustainable development goals

Seven SDGs are relevant to a national strategy on climate change and Australian cities, settlements and built environments, human health and well-being, and Indigenous peoples (see Table 7.5):

1. SDG1 (Poverty eradication);
2. SDG9 (Industry, innovation and infrastructure);
3. SDG10 (Reduced inequalities);
4. SDG11 (Sustainable cities and communities);
5. SDG9 (Industry, innovation and infrastructure);
6. SDG13 (Climate action); and
7. SDG16 (Peace, justice and strong institutions).

Table 7.4 Significant climate risks for cities, settlements and built environments, human health and well-being and potential adaptation strategies

Climate risk	Avoid (transformative)	Accommodate (incremental)	Protect (incremental)	Retreat (transformative)
Sea-level rise, coastal erosion and storm surge risks	Identify future no-build areas in the coastal zone and use planning regulations to prevent new coastal development in areas of high risk now and in the future (based on 1 m sea-level rise and storm surge maps).	Continue to use the land but accommodate change, such as by raising roadbeds, building on piles or relocating essential infrastructure (e.g. auxiliary power supplies, hospitals). Audit of climate-resilient health infrastructure.	Use hard structures such as sea walls, dykes, storm surge barriers, tidal gates; or soft engineering solutions such as dunes and vegetation (or some mixture of the two), to protect the integrity of coastal land. Improved capacity of emergency services and early warning systems.	Withdraw or abandon assets at risk and allow ecosystems to retreat inland (managed retreat). Under high levels of warming and sea-level rise, retreat is likely to be the only feasible long-term strategy for exposed communities.
Riverine and flash flooding risks, extreme rainfall events	Identify future no-build areas in flood plain zone and use planning regulations to prevent new coastal development in areas of high risk now and in the future. Prohibit dams in vulnerable areas in case of catastrophic dam failure.	Continue to use the land but accommodate change, such as by raising land blocks, raising existing buildings, constructing new building on piles or relocating essential infrastructure (e.g. hospitals, shops, schools). Audit of climate-resilient health infrastructure.	Use hard structures such as river levees, raising existing levees to accommodate more frequent and severe floods. Use of soft engineering solutions such as riparian vegetation and rock features to reduce river bank erosion on flood plains. Improved capacity of emergency services, early warning systems.	Withdraw or abandon private assets at risk and convert land to other uses, e.g. parks. Under high levels of warming and more extreme riverine flood events, retreat is likely to be the only feasible long-term strategy for exposed communities, e.g. government buy up and relocation schemes. Deconstruct dams in vulnerable areas in case of catastrophic dam failure.

(Continued)

Table 7.4 (Continued) Significant climate risks for cities, settlements and built environments, human health and well-being and potential adaptation strategies

Climate risk	Avoid (transformative)	Accommodate (incremental)	Protect (incremental)	Retreat (transformative)
Heat stress to human communities	Identify future no-build regions of the country that are likely to be uninhabitable for most of the year by 2070, and prevent future residential development schemes from agreed dates, e.g. parts of Central and Northern Australia (see Chapter 6, Box 6.3).	Investing in improving the thermal comfort of existing dwellings that are unable to protect people from heatwaves is a cost-effective preventative health measure, e.g. air conditioning may be required to reduce the heat stress of occupants. Urban cooling intervention, e.g. green infrastructure and increased albedo (white roofs). Heatwave/fire early warning systems; battery-generators systems for energy security. Education to reduce heat stress.	Strengthen building codes to cost-efficiently improve energy and thermal efficiency performance while protecting occupants from increased heat stress (insulation/cooling). Urban planning and design can significantly influence heat build-up in urban areas. Designing urban spaces to take consideration of heat gain in the first instance is much less costly than retrofitting spaces retrospectively.	Depending on future GHG emissions pathways, later in the century, heat-exposed communities may experience population decline as exposed residents migrate south, e.g. from parts of Northern Australia (see Chapter 6, Box 6.3).
Bushfire risk	Identify future no-build areas in the bushfire risk zone and use planning regulations to prevent new development in areas of high risk now and in the future (see Chapter 5, Box 5.3).	Better resource fire and land management agencies to manage fuel loads and rapidly detect and attack new outbreaks. Add a self-sufficient Australian medium and large aerial firefighting capability to fire services. Create an Indigenous-led National Cultural Fire Strategy.	Strengthen building codes to cost-efficiently improve the bushfire resilience of new buildings while protecting occupants from increased bushfire risk. Improved capacity of emergency services, early warning systems. Strengthening social networks in rural communities.	Withdraw or abandon assets at risk and allow ecosystems to return. Under high levels of warming and more extreme bushfire events, retreat is likely to be the only feasible long-term strategy for exposed communities.

(Continued)

Table 7.4 (*Continued*) Significant climate risks for cities, settlements and built environments, human health and well-being and potential adaptation strategies

Climate risk	Avoid (transformative)	Accommodate (incremental)	Protect (incremental)	Retreat (transformative)
Water scarcity during droughts	Some industries, such as agricultural entities, are already making business decisions to avoid new enterprises in regions that have a high confidence of becoming hotter or drier, e.g. Mediterranean climates and parts of Northern Australia (see Chapter 6, Agricultural sector).	Improved water-use efficiency on irrigated farms and urban areas; water reuse and recycling. On supply side, desalination of water supplies, e.g. Perth. On demand side, consumers are encouraged to reduce their piped water use through the installation of tanks to collect rooftop water and the introduction of drought-tolerant gardens.	Physical or chemical covers for water reservoirs to reduce evaporation; storage of freshwater in underground aquifers. Strengthening social networks in rural communities.	Depending on future GHG emissions pathways, later in the century, drought-exposed communities may experience population decline as residents migrate to wetter regions.

Sources: Moran and Turton (2014), Stock et al. (2017), Steffen et al. (2019), Emergency Leaders for Climate Action & the Climate Council of Australia (2020), Australian Academy of Science (2021), Cresswell et al. (2021), Lawrence et al. (2022), Rice et al. (2022) and Naughtin et al (2022).
Refer also to Chapter 4 (Box 4.1) explanations of active incremental and transformative adaptation strategies, and Table 7.3 for local government planning requirements for effective climate adaptation in Australia.

Table 7.5 Integration of SDGs with Australia's cities, settlements and built environments, human health and well-being, and Indigenous peoples responses to climate change

SDG	Goals	Targets	Indicators
SDG1: Poverty eradication (*economic and social*)	1. End poverty in all its forms everywhere	Build the resilience of lower socio-economic groups and communities (e.g. Indigenous peoples), and those in vulnerable situations and reduce their exposure and vulnerability to climate-related extreme events and natural disasters	• Number of deaths, and missing persons and persons affected by disaster per 100,000 people • Proportion of local governments that adopt and implement local disaster risk reduction strategies in line with national disaster risk reduction strategies
SDG9: Industry, innovation and infrastructure (*economic and social*)	2. Build climate-resilient critical infrastructure	Develop quality, reliable, sustainable and resilient infrastructure, to support economic development and human well-being	• Proportion of the rural population who live within 2 km of an all-season road • Passenger and freight volumes, by mode of transport
SDG10: Reduced inequalities (*economic and social*)	3. Reduce inequalities to climate risk	Empower and promote the social, economic and political inclusion of all, irrespective of age, sex, geographical location, disability, race, ethnicity, origin, religion, economic or other status	• Proportion of people living below 50% median income, by age, sex, geographical location and persons with disabilities
SDG11: Sustainable cities and communities (*social and economic*)	4. Make cities and human settlements inclusive, safe, resilient and sustainable	Ensure access to all to adequate, safe and affordable housing Reduce the number of deaths and number of people affected and substantially decrease the direct economic losses relative to GDP caused by climate-related disasters	• Proportion of population living inadequate housing • Number of deaths, missing persons directly affected to climate-related disasters per 100,000 persons • Direct economic loss in relation to GDP, damage to critical infrastructure and number of disruptions to basic services, attributed to climate-caused natural disasters

(*Continued*)

Table 7.5 (*Continued*) Integration of SDGs with Australia's cities, settlements and built environments, human health and well-being, and Indigenous peoples responses to climate change

SDG	Goals	Targets	Indicators
SDG13: Climate action (*environmental*)	5. Carbon neutral production 6. Resilience and adaptive responses to natural hazards	Reduced global warming Closed loop production (circular economy) and supply chain systems Mitigation of climate change impacts on cities, settlements and built environments Improve education, awareness-raising and human and institutional capacity on climate change mitigation, adaptation, impact reduction and early warning systems	• Reduction in built environments GHG emissions contribution (to zero) • Percentage of LGAs using renewable energy • Percentage of LGAs with a climate action plan • Number of LGAs that have communicated the strengthening of institutional, systemic and individual capacity-building to implement adaptation, mitigation and technology transfer and development actions
SDG16: Peace, justice and strong institutions (*economic and social*)	7. Inclusive society for sustainable development, access to justice for all, and build effective, accountable and inclusive institutions at all levels	Ensure responsive, inclusive, participatory and representative decision-making at all levels on matters related to climate exposure and risk	• Proportion of population who believe climate change decision-making is inclusive and responsive, by sex, age, race, ethnicity, disability and socio-economic group

Refer to Chapter 4 for an explanation of SDG goals, targets and indicators (see also Box 4.5). Triple-bottom line implications for each SDG are shown in italics.

A framework (Table 7.5) shows how these social systems can effectively integrate climate change with goals, targets and indicators for each of the seven relevant SDGs. Climate-resilient development pathways for Australia's vulnerable social systems will require incremental, structural and transformational changes from this decade. If society is to meet (or make progress towards) the 2030 SDG targets, similar changes will need to co-occur, particularly in the energy, transportation, agricultural, manufacturing and construction sectors (see Chapter 6, Tables 6.1, 6.3 and 6.5). A whole-of-economy perspective will be critical in driving innovation and change towards a net-zero world.

SUMMARY

- Appreciating the human geography of Australia is essential to gaining a better understanding of the observed and the likely future impacts of climate change on its social systems. About 90% of Australians live in cities and towns, making the country among the most urbanised in the world. Of those, 85% live within 75 km of the coastline. From a population size, distribution and average age perspective, most Australians exposed to climate change and extreme weather events are 'ageing urban dwellers' living in the continental southeast of the country, centred around the country's three largest cities: Sydney, Melbourne and Brisbane. Many remote regional settlements and Aboriginal and Torres Strait Islander (Indigenous) communities in the north are also highly vulnerable to climate change. LGAs are at the forefront of climate change and face many institutional, financial and logistical challenges when responding to extreme weather events. However, it is from these LGAs that solutions are most likely to emerge.

- Australia's cities, towns and shires (LGAs) are at high risk from worsening climate impacts, such as rising sea levels, floods, bushfires and extreme heatwave. This risk is exacerbated by the high concentration of cities, towns and settlements near the coast. Climate change and extreme weather events also significantly threaten Australia's finance, insurance and property sectors. These climate threats compromise the long-term sustainability and viability of many exposed residential and commercial assets and critical public infrastructure.

- The increasing frequency and intensity of extreme weather events in Australia affect human health and well-being and the livelihoods of exposed communities and their people. An ageing population compounds these adverse effects on public health. Other socially disadvantaged groups are particularly exposed to climate change, including low-income earners living in high climate risk suburbs, people with disabilities, remote settlements and Indigenous communities. For many of these disadvantaged groups, building and contents insurance is often beyond their reach, and their well-being is often neglected in the aftermath of natural disasters. Climate risks and threats to human health and well-being include heatwaves, bushfires, water scarcity, floods and storms.

- Australian Indigenous peoples have distinctive sources of resilience and vulnerability to climate change. Their resilience to climate change is based on their unique 'place-based' knowledge and cultural practices honed over millennia. 'Healthy Country' is all-encompassing to Indigenous peoples' livelihoods, stewardship responsibilities and health and well-being. However, social inequalities and injustices are compounded by climate risk and exposure for Indigenous communities and their people. Indigenous Australians should be crucial stakeholders in climate change decision-making, including the participation of Indigenous culture and science experts as lead authors in future IPCC Assessment Reports.

- There are likely to be cascading, compounding and aggregate climate impacts on Australian cities, settlements, built environments and the health and well-being of their inhabitants. Effects of compound extreme events will also extend to supply chains and services in urban centres and between urban and rural communities. Such events will significantly increase in the future, leading to widespread damage and disruption to human activities. The consequences of extreme compound events will be exacerbated by the highly interdependent and interconnected characteristics of natural, economic and social systems. One of the most significant challenges of climate change is the inability of institutions and governance systems to manage climate risks.

- While Australia's urban areas are significant emitters of GHGs, its cities and towns can deliver about 70% of required GHG reductions, thereby contributing significantly to the country-level pledge to achieve net-zero emissions by 2050. The country's LGAs are well placed to deliver on GHG reductions by embracing renewable energy, energy efficiency and sustainable transport systems.

However, they are often the least equipped to deal with climate risk, particularly those with small rate bases that dominate many climate-exposed regional and rural areas. These LGAs include many remote councils, notably Aboriginal and Torres Strait Islander Councils.

- Climate adaptation will be needed to reduce the exposure of millions of Australians living in cities, towns and shires at risk of significant climate change. Incremental adaptation is already happening in many places, but as time progresses and climate threats increases, transformative adaptation will be required to reduce risks. The lack of national policies and plans to guide state, territory and local governments in developing adaptation strategies to manage climate risks is a significant concern. Climate change is a pervasive national threat, and it covers all jurisdictions. Therefore, having national policies for tackling climate risk and impending threats is needed to ensure uniformity of climate adaptation around the country.

- Addressing the SDGs for Australia's cities, settlements and built environments, human health and well-being, and Indigenous peoples inform incremental, structural and transformative strategies to avoid mal-mitigation and maladaptation pathways in the future.

8

Synthesis

If working apart we are a force powerful enough to destabilise our planet, surely working together we are powerful enough to save it…in my lifetime I've witnessed a terrible decline. In yours, you could and should witness a wonderful recovery.

Sir David Attenborough OM, GHMG, CH, CVO, CBE, FRS, FSA, FRSA, FLS, FZS, FRSFG, FBSB (2021), English broadcaster, biologist, natural historian and author

Current rates of global warming are unprecedented in more than 2,000 years, and temperatures now exceed the warmest multi-century period in more than 100,000 years (Intergovernmental Panel on Climate Change, 2021). The global average surface temperature for the decade from 2011 to 2020 was about 1.1°C higher than from 1850 to 1900 (see Chapter 2, Figure 2.6c). Each of the past four decades has been warmer than any preceding decade since 1850 when reliable weather observations began. Australian land areas have warmed by 1.4°C between 1910 and 2020, while ocean temperatures have increased by about 1°C in that time (see Chapter 3, the 'Historical Changes in Climate' section). The southern movement of the East Australia Current has caused even greater ocean warming in the country's southeast. Background warming of land and sea areas has increased the frequency, severity and duration of terrestrial and marine heatwaves around the country.

The most striking change in climate over the past 50 years has been the general warming and drying of southern Australia, driven mainly by the southward expansion of the dry sub-tropical zone (Turton, 2017). Significant declines in winter wet season rainfall over the southwest of Western Australia and the continental southeast (including the Murray-Darling Basin) have been evident, along with increased agricultural and ecological droughts. Summer wet season rainfall has increased slightly over parts of northern Australia, while high annual variability has masked any discernible rainfall trends for other areas in the north. Warming and drying across southern and central regions of the country have increased extreme fire weather events and extended the bushfire season since the 1950s (see Chapter 3, the 'Historical Changes in Climate' section).

Australia's climate will continue to change, with an increased risk for extreme climate-driven events, such as droughts and riverine and coastal flooding (see Chapter 3, Table 3.2). Even if net-zero global emissions are achieved by 2050, the country will be locked into more climate change, and extreme weather events, over the coming decades. Increases in average air temperatures may be expected, along with more frequent, prolonged and severe heat events. Most areas of southern Australia, particularly the southwest, south and continental southeast, are likely to continue the drying trend for winter and spring rainfall. Changes in rainfall across northern Australia are less certain, but a warming ocean and atmosphere mean a greater risk for more severe tropical cyclones and extreme rain and flood events. The sea level is expected to rise around the country in concert with global trends, with some regional differences. Ocean acidification and deoxygenation rates are likely in the waters around the country. The southward movement of the East Australia Current is expected to continue with the corresponding deoxygenation of waters off southeast Australia.

Australia is considered a global climate change 'hotspot', and climate change risk must therefore be factored into policy and decision-making at all levels of government and across the private sector. Failing to mitigate greenhouse emissions and prepare the country's natural, economic and social systems now will bring a range of significant climate risks to future generations (Naughtin et al., 2022). The latest IPCC Report on Australia (Lawrence et al., 2022) states some of the climate risks for natural systems are close to 'tipping points', and autonomous and planned adaptation may be insufficient to prevent ecosystem collapse (see Chapter 5, Boxes 5.5 and 5.6). Other risks to ecosystems, ecosystem services and biodiversity are likely to be severe (Cresswell et al., 2021). However, human intervention has the potential to minimise some adverse impacts if policymakers and natural resource managers apply active adaptation measures from now on. Many of these adaptation strategies will involve low or no-regrets measures, such as sustainable agriculture and land management. Other strategies will require more profound transformative changes in policy and management actions, such as relocation of vulnerable agribusinesses to new locations (see Chapter 6, see 'Agriculture and Land' section).

The recent IPCC Report also warns that many climate risks have the potential to cascade across economic (see Chapter 6) and human (see Chapter 7) systems with widespread adverse effects on Australian society (Lawrence et al., 2022). There are also concerns that institutions and governance systems may be unable to adapt to climate risk profiles due to rapid rates of global change. The inability to adapt to extreme climate-driven events is already happening, as evidenced by the 2019–2020 catastrophic bushfires and the record-breaking 2022 eastern Australian floods (see Chapter 7, Table 7.4). These and other extreme events (e.g. heatwaves and droughts) have resulted in deaths and injuries. They have affected many households, communities and enterprises through adverse impacts on ecosystems, critical public and private infrastructure, supply chains, health services, food

production and the economy (Lawrence et al., 2022).

What are the significant risks from climate change in Australia over the coming decades? What are the main climate drivers, and what are the likelihoods of system transition or collapse? The latest IPCC Report (Lawrence et al., 2022) has identified nine critical risks for Australia that will be worsened by underlying vulnerabilities and exposures within existing natural, economic and social systems (see Chapter 4, Figure 4.1):

1. *Loss and degradation of tropical shallow coral reefs:* There is *very high confidence* that Australia's tropical shallow coral reefs will transition to a different state due to ocean warming and increases in the frequency, severity and duration of marine heat waves (see Chapter 5, Box 5.6). As these coral reefs transition to more degraded states, the country will lose biodiversity and ecosystem services values (see Chapter 5, Box 5.2). Reef-dependent industries, such as tourism and fisheries, will be adversely impacted, along with the communities and supply chains that depend on them for their livelihoods (see Chapter 6, Box 6.4). At some stage, the Great Barrier Reef World Heritage Area is likely to be listed as a 'world heritage site in danger' due to climate change and other change agents (see Chapter 6, Box 6.7).

2. *Loss of biodiversity in the southeast Australian Alps:* There is *high confidence* that alpine biodiversity will decline due to increasing temperatures and declining snow depths in winter (see Chapter 5, Box 5.5). The main climatic drivers are a combination of declining winter precipitation in southern Australia and rising snow elevations due to warmer winters (see Chapter 3, Table 3.2).

3. *Forest ecosystem collapse or transition to a new state in southern Australia:* There is *high confidence* of a transition or collapse of alpine ash, snowgum woodland, pencil pine and northern jarrah forests in southern Australia (see Chapter 5, Box 5.5). The main climate drivers are declining winter rainfall, increasing average temperatures and more frequent and severe heatwaves and bushfires (see Chapter 3, Table 3.2).

4. *Loss of kelp forests in southern Australia:* There is *high confidence* of a loss of kelp ecosystems

due to ocean warming and overgrazing by southward range extensions of herbivorous fish and sea urchins (see Chapter 5, Box 5.6). The main climate driver is the southern movement and strengthening of the East Australia Current (see Chapter 3, Table 3.2).

5. *Loss of natural and human systems in low-lying areas due to sea level rise:* There is *high confidence* that coastal systems will be adversely affected by sea level rise and associated storm surges and erosion (see Chapter 5, Box 5.6; and Chapter 7, Box 7.1, the 'Cities, Settlements and Built Environments' section). The main climate drivers are rising sea levels and projected increases in the intensity of tropical cyclones and East Coast Lows (see Chapter 3, Table 3.2).

6. *Disruption and decline in agricultural production:* There is *high confidence* of ongoing disruption and decline in agricultural production and increased human stress in rural communities in southwestern, southern and eastern mainland Australia due to hotter and drier climatic conditions (see Chapter 6, the 'Agriculture and Land' section; and Chapter 7, the 'Human Health and Well-Being' section). Declines in runoff and increased drought severity will further stress the already intense competition for water resources (see Chapter 5, Box 5.4). The main climate drivers are warming and drying trends across southern Australia (see Chapter 3, Table 3.2).

7. *Increase in heat-related mortality and morbidity in humans and wildlife:* There is *high confidence* that heatwaves will increase in frequency, severity and duration as the baseline climate warms. Humans and wildlife are highly vulnerable to heatwave events (see Chapter 5, Box 5.5; and Chapter 7, the 'Human Health and Well-Being' section). Increases in the number of hot and very hot days in northern and central Australia are now considered a significant climate risk for this broad region's natural, economic and social systems (see Chapter 6, Box 6.3). The main climate drivers are increases in the frequency, severity and duration of heatwaves across the country (see Chapter 3, Table 3.2).

8. *Cascading, compounding and aggregate impacts:* There is *high confidence* in cascading, compounding and aggregate effects on cities, settlements, infrastructure and supply chains due to bushfires, floods, droughts, heatwaves, storms and sea level rise (see Chapters 6 and 7).

9. *Inability of institutions and governance systems to manage climate risks:* There is *high confidence* that natural, economic and social systems will find it challenging to adapt to rapidly changing climate risk profiles. Resolving barriers to effective climate adaptation across the country's institutions and governance systems requires urgent attention, with sustained input and support from governments, businesses and communities.

This book has shown that projected global heating under current global emissions reduction policies would place many of Australia's natural and social systems at very high risk and beyond adaptation limits (see Chapter 4). If climate conditions extend beyond adaptation limits, there would be significant implications for sectors of the economy that are exposed to climate change (see Chapter 6). Australia must urgently plan and implement climate-resilient development pathways for its natural, economic and social systems. Any delays in implementing climate adaptation and mitigation pathways will impede climate-resilient development (Lawrence et al. 2022), resulting in costly and disruptive maladaptation and mal-mitigation pathways (see Chapter 4, Figure 4.1). Climate-resilient development pathways require significant and rapid GHG emissions reductions to maintain global heating to 1.5°C–2.0°C above pre-industrial levels (see Chapter 4, Box 4.4 for examples). At the same time, active and robust climate adaptation pathways must be in place (see Chapter 4, Figure 4.4). Adaptation pathways will need to overcome any barriers to adaptation and take advantage of any enablers (see Chapter 4, Box 4.2 for examples).

The tenets for reducing GHG emissions and enhancing GHG sinks apply universally in all economies (Australian Academy of Science, 2021). However, their application will differ significantly among countries depending on circumstances, opportunities and preferences (Bataille et al., 2016). Australia is dominated by 'carbon-intensive' industries, notably energy, transportation, agriculture, mining, manufacturing, construction and tourism (see Chapter 6). Having a carbon-intensive economy creates challenges for Australia in meeting its Paris Agreement targets (see Chapter 4, the 'Paris Agreement' section). This national challenge is further hindered by almost a decade of weak climate and energy policies at the federal level. Despite a lack of federal guidance, all its states and territories have driven policy and practice change in climate mitigation. Following an election commitment, in September 2022 the new Australian Labor Government passed legislation through both houses of parliament and has enshrined into law, a commitment to achieve a 43% decline in GHG emissions by 2030 (relative to 2005 levels). This is a positive move, but the science community argues that a more ambitious target is required to ramp up ahead of the significant falls in GHG emissions that must occur after 2030 as the country commits to net zero emissions by 2050. The Australian Academy of Science (Australian Academy of Science, 2021) recommends the following mitigation strategies for the rapid removal of GHG emissions to drive the pathway to 'net zero' by 2050:

- *Remove GHG emissions from electricity generation and distribution*: Achieving an electricity supply with zero emissions will be fundamental for the country by 2050. This target will require investments in renewable energy sources (solar and wind) and storage and a rapid transition from fossil fuel energy generation. Improvements in energy storage are considered a long-term solution to the intermittency in energy supply, with gas likely to play a decreasing role as new renewables technology evolves (see Chapter 6, Box 6.1). The opening-up and regulatory facilitation of renewable energy zones are also required, i.e. areas with high solar and wind resource potential and enabling higher renewable energy penetration through interconnectors, batteries and pumped solar/hydro-energy storage.
- *Electrify the transport sector and stationary energy use*: The transportation sector must seek opportunities to reduce dependence on fossil fuels, e.g. electrification of vehicles and rail systems and green hydrogen-powered heavier vehicles (see Chapter 6, the 'Energy and Transportation' section). Large-scale electric vehicle (EV) adoption will require an expanded charging network and a significant expansion of a renewables-based electricity supply. Commercial and government EV fleets and private EV vehicles also provide a large, conveniently 'decentralised' energy storage capacity

to help balance electricity demand and supply on the grid. However, the aviation sector will rely on offsetting while new aircraft fuel and engine alternatives come into the market. The biggest mitigation challenge to Australia's tourism industry is its high dependence on air travel, so it will continue to rely on offsetting and adopting renewable energy in its infrastructure and operations (see Chapter 6, the 'Tourism' section).

- *Increase energy efficiency and reduce emissions from industrial activities and buildings*: Australia's business sector is well placed to reduce GHG emissions by transitioning to renewable energy sources for its manufacturing and construction industries and supply chains. The decarbonisation of the electricity grid (see above) is considered a key enabler of fuel switching across the economy and within critical sectors. The mining, manufacturing and construction industries will primarily rely on offsetting to mitigate their emissions (see Chapter 6, the 'Mining' and the 'Manufacturing and Construction' sections). However, there are also opportunities to employ renewable energy in their operations, including EVs, hydrogen-powered heavy vehicles and machinery. In the case of remote mine sites and mining communities, there are considerable carbon mitigation benefits in the regulatory facilitation of renewable energy zones, particularly in areas with high wind or solar resource potential. Several renewable energy initiatives are underway across the mining sector. For example, using solar photovoltaic arrays with battery storage, pumped solar/hydropower, utilising existing and abandoned mine water reservoirs, and innovative green solar/hydrogen energy schemes (see Chapter 4, Box 4.4). Energy efficiency incentives are needed to reduce overall load and demand response mechanisms to enable load flexibility. Construction industries must also seek pathways to reduce carbon emissions and effectively manage any 'net-zero' transition risks (see Chapter 4, Box 4.4). At the same time, they need to implement strategies and actions to introduce decarbonisation throughout the construction value chain. For example, the lowering of the carbon intensity of building materials in the upstream production process of materials, the implementation of 'climate-smart', low and clean energy consumption in the use phase of real estate and infrastructure (see Chapter 7, Table 7.3) and the design of more recyclable materials and closed materials flow in the refurbishment and demolition phases. While Australia's urban areas are significant emitters of GHGs, its cities and towns can deliver about 70% of required GHG reductions, thereby contributing significantly to the country-level pledge to achieve net-zero emissions by 2050. Local government areas are well placed to deliver on GHG reductions by embracing renewable energy, energy efficiency and sustainable transport systems (see Chapter 7, Table 7.3 for examples).

- *Reduce non-energy related GHG emissions from industrial processes and agriculture*: Many industrial manufacturing processes in Australia are 'hard-to-abate' industries, as there are few opportunities to substitute emission-free inputs or processes (see Chapter 6, the 'Manufacturing and Construction' section). Reducing GHG emissions from crucial agricultural industries is also challenging, particularly for cattle and sheep production. The red meat sector will have to rely on offsetting in the medium term. However, the agricultural and land sector can significantly contribute to Australia's pathway to net-zero emissions, including manure and fertiliser management and carbon farming initiatives. Including the agriculture and land sectors in any national GHG emissions target is paramount as part of the national target of net zero by 2050. Agri-businesses already use land management strategies that mitigate the effects of climate change, e.g. no-till cultivation before sowing, which retains the previous crop stubble, reduces soil erosion and retains soil carbon. Other GHG emission mitigation strategies in the agricultural sector, such as methane reductions from livestock, are considered medium- to long-term aspirations as new technology is developed (see Chapter 6, Table 6.2 for examples).
- *Implement negative emissions options, through bio-sequestration and technological means*: Australia has great potential to introduce new technologies that remove CO_2 from the atmosphere and store it in living biomass and soils.

For example, biological sequestration through revegetation of degraded landscapes and mangrove and sea grass preservation. Ultimately, large-scale revegetation and soil carbon initiatives will require trade-offs with other land uses, such as farming and urban development (see Chapter 4, Box 4.4). However, using other carbon-sequestering technologies, such as carbon capture, use and storage, remains controversial (see Chapter 6, Box 6.5).

- *Stop deforestation and land degradation, and accelerate revegetation of cleared and degraded land*: State and territory governments should not weaken existing land clearing laws and should keep existing stocks of nature-based carbon at or above current levels. The Australian Government should improve its Emissions Reduction Fund (ERF) by expanding its methods related to agricultural practices (see Chapter 6, the 'Agriculture and Land' section). Notably, there should be no limits on landholders' preferences to perform credible CO_2-removing activities, such as carbon sequestration in agricultural soils under the ERF. Large-scale revegetation of degraded lands will be critical to providing biomass and soil carbon offsets to hard-to-abate industries, such as steel and cement production. At the same time, these high-emitting sectors will need to transition to low or net-zero emissions.
- *Shift energy export industries to zero emissions as a matter of urgency and produce energy-intensive products using renewable energy*: There is much to be gained by expanding and tightening the safeguard mechanism as the primary policy driver to reduce emissions across the manufacturing and construction sectors (see Chapter 6, the 'Manufacturing and Construction' section). The aim is to shift energy export industries to net-zero emissions as quickly as possible and produce energy-intensive products using renewable energy (see Chapter 4, Box 4.4). Australia also has significant opportunities to develop a 'green steel' industry, utilising vast supplies of local iron ore, solar resources and renewable energy-generated hydrogen sources. This innovative process for making steel has the potential for upscaling and would take advantage of substantial amounts of locally produced green hydrogen. There may also be opportunities

to use renewable energy for manufacturing (without hydrogen production), such as by rejuvenating the country's aluminium smelting industry powered entirely by renewable energy.

Over the past decade, there has been limited national leadership on climate adaptation in Australia, although the Australian Government finally released its National Climate Resilience and Adaptation Strategy around the time of the COP26 meeting in Glasgow (Commonwealth of Australia, 2021b). Up until that time, the states and territories have developed and implemented their own climate adaptation strategies. The latest IPCC Report (Lawrence et al., 2022) asserts that a range of incremental and transformative adaptation options and pathways is available for the country's natural, economic and social systems (see Chapter 4, Box 4.1 for examples). There is *high confidence* of their implementation over time, provided enablers are in place to fully implement them (see Chapter 4, Box 4.2 for examples). At the same time, barriers to climate adaptation will have to overcome (see Chapter 4, Box 4.2). However, the IPCC Report also warns responses that lock-in risk by discounting ongoing and changing climate risk can create maladaptation pathways (see Chapter 4, Figures 4.1–4.3) and undermine society's capacity to manage future impacts (Lawrence et al. 2022). This book has identified climate adaptation pathways and opportunities for Australia's natural, economic and social systems. The main conclusions are:

- Australian ecosystems and biodiversity may show some ecological and evolutionary (autonomous) adaptation to climate change. However, the adaptive capacity of ecosystems, habitats and species to rapidly changing climate conditions will vary greatly and may not be enough to avoid irreversible loss of ecosystem functions and many species declines and likely extinctions (see Chapter 5, Boxes 5.5 and 5.6). Both local and Indigenous knowledge and best management principles and practices will be highly relevant to the future management of the country's ecosystems, ecosystem services and biodiversity under climate change (see Chapter 5, Table 5.1). For example, low- or no-regrets actions can build resilience without committing to pathways that may become maladaptive. However, the production of

comprehensive climate adaptation plans for threatened ecosystems has yielded only a few active adaptation strategies to reduce species loss and manage ecosystem transformation in the face of a rapidly changing climate. An ecosystem-based approach to adaptation is needed to deliver multi-benefits for humans and nature while reducing vulnerability and building resilience to climate change (Cresswell et al., 2021). Lastly, the country's ecosystems, ecosystem services and biodiversity are tightly linked to its economy and society (see Chapter 5, Boxes 5.1 and 5.2 for examples).

- Planned climate adaptation is vital across the Australian economy, and all sectors must play their role in responding to the challenges (see Chapter 6). The energy sector is highly exposed to climate change and pressures to meet the net-zero target by 2050. This high exposure has been created by a decade of weak federal energy policy, making it difficult for businesses and industries to invest in renewables and other energy technologies. The most urgent requirement is the rapid construction of a modern, efficient electricity system for the country's two main grids (see Chapter 6, Box 6.1).

- The inclusion of climate change risks in the design, location and rating of future energy, transportation, agricultural, mining, manufacturing, construction and tourism infrastructure will be paramount to building resilience to extreme weather events (see Chapter 6). Agriculture is highly exposed to climate change. As time goes by, incremental adaptation to changing conditions may not be enough, and transformational responses will be required, e.g. relocating to more suitable climates for food and fibre production. In the meantime, 'no regrets' practices (e.g. best practice natural resource management) will remain relevant across the agricultural and land sectors. The mining, manufacturing, construction and tourism industries will also have to invest in climate change adaptation to increase their resilience to extreme weather events. Transformative adaptation strategies will be essential for northern and central Australian industries due to increasing exposure to heat risk and water shortages (see Chapter 6, Box 6.3).

- Climate adaptation will be needed to reduce the exposure of millions of Australians living in cities, towns and shires at risk of significant climate change (see Chapter 7, Table 7.4). Incremental adaptation is already happening in many places, but as time progresses and climate threats increase, transformative adaptation will be required to reduce climate risks. For example, community leaders in the NSW regional city of Lismore are seriously considering relocating parts of their CBD in the aftermath of the devastating February 2022 floods (Vanclay, 2022). The lack of national policies and plans to guide state, territory and local governments in developing adaptation strategies to manage climate risks is a significant concern (see Chapter 7, Box 7.3). Climate change is a pervasive national threat, and it covers all jurisdictions. Therefore, national policies for tackling climate risk and impending threats are paramount to ensure uniformity of climate adaptation around the country.

Climate-resilient development requires mitigation and adaptation strategies to align with the Sustainable Development Goals (SDGs). This book has attempted to integrate climate mitigation and adaptation with the SDGs for Australia's significant natural, economic and social systems (see Chapters 5–7 for examples). Addressing the SDGs for these crucial systems will inform incremental and transformative strategies to avoid mal-mitigation and maladaptation pathways in the future. This integrated perspective will drive innovation and change towards a vibrant and sustainable 'net-zero' Australia.

References

Adept Economics (2022) 'Structural Adjustment Policies Becoming Increasingly Important', https://adepteconomics.com.au/structural-adjustment-policies-becoming-increasingly-important/, accessed 4 March 2022.

Ali, A., Strezov, V., Davies, P.J. and I. Wright, I. (2018) 'River sediment quality assessment using sediment quality indices for the Sydney basin, Australia affected by coal and coal seam gas mining', *Science of The Total Environment*, vol. 616–617, pp. 695–702.

Allen, M.R., Dube, O.P., Solecki, W., et al. (2018) 'Framing and context', in Masson-Delmotte, V., P. Zhai, H.-O. Pörtner, et al. (eds), Global Warming of 1.5°C. An IPCC Special Report on the Impacts of Global Warming of 1.5°C above Pre-Industrial Levels and Related Global Greenhouse Gas Emission Pathways, in the Context of Strengthening the Global Response to the Threat of Climate Change, Sustainable Development, and Efforts to Eradicate Poverty, World Meteorological Organization, Geneva, Switzerland.

Allianz Global Corporate and Specialty (2021) 'Allianz Risk Barometer 2021: Top Global Business Risks', Allianz Global Corporate and Specialty, Munich, Germany, https://www.agcs.allianz.com/news-and-insights/reports/-allianz-risk-barometer.html, accessed 21 October 2021.

Amelung, B. and Nicholls, S. (2014) 'Implications of climate change for tourism in Australia', *Tourism Management*, vol. 41, pp. 228–244.

Andrews, T., Smith, C.J., Myhre, G., et al. (2021) 'Effective radiative forcing in a GCM with fixed surface temperatures', JGR Atmospheres, vol. 126, p. e2020JD033880.

Arias, P.A., Bellouin, N., Coppola, E. et al. (2021) 'Technical summary', in V.P. Masson-Delmotte, A. Pirani, S.L. Connors, et al. (eds), Climate Change 2021: The Physical Science Basis. Contribution of Working Group I to the Sixth Assessment Report of the Intergovernmental Panel on Climate, Cambridge University Press, Cambridge, UK and New York, pp. 33–144.

Ashton, D. and van Dijk, J. (2017) Rice Farms in the Murray-Darling Basin, DAWE and ABARES, Australian Government, Canberra, http://www.agriculture.gov.au:80/abares/-research-topics/ surveys/irrigation/rice, accessed 12 October 2021.

Austin, E.K., Handley, T., Kiem, A.S et al. (2018) 'Drought-related stress among farmers: findings from the Australian Rural Mental Health Study', *Medical Journal of Australia*, vol. 209, pp. 159–165.

Austrade (2013) *Sustainable Mining*, Australia Unlimited, Australian Trade Commission, Australian Government, Canberra.

Austrade (2021) 'Why Australia: Benchmark Report', Australian Government, Canberra, https://www.austrade.gov.au/benchmark-report/resilient-economy, accessed 21 October 2021.

Australasian College for Emergency Medicine (2020) *Heatwave and Heat Health, Policy P59*, Australasian College for Emergency Medicine, Melbourne, Australia.

Australia Institute for Health and Welfare (2022) 'Indigenous health and wellbeing', Australia Institute for Health and Welfare, Australian Government, Canberra, https://www.aihw.gov.au/reports/australias-health/indigenous-health-and-wellbeing, accessed 7 July 2022.

Australian Academy of Science (2021). The Risks to Australia of a 3°C Warmer World, Australian Academy of Science, Canberra.

Australian Bureau of Agricultural and Resource Economics and Sciences (2021) *Snapshot of Australian Agriculture 2021*, DAWE and ABARES, Australian Government, Canberra, https://apo.org.au/sites/default/files/-resource-files/2021-02/apo-nid312618.pdf, accessed 21 October 2021.

Australian Bureau of Statistics (2018) 'Population Projections, Australia', ABS, Australian Government, Canberra, https://www.abs.gov.au/statistics/people/population/-population-projections-australia/latest-release, accessed 12 February 2022.

Australian Bureau of Statistics (2019) Australian National Accounts: Tourism Satellite Account 2018-2019, ABS, Australian Government, Canberra, https://www.abs.gov.au/statistics/economy/national-accounts/australian-national-accounts-tourism-satellite-account/2018-19, accessed 23 October 2021.

Australian Bureau of Statistics (2021a) 'Labour force, Australia, detailed', Australian Government, https://www.abs.gov.au/statistics/labour/employment-and-unemployment/labour-force-australia-detailed/latest-release, accessed 21 October 2021.

Australian Bureau of Statistics (2021b) 'Population', ABS, Australian Government, Canberra, https://www.abs.gov.au/statistics/people/population, accessed 12 February 2022.

Australian Energy Market Operator (2020) Annual Report 2019–2020, AEMO, Melbourne, Australia.

Australian Energy Market Operator (2021) Draft 2022 Integrated System Plan: For the National Electricity Market, Compliance Quarter. AEMO, Melbourne, Australia.

Australian Financial Review (2022) 'Floods crisis could push insurance toll past $3b', *Australian Financial Review*, https://www.afr.com/companies/financial-services/floods-crisis-could-push-insurance-toll-past-3b-20220309-p5a34b, accessed 23 March 2022.

Australian Food and Grocery Council (2021) 'State of the Industry', Canberra, Australia, https://www.afgc.org.au/industry-resources/state-of-the-industry, accessed 21 October 2021.

Australian Industry and Skills Committee (2021) National Industry Insights Report, Department of Education, Skills and Employment, Australian Government, Canberra.

Bataille, C., Waisman, H., Colombier, M., Segafredo, L., Williams, J. and Jotzo, F. (2016) 'The need for national deep decarbonization pathways for effective climate policy,' *Climate Policy*, vol. 16, S7-S26.

Battaglia, M. (2011) 'Greenhouse gas mitigation: sources and sinks in agriculture and forestry' in in H. Cleugh, M. Stafford Smith, M. Battaglia and P. Graham (eds), *Climate Change: Science and Solutions for Australia*, CSIRO Publishing, Melbourne, Australia.

Bauman, D., Fortunel, C., Delhaye, G. et al. (2022) 'Tropical tree mortality has increased with rising atmospheric water stress', Nature, https://doi.org/10.1038/s41586-022-04737-7.

Baylis, K., Fullerton, D. and Karney, D.H. (2014) 'Negative leakage', Journal of the Association of Environmental and Resource Economists, vol. 1, pp. 51–73.

Beck, H.E., Zimmermann, N.E., McVicar, T.R., et al. (2018) 'Present and future Köppen-Geiger climate classification maps at 1km resolution', Scientific Data, vol. 5, p. 180214.

Becken, S. and Mackey, B. (2017) 'What role for offsetting aviation greenhouse gas emissions in a deep-cut carbon world?', *Journal of Air Transport Management*, vol. 63, pp. 71–83.

Benestad, R.E., Nuccitelli, D., Lewandowsky, S., et al. (2015) 'Learning from mistakes in climate research', *Theoretical and Applied Climatology*, vol. 126, pp. 699–703.

Berkes, F. and Folke, C. (eds) (1998) Linking Social and Ecological Systems: Management Practices and Social Mechanisms for Building Resilience, Cambridge University Press, Cambridge, UK.

Blainey, G. (2013) *The Rush that Never Ended A History of Australian Mining, Revised 5th edition*, Melbourne University Publishing, Melbourne, Australia.

Boer, M.M., de Dios, V. R., and Bradstock, R.A. (2020) 'Unprecedented burn area of Australian mega forest fires', *Nature Climate Change*, vol. 10, pp. 170–172.

Bokshi, A.I., Tan, D.K.Y., Thistlethwaite, R.J., Trethowan, R. and Kunz, K. (2021) 'Impact of elevated CO2 and heat stress on wheat pollen viability and grain production', Functional Plant Biology, vol. 48, pp. 503–514.

Bonada, M., Edwards, E.J., McCarthy, M.G., Sepúlveda, G.C. and Petrie, P.R. (2020) 'Impact of low rainfall during dormancy on vine productivity and development', Grape and Wine Research, vol. 26, pp. 325–342.

Booth, T.H., Kirschbaum, M.U.F. and Battaglia, M. (2010) 'Forestry', in C. Stokes and M. Howden (eds) Adapting Agriculture to Climate Change: Preparing Australian Agriculture, Forestry and Fisheries for the Future, CSIRO Publishing, Melbourne, Australia.

Borchers Arriagada, N., Palmer, A.J., Bowman, D.M.J.S., Morgan, G.G., Jalaludin B.B. and Johnston, F.H. (2020) 'Unprecedented smoke-related health burden associated with the 2019–20 bushfires in eastern Australia', Medical Journal of Australia, vol. 213, pp. 282–283.

Boulter, S.L. (2012) 'An assessment of the vulnerability of Australian forests to the impacts of climate change: Synthesis', Contribution of Work Package 5 to the Forest Vulnerability Assessment, National Climate Change Adaptation Research Facility, Gold Coast, Australia.

Bourne

Brewer, T. (2016) 'Climate-adaptive Northern Development', Policy Information Brief 3, National Climate Change Adaptation Research Facility, Gold Coast, Australia.

Buis, A. (2020) 'Milankovitch Cycles', https://climate.nasa.gov/news/2948/milankovitch-orbital-cycles-and-their-role-in-earths-climate/, accessed 7 July 2021.

Büntgen, U., Arseneault, D., Boucher, E. et al. (2020) 'Prominent role of volcanism in common era climate variability and human history', Dendrochronologia, vol. 64, p. 125757.

Bureau of Meteorology (2021) 'Australian Climate Influences', Australian Government, Bureau of Meteorology, http://www.bom.gov.au/watl/about-weather-and-climate/australian-climate-influences.shtml?bookmark=introduction, accessed 12 June 2021.

Bureau of Meteorology and Commonwealth Scientific and Industrial Research Organisation (2020) State of the Climate 2020, Commonwealth of Australia, Melbourne, Australia.

Burke, P.J., Beck, F.J., Aisbett, E. et al. (2022) 'Contributing to regional decarbonization: Australia's potential to supply zero-carbon commodities to the Asia-Pacific', Energy, vol. 248, p. 123563.

Burns, G., Adams, L. and Buckley, G. (2017) Independent review of the extreme weather event South Australia 28 September 5 – October 2016, Report presented to the Premier of South Australia.

Business Council of Australia (2021) Achieving a Net-Zero Economy, Business Council of Australia, Melbourne.

Cai, W., Wang, G., Gan, B., et al. (2018) 'Stabilised frequency of extreme positive Indian Ocean dipole under 1.5 °C warming', Nature Communications, vol. 9, p. 1419.

Cai, W., Ng, B., Wang, G. et al. (2022) 'Increased ENSO sea surface temperature variability under four IPCC emission scenarios', Nature Climate Change, vol. 12, pp. 228–231.

Canadell, J.G., Meyer, C.P., Cook, G.D. et al. (2021) 'Multi-decadal increase of forest burned area in Australia is linked to climate change', Nature Communications, vol. 12, p. 6921.

Capon S., Chambers J., Barmuta L. et al. (2017) National Climate Change Adaptation Research Plan for Freshwater Ecosystems and Biodiversity: Update 2017, National Climate Change Adaptation Research Facility, Gold Coast.

Carlyle, J.C., Charmley, E., Baldock, J.A., Polglase, P.J. and Keating, B. (2010) 'Agricultural greenhouse gases and mitigation options', in C. Stokes and M. Howden (eds) Adapting Agriculture to Climate Change: Preparing Australian Agriculture, Forestry and Fisheries for the Future, CSIRO Publishing, Melbourne, Australia.

Caruana, C. (2010) 'Picking up the pieces: Family functioning in the aftermath of natural disaster', Family Matters, vol. 84, pp. 79–88.

Chen, D., Rojas, M., Samset, B.H, et al. (2021) 'Framing, context, and methods', in V.P.

Masson-Delmotte, A. Pirani, S.L. Connors, et al. (eds), Climate Change 2021: The Physical Science Basis. Contribution of Working Group I to the Sixth Assessment Report of the Intergovernmental Panel on Climate Change, Cambridge University Press, Cambridge, UK and New York, pp. 147–286.

Cities Power Partnership (2022) 'About', Cities Power Partnership, The Climate Council of Australia Limited, Sydney, Australia, https://citiespowerpartnership.org.au/about/, accessed 25 March 2022.

Clean Energy Regulator (2018) '2018 Annual Statement', Australian Government, http://www.cleanenergyregulator.gov.au/About/Pages/Accountability%20and%20reporting/Administrative%20Reports/The%20Renewable%20Energy%20Target%202018%20Administrative%20Report/Annual-statement.aspx, accessed 3 November 2021.

Clean Energy Regulator (2022) 'About the Clean Energy Regulator', Australian Government, http://www.cleanenergyregulator.gov.au/About/Pages/default.aspx, accessed 08 April 2022.

Climate Council (2021a) 'What is Carbon Capture and Storage?', Climate Council of Australia Limited, Sydney, Australia, https://www.climatecouncil.org.au/resources/what-is-carbon-capture-and-storage/, accessed 19 October 2021.

Climate Council (2021b) 'Untouchable Playgrounds: Urban Heat and the Future of Western Sydney', The Climate Council of Australia Limited, Sydney, Australia, https://www.climatecouncil.org.au/urban-heat-island-effect-western-sydney/, accessed 12 February 2022.

Climate Watch (2021) 'Historical GHG Emissions', https://www.climatewatchdata.org/ghg-emissions?end_year=2019&start_year=1990, accessed 09 July 2021.

Climate Works Australia (2021) 'Setting up Industry for Net Zero: Highlights Report: Current State and Future Possibilities', https://climateworkscentre.org/wp-content/uploads/2021/08/Phase-1-Highlights-Report-June-2021.pdf, accessed 7 March 2022.

Clonan, M., McConchie, C., Hall, M., Hearnden, M., Olesen, T. and Sarkhosh, A. (2021) 'Effects of ambient temperatures on floral initiation in Australian mango (Mangifera indica L.) selections', Scientia Horticulturae, vol. 276, p. 109767.

CoastAdapt (2017) Climate Change and Sea-Level Rise Based on Observed Data, National Climate Change Adaptation Research Facility, Gold Coast.

Coates, L., van Leeuwen, J., Browning, S., Gissing, A., Bratchell, J. and Avci, A. (2022) 'Heatwave fatalities in Australia, 2001–2018: An analysis of coronial records', International Journal of Disaster Risk Reduction, vol. 67, 102671.

Commonwealth of Australia (2015a) Our North, Our Future: White Paper on Developing Northern Australia, Australian Government, Canberra.

Commonwealth of Australia (2015b) 2015 Intergenerational Report Australia in 2055, The Treasury, Australian Government, Canberra.

Commonwealth of Australia (2017–2018) Australia State of the Environment 2016, Australian Government, Canberra.

Commonwealth of Australia (2020) Royal Commission into National Natural Disaster Arrangements Report, Australian Government, Canberra.

Commonwealth of Australia (2021a) Australia's Long-term Emissions Reduction Plan: A Whole-of-Economy Plan to Achieve Net Zero Emissions by 2050, Australian Government, Canberra.

Commonwealth of Australia (2021b) National Climate Resilience and Adaptation Strategy 2021 to 2025: Positioning Australia to Better Anticipate, Manage and Adapt to Our Changing Climate, Australian Government, Canberra.

Commonwealth of Australia (2021c) 'The Australian Government's Technology Investment Roadmap', Australian Government, https://www.industry.gov.au/data-and-publications/technology-investment-roadmap, accessed 27/10/2021.

Commonwealth of Australia (2021d) 2021 National Gas Infrastructure Plan, Australian Government, Canberra.

Commonwealth Scientific and Industrial Research Organisation and Energy Networks Australia

(2017) 'Electricity Network Transformation Roadmap Final Report', Energy Networks Australia, http://www.energynetworks.com.au/electricity-network-transformation-roadmap, accessed 07 April 2022.

Commonwealth Scientific and Industrial Research Organisation (2022) 'Cape Grim Greenhouse Gas Data', Australian Government, https://capegrim.csiro.au, accessed 03 April 2022.

Conservation International (2022) 'Biodiversity Hotspots', https://www.conservation.org/priorities/biodiversity-hotspots, accessed 27 March 2022.

Cook, G. (2009) 'Historical perspectives on land use development in northern Australia: with emphasis on the Northern Territory', in P. Stone (ed) Northern Australia Land and Water Taskforce Sustainable Development of Northern Australia: A Report to Government from the Northern Australia Land and Water Taskforce, Department of Infrastructure, Transport, Regional Development and Local Government, Australian Government, Canberra.

Cook, J., Nuccitelli, D., Green, S.A., Richardson, M., Winkler, B., Painting, R., Way, R., Jacobs, P. and Skuce, A. (2013) 'Quantifying the consensus on anthropogenic global warming in the scientific literature', Environmental Research Letters, vol. 10, p. 02402.

Cooper, A. and Lemckert, C.J. (2012) 'Extreme sea-level rise and adaptation options for coastal resort cities: A qualitative assessment from the Gold Coast, Australia', Ocean and Coastal Management, vol. 64, pp. 1–14.

CoreLogic (2022) '$25 billion in Australian residential property exposed to high coastal risk', CoreLogic Australia, Sydney, Australia, https://www.corelogic.com.au/news-research/news/2022/25-billion-in-australian-residential-property-exposed-to-high-coastal-riskaccessed 29 March 2022.

Cotts, B.R.T., Prigmore, J.R. and Graf, K.L (2017) 'HVDC Transmission for renewable energy integration', in B.W. D'Andrade (ed) The Power Grid: Smart, Secure, Green and Reliable, Academic Press, Washington DC.

Cradduck, L., Warren-Myers, G. and Stringer, B. (2020) 'Courts' views on climate change inundation risks for developments: Australian perspectives and considerations for valuers',

Journal of European Real Estate Research, vol. 13, pp. 435–453.

Crausbay, S.D., Ramirez, A.R., Carter, S.L. et al. (2017) 'Defining ecological drought for the twenty-first century', Bulletin of the American Meteorological Society, vol. 98, pp. 2543–2550.

Cresswell, I.D., Janke, T. and Johnston, E.L. (2021) Australia State of the Environment 2021: Overview, independent report to the Australian Government Minister for the Environment, Commonwealth of Australia, Canberra.

Crimp, S., Howden, M., Stokes, C., Schroeter, S. and Keating, B.A. (2014) 'Climate change challenges for low input cropping and grazing systems – Australia', in J. Fuhrer and P. Gregory Climate Change Impact and Adaptation in Agricultural Systems, CABI, Wallingford, United Kingdom.

Critical Ecosystem Partnership Fund (2022). 'Learn more about CEPF', https://www.cepf.net, accessed 27 March 2022.

Daly, J., Anderson, K., Ankeny, R., Harch, B., Hastings, A., Rolfe, J. and Waterhouse, R. (2015) Australia's Agricultural Future, Report for the Australian Council of Learned Academies, Melbourne, Australia.

Davey, S.M. and Sarre, A. (2020) 'Editorial: the 2019/20 Black Summer bushfires'. Australian Forestry, vol. 83, pp. 47–51.

Davidson, B.R. (1965) The Northern Myth: A Study of the Physical and Economic Limits to Agricultural and Pastoral Development in Tropical Australia, Cambridge University Press, London, UK and New York, USA.

Deloitte Access Economics (2020) A New Choice: Australia's Climate for Growth, Deloitte Access Economics, Sydney, Australia.

Department of Agriculture, Water and the Environment (2021a) 'Montreal Protocol', https://www.awe.gov.au/environment/protection/ozone/montreal-protocol, accessed 27 July 2021.

Department of Agriculture, Water and the Environment (2021b) 'Future Drought Fund', DAWE, Australian Government, Canberra, https://www.awe.gov.au/agriculture-land/-farm-food-drought/drought/future-drought-fund, accessed 22 October 2021.

Department of Climate Change, Energy and Efficiency (2011) *Climate Change Risks to Coastal Buildings and Infrastructure: A Supplement to the First Pass National Assessment*, Department of Climate Change, Energy and Efficiency, Australian Government, Canberra.

Department of Industry, Science, Energy and Resources (2020) The King Report, Australian Government, Canberra.

Department of Industry, Science, Energy and Resources (2021a) 'Australian Energy Update 2021', Australian Government, https://www.energy.gov.au/publications/australian-energy-update-2021, accessed 07 November 2021.

Department of Industry, Science, Energy and Resources (2021b) 'National Electricity Market', Australian Government, https://www.energy.gov.au/government-priorities/energy-markets/national-electricity-market-nem, accessed 07 November 2021.

Department of Industry, Science, Energy and Resources (2021c) 'Tracking and Reporting Greenhouse Gas Emissions', Australian National Greenhouse Accounts, Australian Government, https://www.industry.gov.au/policies-and-initiatives/australias-climate-change-strategies/tracking-and-reporting-greenhouse-gas-emissions, accessed 07 November 2021.

Department of Industry, Science, Energy and Resources (2022) 'Safeguard mechanism,' Australian Government, https://www.industry.gov.au/regulations-and-standards/national-greenhouse-and-energy-reporting-scheme/safeguard-mechanism, accessed 08 April 2022.

Deser, C., Trenberth, K. and National Center for Atmospheric Research Staff (eds) (2016), 'The climate data guide: pacific decadal oscillation (PDO): Definition and indices', https://climatedataguide.ucar.edu/climate-data/pacific-decadal-oscillation-pdo-definition-and-indices, accessed 09 July 2021.

Diamond, L., Lorrey, A.M. and Renwick, J.A. (2013) 'A southwest Pacific tropical cyclone climatology and linkages to the El Niño-Southern Oscillation', Journal of Climate, vol. 26, pp. 3–25.

Díaz S., Demissew S., Carabias J., et al. (2015) 'The IPBES Conceptual Framework – connecting nature and people', Current Opinion in Environmental Sustainability, vol. 14, pp. 1–16.

Downham, R. and Gavran, M. (2019) 'Australian Plantation Statistics 2019 Update', *ABARES Technical Report 19.2*, Australian Government, Canberra, Australia.

Dwyer, L., Forsyth, P., Spurr, R. and Hoque, S. (2010) 'Estimating the carbon footprint of Australian tourism', *Journal of Sustainable Tourism*, vol. 18, pp. 355–376.

Electric Vehicle Council (2021) 'State of electric vehicles, August 2021', Electric Vehicle Council Incorporated, Sydney, Australia, https://electricvehiclecouncil.com.au/wp-content/uploads/2021/08/EVC-State-of-EVs-2021-sm.pdfm, accessed 12 April 2022.

Emergency Leaders for Climate Action and the Climate Council of Australia (2020) 'Unpacking the National Bushfire and Climate Summit 2020', The Climate Council of Australia Limited, Sydney, Australia, https://www.climatecouncil.org.au/unpacking-national-bushfire-climate-summit-2020/, accessed 24 March 2022.

Engineers Australia (2020) Climate Change and Transport: Transport Australia Society Discussion Paper, Engineers Australia, Barton, ACT.

Eyring, V., Gillett, N.P., Achuta Rao, K.M., et al. (2021) 'Human influence on the climate system', in V.P. Masson-Delmotte, A. Pirani, S.L. Connors, et al. (eds), Climate Change 2021: The Physical Science Basis. Contribution of Working Group I to the Sixth Assessment Report of the Intergovernmental Panel on Climate Change, Cambridge University Press, Cambridge, UK and New York, pp. 423–552.

Fairman, T.A., Nitschke, C.R. and Bennett, L.T. (2022) 'Carbon stocks and stability are diminished by short-interval wildfires in fire-tolerant eucalypt forests', Forest Ecology and Management, vol. 505, p. 119919.

Filkov, A.I., Ngo, T., Matthews, S., Telfer, S. and Penman, T.D. (2020) 'Impact of Australia's catastrophic 2019/20 bushfire season on communities and environment. Retrospective analysis and current trends',

Journal of Safety Science and Resilience, vol. 1, pp. 44–56.

Finkel, A., Moses, K., Munroe, C., Effeney, T. and O'Kane, M. (2017) Independent Review into the Future Security of the National Electricity Market. Blueprint for the Future, Department of Industry, Science, Energy and Resources, Australian Government, Canberra.

Fisk, G.W. (2017) Climate Risks and Adaptation Pathways for Coastal Transport Infrastructure. Guidelines for Planning and Adaptive Responses, National Climate Change Adaptation Research Facility, Gold Coast.

Fogarty, H.E., Cvitanovic, C., Hobday, A.J. and Pecl, G.T. (2020) 'An assessment of how Australian fisheries management plans account for climate change impacts', *Frontiers in Marine Science*, vol. 7. DOI:10.3389/fmars.2020.591642

Food and Agricultural Organization (2014) 'Agriculture, forestry and other land use emissions by sources and removals by sinks', Working Paper Series ES/14-02, FAO Statistics Division.

Forster, P., Storelvmo, T., Armour, K., et al. (2021) 'The Earth's Energy Budget, Climate Feedbacks, and Climate Sensitivity', in V.P. Masson-Delmotte, A. Pirani, S.L. Connors, et al. (eds) Climate Change 2021: The Physical Science Basis. Contribution of Working Group I to the Sixth Assessment Report of the Intergovernmental Panel on Climate Change, Cambridge University Press, Cambridge, UK and New York, pp. 923–1054.

Friedlingstein, P., O'Sullivan, M., Jones, M. W. et al. (2021) 'Global carbon budget 2020', Earth System Science Data, vol. 2, pp. 3269–3340.

Gallagher, S.J., Greenwood, D.R., Taylor, D., Smith, A.J., Wallace M.W., Holdgate, G.R. (2003) 'The Pliocene climatic and environmental evolution of southeastern Australia: evidence from the marine and terrestrial realm', Paleogeography, Paleoclimatology, Paleoecology, vol. 193, pp. 349–382.

Gardner, J., Dowd, A.-M., Mason, C. and Ashworth, P. (2009) 'A framework for stakeholder engagement on climate adaptation', CSIRO Climate Adaptation National Research Flagship Working Paper No. 3, CSIRO, ACT.

Garnaut, R, (2008) The Garnaut Climate Change Review, Cambridge University Press, Cambridge, UK.

Garnaut, R. (2019) Superpower: Australia's Low-Carbon Future, La Trobe University Press, Melbourne.

Gasbarro, F., Rizzi, F. and Frey, M. (2016) 'Adaptation measures of energy and utility companies to cope with water scarcity induced by climate change', Business Strategy and the Environment, vol. 25, pp. 54–72.

Glaser, M., Krause, B., Ratter, B. and Welp, M. (2008) 'Human/nature interaction in the Anthropocene: potentials of socio-ecological systems analysis', GAIA: Ecological Perspective for Science and Society, vol. 17, pp. 77–80.

Global Carbon Capture Storage Institute (2021) 'Global Status CCS 2021', Global Carbon Capture Storage Institute, Melbourne, Australia, https://www.globalccsinstitute. com/resources/global-status-report/, accessed 21 November 2021.

Golding, B. and Campbell, C. (2009) 'Learning to be drier in the southern Murray-Darling Basin: Setting the scene for this research volume', *Australian Journal of Adult Learning*, vol. 49, pp. 423–450.

Granwal, L. (2022) 'Mining industry gross value-added Australia 2012-2021', *Statistica*, https://www.statista.com/statistics/874024/-australia-gross-value-added-mining-industry/, accessed 12 April 2022.

Great Barrier Reef Marine Park Authority (2022) 'Reef health', GBRMPA, Australian Government, Townsville, https://www2. gbrmpa.gov.au/learn/reef-health, accessed 28 August 2022.

Grech, A., Pressey, R. and Day, J. (2016) 'Coal, cumulative impacts, and the Great Barrier Reef', *Conservation Letters*, vol. 9, pp. 200–207.

Greenville, J., McGilvray, H. and Black, S. (2020) 'Australian Agricultural Trade and the COVID-19 Pandemic', DAWE and ABARES, Australian Government, Canberra, https://daff.ent.sirsidynix.net.

au/client/en_AU/search/asset/1030341/0, accessed 21 October 2021.

Griffith Taylor, T. (1924) 'Geography and Australian national problems', Presidential Address, *Section E. Australasian Association for the Advancement of Science*, vol. 16, pp. 433–487.

Gulev, S.K., Thorne, P.W., Ahn, J., et al. (2021) 'Changing state of the climate system', in V.P. Masson-Delmotte, A. Pirani, S.L. Connors, et al. (eds), Climate Change 2021: The Physical Science Basis. Contribution of Working Group I to the Sixth Assessment Report of the Intergovernmental Panel on Climate Change, Cambridge University Press, Cambridge, UK and New York, pp. 287–422.

Haasnoot, M., Jan H. Kwakkel, J.H., Warren E. Walker, W.E. and ter Maat, J. (2013) 'Dynamic adaptive policy pathways: a method for crafting robust decisions for a deeply uncertain world', Global Environmental Change, vol. 23, pp. 485–498.

Hays, J. D., Imbrie, J. and Shacketon, N.J. (1976) 'Variations in the earth's orbit: pacemaker of the ice ages', Science, vol. 194, pp. 1121–1132.

Heimann, M., Reichstein, M. (2008) 'Terrestrial ecosystem carbon dynamics and climate feedbacks', *Nature*, vol. 451, pp. 289–292.

Henzell, T. (2007) *Australian Agriculture: Its History and Challenges*, CSIRO Publishing, Melbourne, Australia.

Hill, R. and Lyons, P. (2014) 'Adaptation pathways and opportunities for Indigenous peoples' in C. Moran, S.M. Turton and R. Hill (eds) *Adaptation Pathways and Opportunities for the Wet Tropics NRM Cluster Region (Volume 2)*, James Cook University, Cairns, Australia.

Hobday, A.J. and Poloczanska, E.S. (2010) 'Marine fisheries and aquaculture', in C. Stokes and M. Howden (eds) *Adapting Agriculture to Climate Change: Preparing Australian Agriculture, Forestry and Fisheries for the Future*, CSIRO Publishing, Melbourne, Australia.

Hochman, Z., Gobbett, D.L. and Horan, H. (2017) 'Climate trends account for stalled wheat yields in Australia since 1990', *Global Change Biology*, vol. 23, pp. 2071–2081.

Hodges, A., Hoang, A.L., Tsekouras, G. et al. (2022) 'A high-performance capillary-fed electrolysis cell promises more cost-competitive renewable hydrogen', Nature Communications, vol. 13, 1304.

Hodgkinson. J.H., Littleboy. A., Howden, M., Moffat, K. and Loechel, B. (2010) 'Climate adaptation in the Australian mining and exploration industries', *CSIRO Climate Adaptation Flagship Working Paper No. 5*, Canberra, Australia.

Hodgkinson, J.H., Hobday, A.J. and Pinkard, E.A. (2014) 'Climate adaptation in Australia's resource-extraction industries: ready or not?', *Regional Environmental Change*, vol. 14, pp. 1663–1678.

Hodgkinson, J.H and Smith, M.H. (2021) 'Climate change and sustainability as drivers for the next mining and metals boom: The need for climate-smart mining and recycling', *Resources Policy*, vol. 74, p. 101205.

Hoeppner, J.M. and Hughes, L. (2019) 'Climate readiness of recovery plans for threatened Australian species', Conservation Biology, vol. 33, pp. 534–542.

Holgate, C.M., Van Dijk, A.I.J.M., Evans, J.P. and Pitman, A.J. (2020) 'Local and remote drivers of southeast australian drought', Geophysical Research Letters, vol. 47, p. e2020GL090238.

Hovenden, M.J. and Williams, A.L. (2010) 'The impacts of rising CO_2 concentrations on Australian terrestrial species and ecosystems', *Austral Ecology*, vol. 35, pp. 665–684.

Howden, S.M. and Stokes, C.J. (2010) 'Introduction', in C. Stokes and M. Howden (eds) *Adapting Agriculture to Climate Change: Preparing Australian Agriculture, Forestry and Fisheries for the Future*, CSIRO Publishing, Melbourne, Australia.

Howden, S.M., Gifford, R.G. and Meinke, H. (2010) 'Grains', in C. Stokes and M. Howden (eds) *Adapting Agriculture to Climate Change: Preparing Australian Agriculture, Forestry and Fisheries for the Future*, CSIRO Publishing, Melbourne, Australia.

Howlett, B. and Henry, R. (2020) 'Australian agriculture and climate change: a two-way street', Australian Academy of Science, Canberra, Australia, https://www.science.

org.au/curious/policy-features/australian-agriculture-and-climate-change-two-way-street, accessed 23 September 2021.

Hughes, L. (2014) 'Changes to Australian terrestrial biodiversity', in P. Christoff (ed), Four Degrees of Global Warming in a Hot World, Routledge, London.

Hughes, L., Steffen, W., Rice, M. and Pearce, A. (2015) *Feeding a Hungry Nation: Climate Change, Food and Farming in Australia*, Climate Council of Australia Limited, Sydney, Australia.

Hughes, L., Stock, P., Brailsford, L. and Alexander, D. (2018) *Icons At Risk: Climate Change Threatening Australian Tourism*, Climate Council of Australia Limited, Sydney, Australia.

Hughes, L., Steffen, S., Mullins, G., Dean, A., Weisbrot, E. and Rice, M. (2020) Summer of Crisis, Climate Council of Australia Limited, Sydney.

Hughes, L., Steffen, S., Rice, M. and Pearce, A. (2015) Feeding a Hungry Nation: Climate Change, Food and Farming in Australia, Climate Council of Australia Limited, Sydney.

Hughes, N. and Gooday, P. (2021) 'Climate change impacts and adaptation on Australian farms', Department of Agriculture, Water and the Environment, Australian Government, Canberra, https://www.awe.gov.au/abares/products/insights/climate-change-impacts-and-adaptation#future-changes-in-climate-could-make-conditions-tougher-for-australian-farms, accessed 23 September 2021.

Hughes, T.P., Kerry, J.T., Baird, A.H. et al. (2019) 'Global warming impairs stock–recruitment dynamics of corals', Nature, vol. 568, pp. 387–390.

Hughes, T.P., Kerry, J.T., Connolly, S.R., et al. (2021) 'Emergent properties in the responses of tropical corals to recurrent climate extremes', Current Biology, vol. 31, pp. 5393–5399.

Hurlimann, A.C., Warren-Myers, G. and Browne, G. (2019) 'Is the Australian construction industry prepared for climate change?', *Building and Environment*, vol. 153, pp. 128–137.

Infrastructure Australia (2019) Infrastructure Australia Audit 2017, Australian Government, Canberra.

Intergovernmental Panel on Climate Change (2013) in T.F. Stocker, D. Qin, G.-K. Plattner, et al. (eds), Climate Change 2013: The Physical Science Basis. Contribution of Working Group I to the Fifth Assessment Report of the Intergovernmental Panel on Climate Change, Cambridge University Press, Cambridge, United Kingdom and New York.

Intergovernmental Panel on Climate Change (2014a) in O. Edenhofer, R. Pichs-Madruga, Y. Sokona, et al. (eds), Climate Change 2014: Mitigation of Climate Change. Contribution of Working Group III to the Fifth Assessment Report of the Intergovernmental Panel on Climate Change, Cambridge University Press, Cambridge, United Kingdom and New York.

Intergovernmental Panel on Climate Change (2014b) in C.B. Field, V.R. Barros, D.J. Dokken, et al. (eds), Climate Change 2014: Impacts, Adaptation, and Vulnerability. Part A: Global and Sectoral Aspects. Contribution of Working Group II to the Fifth Assessment Report of the Intergovernmental Panel on Climate Change, Cambridge University Press, Cambridge, United Kingdom and New York.

Intergovernmental Panel on Climate Change (2018) 'Summary for policymakers', in Global Warming of 1.5°C. An IPCC Special Report on the Impacts of Global Warming of 1.5°C above Pre-Industrial Levels and Related Global Greenhouse Gas Emission Pathways, in the Context of Strengthening the Global Response to the Threat of Climate Change, Sustainable Development, and Efforts to Eradicate Poverty, V. Masson-Delmotte, P. Zhai, H.-O. Pörtner, et al. (eds), World Meteorological Organization, Geneva, Switzerland.

Intergovernmental Panel on Climate Change (2019a) in H.-O. Pörtner, D.C. Roberts, V. Masson-Delmotte, et al. (eds), IPCC Special Report on the Ocean and Cryosphere in a Changing Climate, Cambridge University Press, Cambridge, United Kingdom and New York.

Intergovernmental Panel on Climate Change (2019b) in P.R. Shukla, J. Skea, E. Calvo Buendia, et al. (eds), Climate Change and Land: An IPCC Special Report on Climate Change, Desertification, Land Degradation, Sustainable Land Management, Food Security, and Greenhouse Gas Fluxes in Terrestrial Ecosystems, Cambridge University Press, Cambridge, United Kingdom and New York.

Intergovernmental Panel on Climate Change (2021) 'Summary for policymakers', in V.P. Masson-Delmotte, A. Pirani, S.L. Connors, et al. (eds), Climate Change 2021: The Physical Science Basis. Contribution of Working Group I to the Sixth Assessment Report of the Intergovernmental Panel on Climate Change, Cambridge University Press, Cambridge, United Kingdom and New York, pp. 3–32.

Intergovernmental Panel on Climate Change (2022a) in P.R. Shukla, J. Skea, R. Slade, et al. (eds), Climate Change 2022 Mitigation of Climate Change. Contribution of Working Group III to the Sixth Assessment Report of the Intergovernmental Panel on Climate Change, Cambridge University Press, Cambridge, United Kingdom and New York.

Intergovernmental Panel on Climate Change (2022b) in H.-O. Pörtner, D.C. Roberts, M. Tignor, et al. (eds), Climate Change 2022: Impacts, Adaptation, and Vulnerability. Contribution of Working Group II to the Sixth Assessment Report of the Intergovernmental Panel on Climate Change, Cambridge University Press, Cambridge, United Kingdom and New York.

Intergovernmental Science-Policy Platform on Biodiversity and Ecosystem Services (2019) in S. Díaz, J. Settele, E. S. Brondízio, et al. (eds), Summary for Policymakers of the Global Assessment Report on Biodiversity and Ecosystem Services of the Intergovernmental Science-Policy Platform on Biodiversity and Ecosystem Services, IPBES Secretariat, Bonn, Germany.

International Energy Agency (2020) Global Energy Review 2020: The Impacts of the Covid-19 Crisis on Global Energy Demand and CO2 Emissions, International Energy Agency, Paris, France.

International Monetary Fund (2021) 'GDP based on PPP, share of world', https://www.imf.org/external/datamapper/PPPSH@WEO/OEMDC/ADVEC/WEOWORLD, accessed 21 October 2021.

International Union for the Conservation of Nature (2020) 'Great Barrier Reef', World Heritage Outlook Report, https://worldheritageoutlook.iucn.org/explore-sites/wdpaid/2571, accessed 21 June 2022.

Jacobs, B., Nelson, R., Kuruppu, N. and Leith, P. (2015) An Adaptive Capacity Guide Book: Assessing, Building and Evaluating the Capacity of Communities to Adapt in a Changing Climate, Southern Slopes Climate Change Adaptation Research Partnership (SCARP), University of Technology Sydney and University of Tasmania. Hobart, Tasmania.

Johnston, F.H., Borchers-Arriagada, N., Morgan, G.G. et al. (2021) 'Unprecedented health costs of smoke-related PM2.5 from the 2019–20 Australian megafires', Nature Sustainability, vol. 4, pp. 42–47.

Jones, R.N. and Mearns, L.O. (2005) 'Assessing climate risks', in B. Lim, E. Spranger-Siegfried, I. Burton, E. Malone and S. Huq (eds) Adaptation Policy Frameworks for Climate Change: Developing Strategies, Policies and Measures, Cambridge University Press, Cambridge, United Kingdom and New York, NY.

Jones, L., Harvey, B., Cochrane, L. et al. (2018) 'Designing the next generation of climate adaptation research for development', Regional Environmental Change, vol. 18, pp. 297–304.

Jouzel, J. and Masson-Delmotte, V. (2010) 'Deep ice cores: the need for going back in time', Quaternary Science Reviews, vol. 29, pp. 3683–3689.

Kayrros (2021) 'Methane Emissions from Australia's Bowen Basin', Kayrros, Paris, France, https://www.kayrros.com/blog/-methane-emissions-from-australias-bowen-basin/, accessed 21 October 2021.

Keogh, M. (2012) 'Including risk in enterprise decisions in Australia's riskiest businesses', Australian Agricultural and Resource Economics Society, 2012 Conference (56th), February 7-10, 2012, Fremantle, Australia,

124202, https://ideas.repec.org/p/ags/ aare12/124202.html, accessed March 23 2021.

Kienberger, S., Hagenlocher, M., Delmelle, E. and Casas, I. (2013) 'A WebGIS tool for visualizing and exploring socioeconomic vulnerability to dengue fever in Cali, Colombia', Geospatial Health, vol. 8, pp. 313–316.

Kompas, T. (2020) Cited in B. Silvester 'Trillions up in smoke: The staggering economic cost of climate change inaction', New Daily, https://thenewdaily.com. au/news/national/ 2020/09/10/economic-cost-climate- change/, accessed 5 March 2022.

Köppen, W. (1936) Das geographische System der Klimate, Verlag von Gebrüder Borntraeger, Berlin, Germany.

Kossin, J. P. (2018) 'A global slowdown of tropical cyclone translation speed', Nature, vol. 558, pp. 104–107.

Kotey, B. (2015) 'Demographic and economic changes in remote Australia', Australian Geographer, vol. 46, pp. 183–201.

Kuleshov, Y., Qi, L., Fawcett, R. and Jones, D. (2008) 'On tropical cyclone activity in the Southern Hemisphere: Trends and the ENSO connection', Geophysical Research Letters, vol. 35, p. 14.

Kuruppu, N., Murta, J.P., Mukheibir, P., Chong, J. and Brennan, T. (2013) Understanding the Adaptive Capacity of Australian Small-To-Medium Enterprises to Climate Change and Variability, National Climate Change Adaptation Research Facility, Gold Coast, Australia.

Lambeck, K., Rouby, H., Purcell, A., Sun, Y. and Sambridge, M. (2014) 'Sea level and global ice volumes from the last glacial maximum to the holocene', PNAS Earth, Atmospheric, and Planetary Sciences, vol. 111, pp. 15296–15303.

Langcake, S. (2016) 'Conditions in the Manufacturing Sector', Bulletin June Quarter 2016, Reserve Bank of Australia, Sydney, Australia, https://www.rba.gov. au/publications/bulletin/2016/jun/pdf/- bu-0616-4.pdf, accessed 22 October 2021.

Laurance, W.F., Dell, B., Turton, S.M., et al. (2011). 'The ten Australian ecosystems most vulnerable to tipping points', Biological Conservation, vol. 144, pp. 1472–1480.

Lawrence, J., Mackey, B., Chiew, B.F., et al. (2022) 'Australasia', in H.-O. Pörtner, D.C. Roberts, M. Tignor, et al. (eds) Climate Change 2022: Impacts, Adaptation, and Vulnerability. Contribution of Working Group II to the Sixth Assessment Report of the Intergovernmental Panel on Climate Change, Cambridge University Press, Cambridge, United Kingdom and New York.

Lee, J.-Y., Marotzke, J., Bala, G., et al. (2021) 'Future global climate: scenario-based projections and near-term information', in V.P. Masson-Delmotte, A. Pirani, S.L. Connors, et al. (eds) Climate Change 2021: The Physical Science Basis. Contribution of Working Group I to the Sixth Assessment Report of the Intergovernmental Panel on Climate Change, Cambridge University Press, Cambridge, United Kingdom and New York.

Lenzen, M., Sun, Y.Y., Faturay, F. et al. (2018) 'The carbon footprint of global tourism', Nature Climate Change, vol. 8, pp. 522–528.

Li, S., Wu, L., Yang, Y., et al. (2019) 'The Pacific Decadal Oscillation less predictable under greenhouse warming', Nature Climate Change, vol. 10, pp. 30–34.

Liedloff, A.C., Woodward, E.L., Harrington, G.A. and Jackson, S. (2013) 'Integrating indigenous ecological and scientific hydro-geological knowledge using a Bayesian Network in the context of water resource development', Journal of Hydrology, vol. 499, pp. 177–187.

Lim, E.-P., Hendon, H.H., Arblaster, J.M., et al. (2016) 'The impact of the Southern Annular Mode on future changes in Southern Hemisphere rainfall', Geophysical Research Letters, vol. 43, pp. 7160–7167.

Liu, P.R. and Raftery, A.E. (2021) 'Country-based rate of emissions reductions should increase by 80% beyond nationally determined contributions to meet the 2°C target', Communications Earth and Environment, vol. 2, p. 29.

Liverman, D.M. (2018) 'Geographic perspectives on development goals: constructive engagements and critical perspectives on the MDGs and the SDGs', Dialogues in Human Geography, vol. 8, pp. 168–185.

Loechel, B., Hodgkinson, J. and Moffat, K. (2013) 'Climate change adaptation in Australian mining communities: comparing mining company and local government views and activities', *Climatic Change*, vol. 119, pp. 465–477.

Loginova, J., and Batterbury, S. (2019) 'Incremental, transitional and transformational adaptation to climate change in resource extraction regions', *Global Sustainability*, vol. 2, E17.

Lucas, C., Booth, K.I. and Garcia, C. (2021) 'Insuring homes against extreme weather events: a systematic review of the research', *Climatic Change*, vol. 165, p. 61.

Ludt, W.B. and Rocha, L.A. (2014) 'Shifting seas: the impacts of Pleistocene sea-level fluctuations on the evolution of tropical marine taxa', Journal of Biogeography, vol. 42, pp. 25–38.

Lüthi, D., Le Floch, M., and Bereiter, B. (2008b) 'EPICA Dome C Ice Core 800KYr Carbon Dioxide Data', IGBP PAGES/World Data Center for Paleoclimatology Data Contribution Series # 2008–055, NOAA/NCDC Paleoclimatology Program, Boulder CO.

Lüthi, D., Le Floch, M., Bereiter, B., et al. (2008a) 'High-resolution carbon dioxide concentration record 650,000–800,000 years before present', Nature, vol. 453, pp. 379–382.

Lynas M., Houlton, B.J. and Perry, S. (2021) 'Greater than 99% consensus on human caused climate change in the peer-reviewed scientific literature', Environmental Research Letters, vol. 16, 114005.

Ma, S. and Kirilenko, A.P. (2019) 'Climate change and tourism in English-language newspaper publications', *Journal of Travel Research*, vol. 59, pp. 1–15.

Macintosh, A., Butler, D., Ansell, D. and Waschka, M. (2022) 'The Emissions Reduction Fund (ERF): problems and solutions', Australian National University, https://law.anu.edu.au/sites/all/files/erf_-_problems_and_solutions_final_6_april_2022.pdf, accessed 08/04/2022.

Madden, R.A. and Julian, P.R (1971) 'Detection of a 40–50 Day Oscillation in the Zonal Wind in the Tropical Pacific', *Journal of Atmospheric Sciences*, vol. 28, pp. 702–708.

Madeddu, S., Ueckerdt, F., Pehl, M. et al. (2020) 'The CO2 reduction potential for the European industry via direct electrification of heat supply (power-to-heat)', *Environmental Research Letters*, vol. 15, 124004.

Mannocchi, F., Todisco, F. and Vergni, L. (2004) 'Agricultural drought: indices, definition and analysis', The Basis of Civilization – Water Science? Proceedings of the UNESCO/IAHS/IWHA symposium held in Rome, December 2003, IAHS Publication 286.

Mantua, N.J. and Hare, S.R. (2002) 'The pacific decadal oscillation', Journal of Oceanography, vol. 58, pp. 35–44.

Mao, X and Retallack, G. (2019) 'Late Miocene drying of central Australia', Palaeogeography, Palaeoclimatology, Palaeoecology, vol. 514, pp. 292–304.

Marshall, N.A. (2015) 'Adaptive capacity on the northern Australian rangelands', *The Rangeland Journal*, vol. 37, pp. 617–622.

Martek, I. and Hosseini, M.R. (2019) 'Buildings produce 25% of Australia's emissions. What will it take to make them 'green' – and who'll pay?', *The Conversation*, https://theconversation.com/buildings-produce-25-of-australias-emissions-what-will-it-take-to-make-them-green-and-wholl-pay-105652, accessed 23 October 2021.

Martin, H.A. (2006) 'Cenozoic climatic change and the development of the arid vegetation in Australia', Journal of Arid Environments, vol. 66, pp. 533–563.

Martin, P., Randall, L. and Jackson, T. (2020) *Labour Use in Australian Agriculture, Research Report 20.20*, DAWE and ABARES, Australian Government, Canberra.

Mason, L. and Giurco, D. (2013) *Climate change adaptation for Australian minerals industry professionals, Final Report*, National Climate Change Adaptation Research Facility, Gold Coast Australia.

Mason, J. (2020) 'The history of climate science', Skeptical Science, https://skepticalscience.com/history-climate-science.html, accessed 23 August 2022.

Matthews, V., Atkinson, A.-R., Lee, G., Vine, K. and Longman, J. (2021) *Climate Change and Aboriginal and Torres Strait Islander Health, Climate Change and Aboriginal and Torres Strait Islander Health*, Discussion Paper, Lowitja Institute, Melbourne, Australia.

Mavrommatis, E., Damigos, D, and Mirasgedis, S. (2019) 'Towards a comprehensive framework for climate change multi-risk assessment in the mining industry', *Infrastructures*, vol. 4, p. 38.

McArthur, A.G. (1967) 'Fire Behaviour in Eucalypt Forests', Leaflet 107, Department of National Development Forestry and Timber Bureau, Canberra.

McRobert, K., Admassu, S., Fox, T. and Heath, R. (2019) *Change in the Air: Defining the Need for an Australian Agricultural Climate Change Strategy, Research Report*, Australian Farm Institute, Surry Hills, New South Wales, Australia.

McTernan, W.P., Dollard, M.F., Tuckey, M.R. and Vandenberg, R.J. (2016) 'Beneath the surface: an exploration of remoteness and work stress in the mines', in A. Shimazu, R. Bin Nordin, M. Dollard and J. Oakman, J. (eds) *Psychosocial Factors at Work in the Asia Pacific*, Springer, Cham, Switzerland.

Meat and Livestock Australia (2017) 2017 *Framework Report, Australian Beef Sustainability Framework*, Sydney, Australia, https://www.sustainableaustralianbeef. com.au/resources/news/carbon-neutral-announcement-supported-by-australian-beef-sustainability-framework/, accessed 12 October 2021.

Meert, J.G. (2011) 'Gondwanaland, formation', in J. Reitner and V. Thiel (eds) Encyclopedia of Geobiology. Encyclopedia of Earth Sciences Series, Springer, Dordrecht.

Meinshausen, M., Lewis, J., McGlade, C. et al. (2022) 'Realization of Paris agreement pledges may limit warming just below 2°C', Nature, vol. 604, pp. 304–309.

Mieczkowski, Z. (1985) 'The tourism climatic index: a method for evaluating world climates for tourism', *The Canadian Geographer*, vol. 29, pp. 220–233.

Millán, L., Santee, M.L., Lambert, A., et al. (2022) 'The Hunga Tonga-Hunga Ha'apai Hydration of the Stratosphere'. *Geophysical Research Letters*, vol. 49, e2022GL099381.

Minerals Council of Australia (2021) 'Land use', Minerals Council of Australia, Canberra, https://www.minerals.org.au/land-use, accessed 21 October 2021.

Mobsby, D. and Curtolli, R. (2018) Snapshot of Australia's Commercial Fisheries and Aquaculture, Department of Agriculture, Water and the Environment, Australian Government, Canberra.

Moggridge, B.J., Peci, G., Lansbury, N., Creamer, S. and Mosby, V. (2022) 'IPCC reports still exclude Indigenous voices. Come join us at our sacred fires to find answers to climate change', *The Conversation*, https://theconversation. com/ipcc-reports-still-exclude-indigenous-voices-come-join-us-at-our-sacred-fires-to-find-answers-to-climate-change-178045, accessed 20 March 2022.

Moran, C. and Turton, S.M. (2014) 'Adaptation pathways and opportunities for infrastructure', in C. Moran, S.M. Turton and R. Hill (eds), Adaptation Pathways and Opportunities for the Wet Tropics NRM Cluster Region (Volume 2), James Cook University, Cairns.

Morton, S. and Hill, R. (2014) 'What is biodiversity, and why is it important?' in S. Morton, M. Lonsdale and A. Sheppard (eds) Biodiversity: Science and Solutions for Australia, CSIRO Publishing, Melbourne.

Mountain, B. (2022) 'Labor's plan to green the Kurri Kurri gas power plant makes no sense', The Conversation, https://theconversation. com/labors-plan-to-green-the-kurri-kurri-gas-power-plant-makes-no-sense-176157, accessed 4 March 2022.

Moyle, C.J., Moyle, B.D., Chai, A., Hales, R., Banhalmi-Zakar, Z. and Bec, A. (2018) 'Have Australia's tourism strategies incorporated climate change?', *Journal of Sustainable Tourism*, vol. 26, pp. 703–721.

Müller, M. (2020) 'Some systems perspectives on demand management during Cape Town's 2015–2018 water crisis', *International Journal of Water Resources Development*, vol. 36, pp. 1054–1072.

Myers, N. (1988) 'Threatened biotas: 'hotspots' in tropical forests', Environmentalist, vol. 8, pp. 87–208.

Myers, N., Mittermeier, R., Mittermeier, C. et al. (2000) 'Biodiversity hotspots for conservation priorities', Nature, vol. 403, pp. 853–858.

National Climate Change Adaptation Research Facility (2013a) 'Adapting ecosystems to climate change', Policy Guidance Brief 8, NCCARF, Gold Coast, Australia.

National Climate Change Adaptation Research Facility (2013b) 'Ensuring business and industry are ready for climate change', Policy Guidance Brief 11, NCCARF, Gold Coast, Australia.

National Climate Change Adaptation Research Facility (2013c) 'Adapting agriculture to climate change', Policy Guidance Brief 4, NCCARF, Gold Coast, Australia.

National Climate Change Adaptation Research Facility (2016) 'Terrestrial ecosystems', Synthesis Summary 6, NCCARF, Gold Coast, Australia.

National Climate Change Adaptation Research Facility (2017) 'Marine biodiversity', Synthesis Summary 7, NCCARF, Gold Coast, Australia.

National Electricity Market (2022) 'Government Priorities', Australian Government, https://www.energy.gov.au/government-priorities/energy-markets/national-electricity-market-nem, accessed 3 March 2022.

National Farmers Federation (2020) Climate Change Policy, National Farmers Federation, Canberra, Australia, https://nff.org.au/wp-content/uploads/2020/08/2020.08.06_Policy_NRM_Climate_Change.pdf, accessed 12 October 2021.

Naughtin, C., Hajkowicz, S., Schleiger, E., Bratanova, A., Cameron, A., Zamin, T. and Dutta, A (2022) Our Future World: Global Megatrends Impacting the Way We Live Over Coming Decades, CSIRO, Brisbane, Australia.

Nelson, J. and Schuchard, R. (2011) Adapting to Climate Change: A Guide for the Mining Industry, Business for Social Responsibility, Industry Series, Asia-Pacific, accessed 21 October 2021.

Nguyen, P.-L., Min, S.-K. and Kim, Y.-H. (2021) 'Combined impacts of the El Niño-Southern Oscillation and Pacific Decadal Oscillation on global droughts assessed using the standardized precipitation evapotranspiration index', International Journal of Climatology, vol. 41, pp. E1645–E1662.

Nicholls, N., Lavery, B., Frederiksen, C., Drosdowsky, W. and Torok, S. (1996) 'Recent changes in relationships between the El Niño–Southern Oscillation and Australian rainfall and temperature', Geophysical Research Letters, vol. 3, pp. 3357–3360.

Norman, B. (2022) 'The floods have killed at least 21 Australians. Adapting to a harsher climate is now a life-or-death matter', The Conversation, https://theconversation.com/the-floods-have-killed-at-least-21-australians-adapting-to-a-harsher-climate-is-now-a-life-or-death-matter-178761, accessed 12 March 2022.

Nunn, P. and N. Reid, N. (2016) 'Aboriginal memories of inundation of the Australian coast dating from more than 7000 years ago', Australian Geographer, vol. 47, pp. 11–47.

O'Brien, J., Pradeep, P. and Rayward-Smith, W. (2020) Connecting the Green Transition: Connected Future | Future Connected, Deloitte Touche Tohmatsu, Sydney.

Odell, S.D, Bebbington, A. and Frey, K.E. (2018) 'Mining and climate change: A review and framework for analysis', The Extractive Industries and Society, vol. 5, pp. 201–214.

Oke, T.R. (1987) Boundary Layer Climates, Second Edition, Routledge, Taylor and Francis e-Library 2002, London, UK.

Olsson, L., Opondo, M., Tschakert, P. et al. (2014) 'Livelihoods and poverty', in C. B. Field, V. R. Barros, D. J. Dokken, K. J. Mach, M. D. Mastrandrea, T. E. Bilir, M. Chatterjee, K. L. Ebi, Y. O. Estrada, R. C. Genova, B. Girma, E. S. Kissel, A. N. Levy, S. MacCracken, P. R. Mastrandrea, and L. L. White (eds) Climate change 2014: impacts, adaptation and vulnerability. Part A: global and sectoral aspects. Contribution of working Group II to the Fifth Assessment Report of the Intergovernmental Panel on Climate Change. Cambridge University Press, Cambridge, UK.

Organisation for Economic Co-operation and Development (2021) 'Who we are', https://www.oecd.org/about/, accessed 12 August 2021.

Paice, R. and Chambers, J. (2016) 'Climate change adaptation planning for protection of coastal ecosystems', CoastAdapt Information Manual 10, National Climate Change Adaptation Research Facility, Gold Coast, Australia.

Parker, T.J., Berry, G.J. and Reeder, M.J. (2014) 'Variability in severe coastal flooding, associated storms, and death tolls in Southeastern Australia since the mid–nineteenth century', Journal of Climate, vol. 27, pp. 5768–578.

Parkinson, G. (2021) 'Monumental: why AEMO's stunning Hydrogen Superpower scenario can't be ruled out Hydrogen Superpower option', Renew Economy, https://reneweconomy.com.au/monumental-why-aemos-stunning-hydrogen-superpower-scenario-cant-be-ruled-out/, accessed 4 March 2022.

Pascoe, C., Lawrence, B.N., Guilyardi, E., Juckes, M. and Taylor, K.E. (2020) 'Documenting numerical experiments in support of the Coupled Model Intercomparison Project Phase 6 (CMIP6)', Geoscientific Model Development, vol. 13, pp. 2149–2167.

Peel, M.C., Finlayson, B.L. and McMahon, T.A. (2007) 'Updated world map of the Köppen-Geiger climate classification', Hydrology and Earth System Sciences, vol. 11, pp. 1633–1644.

Perkins, S.E. and Alexander, L.V. (2013) 'On the measurement of heat waves', Journal of Climate, vol. 26, pp. 4500–4517.

Pepler, A.S., Trewin, B. and Ganter, C. (2015) 'The influences of climate drivers on the Australian snow season', Australian Meteorological and Oceanographic Journal, vol. 65, pp. 195–205.

Petit J.R., Jouzel J., Raynaud D., et al. (1999) 'Climate and atmospheric history of the past 420,000 years from the Vostok Ice Core, Antarctica', Nature, vol. 399, pp. 29–436.

Phillips, J. (2016) 'Climate change and surface mining: A review of environment-human interactions & their spatial dynamics', Applied Geography, vol. 74, pp. 95–108.

Physick, W., Cope, M. and Lee, S. (2014) 'The impact of climate change on ozone-related mortality in Sydney', Environmental Research and Public Health, vol. 11, pp. 1034–1048.

Planning Institute of Australia, Queensland Division (2021) 'Planning to tackle climate change: 10 actions for a climate-conscious planning system in Australia', Planning Institute of Australia, Queensland Division, https://www.planning.org.au/documents/item/11362, accessed 24 March 2022.

Porfirio, L.L., Newth, D., Finnigan, J.J. and Cai, Y. (2018) 'Economic shifts in agricultural production and trade due to climate change', Palgrave Communications, vol. 4, p. 111.

Power, S.B. and Callaghan, J. (2016) 'Variability in severe coastal flooding, associated storms, and death tolls in southeastern australia since the mid–nineteenth century', Journal of Applied Meteorology and Climatology, vol. 55, pp. 1139–1149.

Price Waterhouse Coopers Australia (2019) 'Australian Findings', Price Waterhouse Coopers Australia, Southbank, Victoria, Australia, https://www.pwc.com.au/ceo-agenda/ceo-survey/2019.html, accessed 23 October 2021.

Ramieri, E., Hartley, A., Barbanti A., et al. (2011) 'Methods for assessing coastal vulnerability to climate change', ETC CCA Technical paper 1/2011, European Environment Agency.

Ramsar (2022) 'About the Ramsar Convention on Wetlands', https://www.ramsar.org/about-the-convention-on-wetlands-0, accessed 28 March 2022.

Reside, A.E, Ceccarelli, D.M, Isaac, J.L. et al. (2014) 'Biodiversity – adaptation pathways and opportunities', in C. Moran, S.M. Turton and R. Hill (eds), Adaptation Pathways and Opportunities for the Wet Tropics NRM Cluster Region (Volume 1), James Cook University, Cairns.

Riahi, K., van Vuuren, D.P., Kriegler, E. et al. (2017) 'The shared socioeconomic pathways and their energy, land use, and greenhouse gas emissions implications: an overview', Global Environmental Change, vol. 42, pp. 153–168.

Rice, M., Hughes, L., Steffen, W., et al. (2022) A Supercharged Climate: Rain bombs, Flash Flooding and Destruction, Climate Council of Australia Limited, Sydney.

Ridgway, K. and Hill, K. (2009) 'The East Australian current', in E.S Poloczanska, A.J. Hobday, and A.J. Richardson (eds) A Marine Climate Change Impacts and Adaptation Report Card for Australia 2009, National Climate Change Adaptation Research Facility Publication 05/09.

Rickards L. and Howden S.M. (2012) 'Transformational adaptation: agriculture and climate change', *Crop and Pasture Science*, vol. 63, pp. 240–250.

Röck, M., Ruschi, M., Saade, M. et al. (2020) 'Embodied GHG emissions of buildings – The hidden challenge for effective climate change mitigation', *Applied Energy*, vol. 258, 114107.

Roger, E., Duursma, D.E., Downey, P.O., Gallagher, R.V., Hughes, L., Steel, J., Johnson, S.B. and Leishman, M.R. (2015) 'A tool to assess potential for alien plant establishment and expansion under climate change', *Journal of Environmental Management*, vol. 159, pp. 121–127.

Rohling, E.J., Hibbert, F.D., Grant, K.M. et al. (2019) 'Asynchronous Antarctic and Greenland ice-volume contributions to the last interglacial sea-level highstand', Nature Communications, vol. 10, 5040.

Roque, B.M., Venegas, M., Kinley, R.D., et al. (2021) 'Red seaweed (Asparagopsis taxiformis) supplementation reduces enteric methane by over 80 percent in beef steers' PLoS One, vol. 16, e0247820.

Rosenbloom, D. (2017) 'Pathways: an emerging concept for the theory and governance of low-carbon transitions', Global Environmental Change, vol. 43, pp. 37–50.

Sadavarte, P., Pandey, S., Maasakkers, J.D. et al. (2021) 'Methane Emissions from Superemitting Coal Mines in Australia Quantified Using TROPOMI Satellite Observations', *Environmental Science and Technology*, vol. 55, pp. 16573–16580.

Saddler, H. (2021) *Back of the Pack: An Assessment of Australia's Energy Transition*, The Australia Institute, Canberra, Australia.

Sánchez, B., Rasmussen, A. and Porter, J.R. (2014) 'Temperatures and the growth and development of maize and rice: a review', *Global Change Biology*, vol. 20, pp. 408–417.

Scarano, F.R. (2017) 'Ecosystem-based adaptation to climate change: concept, scalability and a role for conservation science', Perspectives in Ecology and Conservation, vol. 15, pp. 65–73.

Scheiter, S., Higgins, S.I., Beringer, J. and Hutley, L.B. (2015) 'Climate change and long-term fire management impacts on Australian savannas', *New Phytologist*, vol. 205, pp. 1211–1226.

Schiller, B. (2010) The Micro Economy Today, McGraw-Hill/Irwin, New York.

Schipper, E.L.F., Revi, A., Preston, B.L. et al. (2022) 'Climate resilient development pathways', in H.-O. Pörtner, D.C. Roberts, M. Tignor, et al. (eds), Climate Change 2022: Impacts, Adaptation, and Vulnerability. Contribution of Working Group II to the Sixth Assessment Report of the Intergovernmental Panel on Climate Change, Cambridge University Press, Cambridge, Cambridge, United Kingdom and New York.

Schweinsberg, S., Darcy, S. and Beirman, D. (2020) 'Climate crisis' and 'bushfire disaster: Implications for tourism from the involvement of social media in the 2019–2020 Australian bushfires', *Journal of Hospitality and Tourism Management*, vol. 43, pp. 294–297.

Scott, D., Hall, C.M. and Gössling, S. (2019) 'Global tourism vulnerability to climate change', *Annals of Tourism Research*, vol. 77, pp. 49–61.

Shabani, F. and Kotey, B. (2016) 'Future distribution of cotton and wheat in Australia under potential climate change', *The Journal of Agricultural Science*, vol. 154, pp. 175–185.

Sharma, V., van de Graaff, S., Loechel, B., Franks, D.M. (2013) *Extractive Resource Development in a Changing Climate: Learning the Lessons from Extreme Weather Events in Queensland, Australia*, National Climate Change Adaptation Research Facility, Gold Coast, Australia.

Sharmila, S. and Walsh, K.J.E. (2018) 'Recent poleward shift of tropical cyclone formation linked to Hadley cell expansion', Nature Climate Change, vol. 8, pp. 730–736.

Sloan, L.C. and Rea, D.K. (1996) 'Atmospheric carbon dioxide and early Eocene climate: a general circulation modeling sensitivity study', Palaeogeography, Palaeoclimatology, Palaeoecology, vol. 119, pp. 275–292.

Smith, M.H. (2013) *Assessing climate change risks and opportunities for investors: mining and minerals processing sector*, Australian National University, Canberra, Australia.

Stafford Smith, M. and Ash, A. (2011) 'Adaptation: reducing risk, gaining opportunity', in H. Cleugh, M. Stafford Smith, M. Battaglia and P. Graham (eds), Climate Change:

Science and Solutions for Australia, CSIRO Publishing, Melbourne, Australia.

Stanke, C., Kerac, M., Prudhomme, C., Medlock, J. and Murray, V. (2013) 'Health effects of drought: a systematic review of the evidence' *PLoS Currents*, vol. 5, PMC3682759.

Steffen, W., Mallon, K., Kompas, T., Dean, A. and Rice, M. (2019) Compound Costs: How Climate Change is Damaging Australia's Economy, Climate Council of Australia Limited, Sydney.

Stock, P., Rice, M., Hughes, L., Steffen, W., Pearce, A., Hussey, K. and Flannery, T. (2017) *Local Leadership: Tracking Local Government Progress on Climate Change*, The Climate Council of Australia Limited, Sydney, Australia.

Stock, A., Stock, P., Bourne, G. and Brailsford, L. (2018a) Clean and Reliable Power: Roadmap to a Renewable Future, Climate Council of Australia Limited, Sydney.

Stock, P., Steffen, W., Bourne, G. and Brailsford, L. (2018b) Waiting for the Green Light: Transport Solutions to Climate Change, Climate Council of Australia Limited, Sydney.

Stokes, C.J. and Howden, S.M. (2010) 'Summary', in C. Stokes and M. Howden (eds) *Adapting Agriculture to Climate Change: Preparing Australian Agriculture, Forestry and Fisheries for the Future*, CSIRO Publishing, Melbourne, Australia.

Stork, N.E. and Turton, S.M. (eds) (2008) Living in a Dynamic Tropical Forest Landscape, Wiley-Blackwell Publishing, Oxford, United Kingdom.

Sturman, A.P. and Tapper, N.J. (2006) The Weather and Climate of Australia and New Zealand (2nd Edition), Oxford University Press, Melbourne, Australia.

Taschetto, A.S. and England, M.H. (2009) 'El Niño Modoki impacts on Australian rainfall', Journal of Climate, vol. 22, pp. 3167–3174.

Termeer, C.J.A.M., Art Dewulf, A. and Biesbroek, G.R. (2017) 'Transformational change: governance interventions for climate change adaptation from a continuous change perspective', Journal of Environmental Planning and Management, vol. 60, pp. 558–576.

The Australian Climate Roundtable (2021). 'Australian climate roundtable: joint principles for climate policy', https://www.australianclimateroundtable. org.au/wp-content/uploads/2020/11/ Climate_roundtable_joint_principles-Updated_November_2020.pdf, accessed 7 November 2021.

The National Construction Code of Australia (NCC) (2019) 'National Construction Code', Australian Building Codes Board, DISER, Australian Government, Canberra, https:// ncc.abcb.gov.au, accessed 23 October 2021.

The Planning Institute of Australia (2021) 'Planning in a changing climate: position statement', PIA Climate Series, Canberra, Australia.

Tourism Australia (2022) 'Sustainable Tourism', Tourism Australia, Sydney, Australia, https:// www.tourism.australia.com/en/about/our-organisation/sustainable-tourism-statement. html, accessed 3 February 2022.

Tourism Research Australia (2020) 'The Economic Importance of Tourism 2018-2019', Tourism Research Australia, Australian Government, Austrade, Canberra, https://www.tourism. australia.com/en/markets-and-stats/tourism-statistics/the-economic-importance-of-tourism.html, accessed 23 October 2021.

Turton, S.M., Hadwen, W. and Wilson, R. (Eds) (2009). *The Impacts of Climate Change on Australian Tourism Destinations: Developing Adaptation and Response Strategies – A Scoping Study*, Cooperative Research Centre for Sustainable Tourism, Gold Coast, Australia.

Turton, S., Dickson, T., Hadwen, W., Jorgensen, B., Pham, T., Simmons, D., Tremblay, P. and Wilson, R. (2010) 'Developing an approach for tourism climate change assessment: evidence from four contrasting Australian case studies', *Journal of Sustainable Tourism*, vol. 18, pp. 429–447.

Turton, S.M. (2012) 'Securing landscape resilience to tropical cyclones in Australia's wet tropics under a changing climate: Lessons from Cyclones Larry (and Yasi)', *Geographical Research*, vol. 50, pp. 15–30.

Turton, S.M. (2014) 'Climate change and rainforest tourism in Australia', in B. Prideaux (ed) *Rainforest Tourism, Conservation and Management: Challenges for Sustainable Development*, Routledge, New York.

Turton, S.M. (2017) 'Expansion of the tropics: revisiting frontiers of geographical knowledge', Geographical Research, vol. 55, pp. 3–12.

Turton, S.M. (2019a) 'Reef-to-ridge ecological perspectives of high-energy storm events in northeast Australia', Ecosphere, vol. 10, pp. 1–20.

Turton, S.M. (2019b) 'How climate change can make catastrophic weather systems linger for longer', The Conversation, https://theconversation.com/how-climate-change-can-make-catastrophic-weather-systems-linger-for-longer-111832, accessed 23 March 2022.

Turton, S.M. (2020) 'Ecosystems services and integrity trend', in W. Leal Filho, A. Azul, L. Brandli, A. Lange Salvia and T. Wall T (eds) Life on Land. Encyclopedia of the UN Sustainable Development Goals, Springer, Cham.

Turton, S.M. (2022) 'The '97% climate consensus' is over. Now it's well above 99% (and the evidence is even stronger than that)', The Conversation, https://theconversation.com/the-97-climate-consensus-is-over-now-its-well-above-99-and-the-evidence-is-even-stronger-than-that-170370, accessed 21 June 2022.

Uluru Statement (2017) The Uluru Statement from the Heart, The Uluru Statement, https://ulurustatement.org, accessed 30 March 2022.

United Nations (2015) 'Sustainable Development Goals'. https://www.un.org/sustainabledevelopment/sustainable-development-goals/, accessed 05 May 2021.

United Nations (2019) 'Sustainable Development', https://sdgs.un.org, accessed 05 May 2021.

United Nations (2020a) The Sustainable Development Goals Report, 2020, Department of Economic and Social Affairs, United Nations, Statistics Division, New York.

United Nations (2020b) 'Progress in the Process to Formulate and Implement National Adaptation Plans', Document FCCC/SBI/2020/INF.13, United Nations, New York.

United Nations (2021) 'Inter-Agency Task Force On Financing for Development, Financing for Sustainable Development Report 2021', United Nations, New York.

United Nations Environment Program (2021a) 'About the Montreal Protocol', https://www.unep.org/ozonaction/who-we-are/about-montreal-protocol, accessed 07 July 2021.

United Nations Environment Program (2021b) Emissions Gap Report 2021: The Heat Is On – A World of Climate Promises Not Yet Delivered, UNEP, Nairobi, Kenya.

United Nations Environment Program/GRID-Arendal (2005) 'The Greenhouse Effect', https://www.grida.no/resources/6888, accessed 07 July 2021.

United Nations Framework Convention on Climate Change (1992) 'United Nations Framework Convention on Climate Change', https://unfccc.int/files/essential_background/background_publications_htmlpdf/application/pdf/conveng.pdf, accessed 07 July 2021.

United Nations Framework Convention on Climate Change (2021a) 'The Doha Amendment', https://unfccc.int/process/the-kyoto-protocol/the-doha-amendment, accessed 09 August 2021.

United Nations Framework Convention on Climate Change (2021b) 'The Paris Agreement', https://unfccc.int/process-and-meetings/the-paris-agreement/the-paris-agreement, accessed 09 August 2021.

United Nations Human Rights, Office of the High Commissioner (1966a), 'International Covenant on Civil and Political Rights', General Assembly resolution 2200A (XXI), United Nations, New York.

United Nations Human Rights, Office of the High Commissioner (1966b) 'International Covenant on Economic, Social and Cultural Rights', General Assembly resolution 2200A (XXI), United Nations, New York

Valentine, P.S. (2019) World Heritage Sites of Australia, National Library of Australia, Canberra, Australia.

Vanclay, J. (2022) 'It's time to come clean on Lismore's future. People and businesses have to relocate away from the floodplains', The Conversation, https://theconversation.com/its-time-to-come-clean-on-lismores-future-people-and-businesses-have-to-relocate-away-from-the-floodplains-184636, accessed 15 June 2022.

Van der Heijden, J. (2018) 'From leaders to majority: a frontrunner paradox in built-environment climate governance experimentation', Journal of Environmental Planning and Management, vol. 61, pp. 1383–1401.

Vardoulakis, S., Jalaludin, B.B., Morgan, G.G., Hanigan, I.C. and Johnson, F.H. (2020) 'Bushfire smoke: Urgent need for a national health protection strategy', *Medical Journal of Australia*, vol. 212, pp. 349–353.

Varghese, B.M., Hansen, A.L., Williams, S. et al. (2020) 'Determinants of heat-related injuries in Australian workplaces: Perceptions of health and safety professionals', *Science of The Total Environment*, vol. 718, 137138.

Venkataraman, M., Csereklyei, Z., Aisbett, E., Rahbari, A., Jotzo, F., Lord, M. and Pye, J. (2022) 'Zero-carbon steel production: The opportunities and role for Australia', *Energy Policy*, vol. 163, 112811.

Vicedo-Cabrera, A.M., Scovronick, N., Sera, F. et al. (2020) 'The burden of heat-related mortality attributable to recent human-induced climate change', *Nature Climate Change*, vol. 11, pp. 492–500.

Vogado N.O., Engert J.E., Linde T.L., Campbell M.J., Laurance W.F. and Liddell M.J. (2022) 'Climate change affects reproductive phenology in lianas of Australia's Wet Tropics', Frontiers in Forests and Global Change, vol. 5, p. 787950.

Wang, G. and Hendon, H.H. (2007) 'Sensitivity of Australian rainfall to inter–El Niño variations', Journal of Climate, vol. 20, pp. 4211–4226.

Wang, G., Cai, W., Santoso, A. et al. (2022) 'Future Southern Ocean warming linked to projected ENSO variability', *Nature Climate Change*, vol. 12, pp. 649–654.

Wardell-Johnson, G.W., Calver, M., Burrows, N. and Di Virgilio, G. (2015) 'Integrating rehabilitation, restoration and conservation for a sustainable jarrah forest future during climate disruption', *Pacific Conservation Biology*, vol. 21, pp. 175–185.

Waters, C.N., Zalasiewicz, J., Summerhayes, C. et al. (2016) 'The Anthropocene is functionally and stratigraphically distinct from the Holocene', Science, vol. 351, p. aad2622.

Weart, S.R. (2008). *The Discovery of Global Warming*, Second Edition, Harvard University Press, Cambridge, MA.

West, J. and Brereton, D. (2013). *Climate Change Adaptation in Industry and Business: Framework for Best Practice in Financial Risk Assessment, Governance and Disclosure*, National Climate Change Adaptation Research Facility, Gold Coast, Australia.

White, I., Wade, A., Worthy, M., Mueller, N., Daniell, T. and Wasson, R. (2006) 'The vulnerability of water supply catchments to bushfires: impacts of the January 2003 wildfires on the Australian Capital Territory', *Australasian Journal of Water Resources*, vol. 10, pp. 179–194.

Wise, R.M, Fazey, I., Stafford Smith, M., Park, S.E., Eakin, H.C., Archer Van Garderen, E.R.M. and Campbell, B. (2014) 'Reconceptualising adaptation to climate change as part of pathways of change and response', *Global Environmental Change*, vol. 28, pp. 325–336.

Williams, A., White, N., Mushtaq, S. et al. (2015) 'Quantifying the response of cotton production in eastern Australia to climate change', *Climatic Change*, vol. 129, pp. 183–196.

Williams, K.J., Ford, A. Rosauer, D.F. et al. (2011) 'Forests of East Australia: the 35th biodiversity hotspot', in F. Zachos and J. Habel (eds) Biodiversity Hotspots, Springer, Berlin, Heidelberg.

Williams, S. E., Falconi L., Lowe A. et al. (2017) National Climate Change Adaptation Research Plan for terrestrial biodiversity: Update 2017, National Climate Change Adaptation Research Facility, Gold Coast.

Wise, R.M, Fazey, I., Stafford Smith, M., et al. (2014) 'Reconceptualising adaptation to climate change as part of pathways of change and response', Global Environmental Change, vol. 28, pp. 325–336.

Wolfe, J. (2020) 'Volcanoes and Climate Change', https://www.earthdata.nasa.gov/learn/sensing-our-planet/volcanoes-and-climate-change, accessed 07 July 2021.

Wood, T., Reeve, A. and Ha. J. (2021) Towards Net Zero: A Practical Plan for Australia's Governments, Grattan Institute, Melbourne.

World Bank (2020) 'Manufacturing, value added (% of GDP) – Australia', World Bank, Washington, DC, https://data.worldbank.org/indicator/NV.IND.MANF.ZS?end=2021&locations=AU&start=2004, accessed 23 October 2021.

World Meteorological Organization (2019) 'WMO Greenhouse Gas Bulletin', https://ig3is.wmo.int/en/news/wmo-greenhouse-gas-bulletin-2019, accessed 07 July 2021.

Zhang, Y., Wallace, J.M. and Battisti, D.S. (1997) 'ENSO-like interdecadal variability', Journal of Climate, vol. 10, pp. 1004–1020.

Index

A

acid sulphate soils *see* land
adaptation *see* climate
adaptation pathways *see* climate, adaptation
adaptive capacity 59, 67, 76, 88, 98, 112, 118, 123, 139, 149, 153, 154, 172, 181, 206
AEMO *see* Australian Energy Market Operator
aerosols *see* forcings, volcanic
agri-business *see* agriculture
agriculture xii, 1, 10, 11, 12–14, 42, 50, 52, 58–59, 64, 65, 67, 72, 73, 74, 85, 90, 91, 95, 107, 113, 116, 117, 119, 121–135, 136, 147, 150, 162, 180, 185, 186, 202–203, 205, 206
  climate impacts and risks 122–127
  climate-resilient development pathways 127–135, 202
    adaptation pathways 133–135, 202
    mitigation pathways 130–133
  emissions 14, 127–128, 130, 205
  employment 121
  exports 121
  field crops 121–123, 125
    cotton 125, 127
    grains (cereals) 125
    maize 126
    pasture 125
    rice 12, 121, 125–126
    sugarcane 121, 123, 126–127
    wheat 121, 125–126
  footprint 119
  frost 42, 48, 125, 126–127
  geographical distribution 127
  grazing 12, 84, 88, 96, 99, 121, 125, 128, 130, 132, 205
  groundwater management 86, 91, 101, 124, 126, 134, 140, 158, 184
  horticulture 107, 121, 126, 129, 134, 137
    mangoes 126
    tree crops 123–124, 126
  irrigation 17, 86, 91, 124, 125, 126, 131, 133–134
  livestock 88, 121, 126, 130, 134, 205
    dairy 65, 100, 121, 126, 127, 131, 134
  managing risk 121–122
  no regrets practices 99, 128, 131, 133, 202, 205–206
  production 121
  relocation 127, 133
  renewables 134
  security 59, 202
  viticulture 121, 126–127, 128, 134, 137
  yields 60, 105, 125, 127
alpine zones *see* ecosystems, terrestrial
aluminium ore *see* mining, metals and minerals
aluminium production *see* manufacturing
amphibians *see* biodiversity
animals *see* biodiversity; biosecurity

Annex B countries *see* Kyoto Protocol
anthropocene 1–2, 20–21
anthropogenic forcing *see* forcings
aquaculture *see* fisheries
ARENA *see* Australian Renewable Energy Agency
Australian Energy Market Operator (AEMO) 112, 113–114
Australian Renewable Energy Agency (ARENA) 116, 132, 143
autonomous adaptation *see* climate, adaptation
aviation *see* transportation

B

barriers to climate adoption *see* climate, adaptation
bauxite *see* aluminium ore
biodiversity xii, 6, 58, 59, 65, 73, 75, 83–84, 85, 86, 87, 88, 90, 93, 94, 96, 97, 98, 99, 100–101, 102, 103, 134, 136–137, 153, 158, 182, 183, 184, 202, 206
  amphibians 83
  animals 96
  birds 83, 93
  climate impacts and risks 88, 90, 93
  climate-resilient development pathways 98–99, 202
    adaptation strategies 98, 100–101, 202
    mitigation strategies 98, 100–101

declines 88, 90, 206

endemic 83–84, 86, 92, 94, 100, 160

extinctions 39, 83, 93, 94–95, 96, 98, 160–161, 206

fishes 86, 93, 97

geographical distribution 88, 92, 97

hotspots 83–86

forests of East Australia 85–86

Southwest Australia 84–85, 91

introduced grazing animals (*see* agriculture)

invertebrates 97

loss 88, 90, 98

mammals 83, 93

plants 83, 93

reptiles 83, 96

sea urchins 96, 97, 101, 203

species

aquatic 86, 90, 93, 94–95

endangered 85

invasive (*see* biosecurity)

marine 28–29, 86, 96, 97

range shifts 28–29, 85, 88, 90, 92, 97

rare 100

recovery plans 99–100

threatened 84, 85, 92, 94, 99, 100, 184

translocation 99, 100–101

vulnerable 96, 99

values 83, 134

biosecurity 83, 93, 98–99, 122, 123, 126, 134

animals 85, 86, 90

cattle ticks 126

climate impacts 84

diseases 84, 85, 90, 98, 126, 129

pathogens 84, 86, 90, 99, 122, 125, 127

pests 85, 90, 95, 100, 125–126

plants 84, 86, 90, 95, 98

woody weeds 90, 100, 126

birds *see* biodiversity

black coal *see* mining, metals and minerals

blue carbon *see* carbon

blue hydrogen *see* energy, hydrogen

brown coal *see* energy

built environments 1, 54, 58, 61, 73, 75, 77, 109, 111, 147, 161, 172, 204–205, 207

cities 1, 46, 54, 58, 61, 73, 75, 77, 91, 109, 111, 147, 169–172, 177, 207

capital 46, 109, 169–171, 174–175

regional 109, 147, 169–171

climate impacts, hazards and risks 172–178, 184–185, 186, 193–195

climate-resilient development pathways 186–192, 193–195

adaptation pathways 188–192, 193–195

mitigation pathways 186–188

critical infrastructure 58–59, 105, 111–112, 152, 172, 175, 176, 180, 202

emissions 14, 150, 152, 186, 205

local government areas (LGAs) 54, 64, 116, 154, 172, 186–192, 195, 205

remote communities 141, 147, 171

settlements (communities) 54, 58, 73, 109, 111, 172, 177, 193–195, 207

coastal 59, 86, 93, 97–98, 103, 169–172, 177–178, 180, 188–189

regional 122, 147, 171

rural 122

urban environments 174–175, 177, 207

heat island effect 175

vulnerability 172, 175, 177–178, 207

exposure 172, 175, 178, 207

bushfires *see* fire

business 66, 78, 106–107, 112–114, 116, 118–119, 122, 139–140, 141, 143, 146–147, 149–150, 152–154, 156, 159, 162–164, 189, 204

C

carbon

abatement 70, 73, 78, 106, 128, 130, 143, 150; *see also* carbon, offsets

blue 143–144

capture and storage (CCS) 78, 143–145

capture use and storage (CCUS) 78, 144–145, 151

credits 70, 78, 131

cycle 3, 9, 33

dioxide (CO_2) concentrations 1, 4, 19–20, 26–27, 39

dioxide (CO_2) emissions 1–4, 11–14, 26–27, 31–32

dioxide equivalent (CO_2-e) emissions 11–14, 143 (*see also* greenhouse, gases)

dioxide (CO_2) fertiliser effect 88, 97, 98, 125, 141

farming 78, 134, 137, 205

footprint 151, 159, 162, 186

markets 75, 78, 134

offsets 78, 106, 139, 205 (*see also* carbon, abatement)

price 116, 139

sequestration 10, 97, 103, 128, 130, 134, 162, 184, 205

sinks 11–14, 61–62, 130–131, 143

land 12–13, 33

ocean 12–13, 33

sources 61–62

cattle ticks *see* biosecurity, pests

CCS *see* carbon

CCUS *see* carbon

cement production *see* manufacturing

chemical production *see* manufacturing

circular economy *see* economy

cities *see* built environments

Clean Energy Regulator 78, 108, 162

renewable energy target 108, 113

safeguard crediting mechanism 143

safeguard mechanism 78, 106, 119, 143, 151, 205

clean hydrogen *see* blue hydrogen; green hydrogen
climate
 adaptation 57–58, 206–207
  autonomous 1–2, 58–59, 60, 61, 206
  barriers 64–67
  enablers 64–67
  incremental 61, 63, 206–207
  maladaptation 58, 60, 61, 63–64, 67
  pathways 62–64
  planned 1–2, 58, 60–61
  strategies xii, 61, 65, 207
  structural 61, 113
  transformational (transformative) 61, 63–64, 206–207
 analogues 21
 change 57
  anthropogenic 4–5, 9, 57–58
  hotspot xi, 202
 crisis xii
 driven events (*see* extreme events)
 drivers (*see* forcings)
 hazards 60
 impacts
  cultural 59
  environmental 59
  economic 59
  negative 59
  observed 61
  positive 59
  social 59
  potential 58, 61
 mitigation 57–58, 61–62, 207
  mal-mitigation 58, 62, 67
  pathways 62, 63
  strategies xii, 73, 207
 models 30–31, 51–51
 -resilient development pathways xii, 5, 62–63, 65–68, 207
 risks 60
 scenarios 62
 types 45–47, 52, 159
 variability 1–2, 9, 15–17, 57
 vulnerability xi, 1–2, 58, 60
  exposure 58, 59
  sensitivity 58, 59

climate-resilient development pathways *see* climate
climate smart buildings *see* construction
coal *see* energy, non-renewable; mining, metals and minerals
coastal *see* built environments; ecosystems
construction 1, 58, 66, 74, 107, 112, 116, 140, 142, 145, 147–153, 155, 159, 186, 187, 189, 191, 203–205, 206
 biodiversity enablement 153, 188 (*see also* biodiversity)
 carbon intensity 150, 152, 204–205
 climate impacts and risks 149–150
 climate-resilient development pathways 150–153
  adaptation pathways 152–153
  mitigation pathways 150–152
 climate smart buildings 152, 186, 188, 189, 205
 decommissioning 152
 emissions 150, 152, 204–205
 employment 147
 green buildings 61, 152–153, 188, 189, 205
 managing risk 149, 152
 materials 149, 152–153
 recycling 152, 205
 renewables 152, 204–205
 sustainable buildings 152–153, 205
cooling *see* climate; temperature
coping
 capacity 58, 60
 ranges 1–2, 58–59
CO_2 *see* carbon (dioxide)
CO_2-e *see* carbon, dioxide equivalent
coral reefs *see* ecosystems, marine
cotton *see* agriculture, field crops
critical infrastructure *see* built environments
critical minerals *see* mining, metals and minerals

cropping *see* agriculture
crops *see* agriculture
cryosphere 28, 35
 ice sheets 28, 34, 35
 land glaciers 28, 35
 sea ice 28, 35
cyclones
 extra-tropical 29–30, 33–34, 35
 tropical 29, 35, 42–43, 50, 53–54, 93, 97, 111, 123, 139–140, 149

D

Development of Northern Australia Agenda 124–125, 132–133, 141, 150, 174–175, 206–207
diseases *see* biosecurity; human health
Doha Amendment *see* Kyoto Protocol
droughts xi, 2, 5, 28, 30, 32, 35, 41, 51, 53, 88, 90, 91, 92–93, 94–95, 111, 122, 126, 127, 137, 140, 175, 180, 181, 183, 185, 190, 195, 201, 202, 203
 agricultural 27, 35, 50–51, 53, 122, 180, 201
 ecological 28, 35, 50–51, 53, 90, 92, 93, 201
 hydrological 35, 91, 149
 meteorological 35, 53

E

East Australian Current 49, 52–53, 86, 96, 97, 129, 201–202
east coast lows 44, 54, 97, 149, 158, 178, 185, 203; *see also* cyclones, extra-tropical
economic systems *see* agriculture; construction; energy; land; mining; manufacturing; transportation; tourism
economy 105, 144, 149, 154, 185, 202, 206
 circular 144, 152, 205

ecosystems 13, 29, 42, 50, 52, 58–59, 65, 66, 68, 72, 79, 83–84, 85, 86, 87, 99, 100–101, 102, 202, 206
  climate impacts and risks 88, 93, 202
    coastal 88, 93
    estuarine 88, 93, 97
    freshwater 88
    marine 88, 93
    terrestrial 88
  climate-resilient development pathways 88, 98–103, 202
    adaptation strategies 98, 100–102, 202
    mitigation strategies 98, 100–102
  coastal 58, 86, 88, 93, 96–97, 102, 178
    mangroves 61, 65, 92, 97, 102, 103, 143, 178
  estuarine 58, 97, 101, 178
  freshwater 58, 88, 90–91, 93–95, 101, 176, 178 (see also hydrology)
    Ramsar sites 86, 90, 92, 122
    streams 86, 88, 90, 91, 101
    wetlands 33, 61, 86, 87, 88, 90, 91, 95–96, 97, 102, 122, 161
  marine 13, 29, 72, 87–88, 93, 95–98, 100–101
    coral reefs xi-xii, 49, 53, 87, 90, 95–96, 98, 99, 101, 164, 202
    kelp forests 49, 53, 96, 97, 101, 143, 202
    seagrass meadows 49, 96, 97–98, 99, 143, 161, 178
  services 6, 28, 58, 59, 73, 75, 84, 86, 87, 88, 93, 97–99, 100–102, 103, 123, 158, 186, 202, 206
  terrestrial 58, 88, 90, 92–95, 100
    alpine 92, 94, 100, 202
    forests (sclerophyll) xi, 39, 84, 85–86, 87, 89, 93, 100, 102, 129, 137, 156, 185, 202
    rainforests 39, 60, 84, 86, 89, 93, 94, 100, 156, 160–161

    shrublands 84, 86
    woodlands 84, 89, 90, 92
  threatened 84, 90, 99, 190, 206
  transformation 99
effective radiative forcing (ERF) see forcings
electric vehicles (EVs) see energy, electricity
electricity see energy
electricity transmission grid see energy
El Niño-Southern Oscillation (ENSO) 1, 15, 18, 34, 41–43, 48, 51, 52
  El Niño 14, 18–19, 41–43, 44, 48, 49, 50, 96
  La Niña 18–19, 41–43, 44, 48, 50, 96
  Walker Circulation 18–19, 41
emissions; see also carbon; greenhouse gases (GHGs); Paris Agreement
  net-zero 72, 77–78, 79, 106–107, 113, 132, 138–139, 149, 153, 164, 186, 206–207
  Reduction Fund (ERF) 78, 128, 132, 143, 162, 205
  targets 77, 204
    2030 77, 106, 204
    2050 77, 106, 113, 128, 138–139, 142, 149, 186, 204–205, 206
Emissions Reduction Fund (ERF) see emissions
enablers to climate adoption see climate, adaptation
endemic see biodiversity
energy 13–14, 58, 105, 107, 120, 139, 142, 149, 159, 204–205, 206
  budget (see radiation balance)
  climate impacts and risks 111–112, 175
  climate-resilient development pathways 112–118
    adaptation pathways 117
    mitigation pathways 113–115
  consumption 108
  efficiency 73, 114, 117, 120, 114, 148, 155, 162, 186, 187, 204

electricity 13–14, 72–73, 107–109, 115, 204
  micro-grid 151, 163
  powered vehicles (EVs) 115–116, 143, 151, 162, 188, 204
  supply 113, 115, 143, 175, 204
  transmission grid 108, 109, 110–112, 113, 114–115, 116–117, 150, 162, 204
  emissions 14, 112–113, 150
  employment 110
  hydrogen 73, 78, 114, 116, 117, 139, 144, 145, 151–152, 156, 162, 204, 205–206
    blue 78, 114, 151
    clean 78, 116, 151
    fuel cell 116
    green 78, 114, 116–117, 139, 143–144, 151, 156, 162, 204–206
    powered vehicles 117, 143, 156, 162, 204
  managing risk 110–111
  non-renewable 144, 150, 162 (see also mining)
    black coal 73, 91, 108, 110, 111, 113–115, 135, 138, 143, 175
    brown coal 108, 111, 113, 175
    diesel 14, 72, 108, 111, 113, 114, 132, 139, 140, 162
    gas (natural) 12, 13, 108, 110, 111, 113–114, 145, 150, 152
    oil 14, 108, 117, 144
    petroleum 14, 111, 112, 113, 147, 149, 150
  production 108, 113, 117, 124, 134, 143
  renewable 108, 113, 117, 144, 150, 152, 186, 187–188, 204–206
    battery storage 78, 110, 114–117, 145, 156, 194, 204
    biomass (biofuels) 73, 114, 163
    hydro 86, 91, 112, 114, 115–117, 144, 162, 204
    solar photovoltaic (PV) 108, 114–115, 117, 144, 151, 162–163, 187, 204

wind 108–109, 110, 112, 113–114, 115, 117, 125, 144, 145, 151, 162, 204
security 59, 194
trade 108, 109
use 108, 120, 136, 148, 155, 157, 165, 172, 204
ENSO *see* El Niño-Southern Oscillation
ERF, Effective radiative forcing *see* forcings
ERF *see* emissions
EVs *see* electric vehicles
extinction *see* biodiversity
extractive industries *see* mining
extra-tropical cyclones *see* cyclones
extreme events 1–2, 34–35, 52–53, 88, 111, 172, 178; *see also* droughts; fires; floods
cold days 29
compound 30, 35, 50–51, 53, 92, 178, 184–185
heatwaves 2, 34–35, 42, 49–51, 54, 88, 93, 111, 122, 139–141, 149, 174–175
marine xii, 29, 35, 50, 159, 201
terrestrial 34, 50, 54, 152, 159, 179, 182, 201
hot days 29, 34–35, 42, 54, 157–158, 174–175
storm surges 2, 53, 54, 93, 96, 149, 158–159, 177–178
weather 52, 140–141, 149, 152–153, 157, 163, 172, 182, 201

F

fertiliser production *see* manufacturing
FFDI *see* fire
fibre *see* agriculture, field crops
finance 58; *see also* insurance
fire
bushfires xi, 2, 43, 49, 51, 53, 54, 88–89, 90–93, 100, 111, 121, 139–141, 149, 163, 179–180, 202
Forest Fire Danger Index (FFDI) 49, 54, 89, 176

free-refugia 100
Indigenous management 184, 186 (*see also* Indigenous, Caring for Country)
management 99, 130, 184, 186
management plans 146
regimes 86, 88, 90, 99
weather 6, 30, 43, 49, 51–52, 53, 89, 90, 100, 152, 159, 176, 201
fisheries 29, 58, 86–87, 97, 103, 107, 121, 127–128, 129–130
aquaculture 86–87, 127, 129
commercial 86–87, 93, 98, 127, 129–130
fishes *see* biodiversity
floods xii, 2, 5, 33, 42, 53, 65, 87, 88, 90, 91, 93, 111, 112, 118, 122, 123, 127, 137, 139, 140, 141, 149, 152, 175–176, 179, 180–181, 182, 183, 185, 189, 193, 202, 203, 207
coastal 50, 53, 65, 93, 177–178, 181
flash 53, 175, 178, 180–181
riverine 53, 65, 93, 175, 178, 180–181, 207
food and beverage industries (*see* manufacturing)
food security (*see* agriculture)
forestry 14, 58–59, 70, 72, 73, 74, 107, 121, 127–128, 129–130
native forests 127, 129
plantations 127, 129–130
reforestation 143, 162
forests *see* ecosystems, terrestrial
forcings
aerosols 10, 20, 27, 30, 32
anthropogenic 6, 9, 15, 19–20, 32, 37, 57
effective radiative (ERF) 17–18, 20
internal 15, 18–19, 51–52
natural 1, 15, 40, 57
orbital (Milankovitch cycles) 1–3, 15–17, 21, 24, 39
radiative (RF) 15, 19, 24, 30, 123
solar 1, 15, 17

volcanic 1, 11, 15, 18, 20
aerosols 11, 17, 18, 20
Forest Fire Danger Index (FFDI) *see* fires
fossil fuels *see* energy, non-renewable
freshwater *see* ecosystems; hydrology

G

gas *see* energy, non-renewable
GBR *see* Great Barrier Reef
GCM *see* climate, models
GDP *see* gross domestic product
general circulation model (GCM) *see* climate, models
GHGs *see* greenhouse, gases
glaciers *see* cryosphere
global cooling *see* climate; temperature
global heating *see* climate; temperature
global warming *see* climate; temperature
GMST *see* temperature
gondwanaland 37, 39
governance *see* institutions and governance systems
grains *see* agriculture, field crops
grazing *see* agriculture
Great Australia Bight 86
Great Barrier Reef (GBR) xi, 49, 86, 95–96, 98, 156, 159–60, 163, 169, 202 *see also* World heritage sites
green buildings *see* construction
green hydrogen *see* energy, hydrogen
green steel *see* manufacturing
greenhouse
effect 2–3, 5, 9–11
gases (GHGs) 1–4, 9–14, 17, 19–20, 23, 24, 26–27, 30, 31, 57, 61, 69, 98, 145, 172, 186, 187, 205
gross domestic product (GDP) 105, 121, 138, 147, 154
GSAT *see* temperature

H

habitats
  climate refugia 39, 98, 100–101
  connectivity 60, 90, 98, 100–101
  degradation 65, 66, 75, 84, 87,
      88, 94, 95, 100–101, 102,
      182, 183, 202, 205
  destruction 83, 88, 90, 93, 205
  fragmentation 83, 86, 88, 90
  provision 99, 103
  restoration 99, 103, 141, 162
  revegetation 99, 141, 143, 162,
      205
heat island effect *see* built
      environments, urban
      environments
heating *see* climate; temperature
heatwaves *see* extreme events
holocene 1–2, 11, 21, 23–24, 40, 181
horticulture *see* agriculture
human health 58, 59, 123, 141, 145,
      172, 178–179, 180–181,
      183, 192, 193–194,
      195–197, 203
  climate impacts, hazards
      and risks 141, 178–181,
      184–185, 193–195, 202
    morbidity 178–179, 180, 181,
      202
    mortality 178, 179, 180, 181, 202
  climate-resilient development
      pathways 186–192,
      193–195
    adaptation pathways
      188–192, 193–195
    mitigation pathways 186–188
  diseases 158, 179, 180, 182, 183
  emergency management and
      services 78, 89, 146, 179,
      183, 189, 193–194
  pathogens 179, 180
  pollutants 179, 180
  vulnerability 178–179, 181, 190,
      192
    exposure 178–179, 181, 192
  well-being 58, 59, 68, 75, 83,
      87, 103, 123, 141, 157,
      172, 175, 178–181, 182,
      183, 184–185, 190, 192,
      193–195, 196–197, 203

hydrogen *see* energy
hydrogen-powered vehicles *see*
      energy, hydrogen
hydrology
  freshwater 6, 49, 53, 58, 86,
      88–90, 91, 92–93, 94–95,
      96, 99, 100–102, 111, 120,
      121, 136, 139, 148, 155,
      159, 161, 165, 176, 178, 195
    acidity (acidification) 88, 91,
      93, 123
    deoxygenation 88, 93, 97
    quality 87, 90, 91, 95, 97, 120,
      131, 136, 141, 148, 155,
      159, 165
    salinity 90, 97, 125
    security (scarcity) 59, 91, 123,
      127, 132, 141, 149, 158,
      175–176, 179, 180, 182,
      184, 190, 195
    temperature 88, 90, 91, 95, 97
    use 91, 120, 125, 130,
      140–141, 136, 144, 146,
      148, 155, 165
  hydrodynamics 90, 93, 97,
      175–176
  streamflow 34, 49, 53, 90, 91, 93,
      95–96, 175–176, 184, 195
  water cycle 32, 33–34, 91

I

ice sheets *see* cryosphere
ice cores 3–4, 15–16, 19, 21, 39
impacts *see* climate
incremental adaptation *see* climate,
      adaptation
Indian Ocean Dipole (IOD) 43–44,
      48, 49, 51
Indigenous
  climate impacts and risks 175,
      179, 181, 182–184, 192,
      193–195
  climate-resilient development
      pathways 184, 192,
      193–195
    climate adaptation 184, 192,
      193–195
    climate mitigation 184, 186
  Caring for Country 181, 182,
      184, 186, 189, 206

  cultural flows 184 (*see also*
      hydrology, freshwater)
  land management 98, 181,
      184, 186, 206 (*see also* fire)
  knowledge 98, 181, 184, 186,
      206
  stewardship 181–182, 184,
      186, 206
  communities 58, 65, 67, 125, 141,
      146–147, 171–172, 179,
      181–182, 184, 189–190
  fire management (*see* fire)
  health 172, 181–182 (*see also*
      human health)
  peoples 67, 84, 94, 141,
      172, 186
    Aboriginal and Torres Strait
      Islander 171, 181–182,
      183–184, 186, 192, 196
  population 171–172
  rights 182
  services 172
  Uluru Statement 181
  vulnerability 172, 175, 182
  well-being 172, 181–184, 190
      (*see also* human health,
      well-being)
Industrial revolution 1, 5
infrastructure *see* built
      environment
institutions and governance
      systems 58–59, 122, 149,
      185–186, 202
insurance 58, 66, 105, 112, 117, 122,
      140, 146, 149, 172–174,
      175, 176, 178–179,
      180–181, 185, 190
Intergovernmental Panel on
      Climate Change (IPCC)
      68–69, 202
  assessment reports xi, 3–4, 24,
      26, 69, 202, 206
Intergovernmental Science-Policy
      Platform on Biodiversity
      and Ecosystem Services
      (IPBES) 6, 57, 75
internal forcing *see* forcings
invasives *see* biosecurity
invertebrates *see* biodiversity;
      biosecurity, pests
IOD *see* Indian Ocean Dipole

iron ore *see* mining, metals and minerals
irrigation *see* agriculture

K

Kakadu National Park 83, 90, 92, 95, 156, 157, 160–161, 183; *see also* World heritage sites
Keeling curve 3–4, 9, 19–20
kelp forests *see* ecosystems, marine
Kyoto Protocol 12, 14, 69–71, 74, 128
    Annex B countries 69
    Doha Amendment 69
    market-based mechanisms (MBMs) 70–71

L

land 119, 136–137, 205–206
    acid sulphate soils 123 (*see also* soils, degradation)
    climate impacts and risks 122–127
    climate-resilient development pathways 127–135
        adaptation pathways 133–135
        mitigation pathways 130–133
    degradation 122–123, 132, 205 (*see also* soils, degradation)
    emissions 128, 205
    natural capital 59, 102, 123, 126, 132, 136
    salinisation 123
    saltwater intrusion 65, 90, 95, 127, 161, 183
    sustainable management 119, 128, 132–133, 206
    use and change 5, 14, 17, 30, 32, 57, 65, 66, 70, 72, 73, 74, 100, 119, 128, 133–134, 153, 187, 188, 189, 190, 205
La Niña *see* El Niño-Southern Oscillation (ENSO)
LGA *see* local government area
land glaciers *see* cryosphere
lithium *see* mining, metals and minerals
livestock *see* agriculture

local government area *see* built environments
Long-Term Emissions Reduction Plan (Australia) 77–78, 113, 151

M

Madden-Julian Oscillation (MJO) 44–45
maize *see* agriculture, field crops
maladaptation *see* climate, adaptation
mal-mitigation *see* climate, mitigation
mammals *see* biodiversity
mangoes *see* agriculture, horticulture
mangroves *see* ecosystems, coastal
manufacturing 13, 58, 74, 78, 107, 111–112, 113, 117, 119, 121, 142–143, 147, 149–152, 153, 155, 159, 172, 186, 198, 203, 204–206
    aluminium 13, 152, 205–206
    cement 78, 140, 145, 150, 151, 205
    chemicals 147, 149, 150, 151
    climate impacts and risks 149–150
    climate-resilient development pathways 150–153
        adaptation pathways 152–153
        mitigation pathways 150–152
    emissions 14, 150, 205
    employment 147
    equipment and machinery 147
    fertiliser 12, 78, 131, 132, 145, 150, 205
    food and beverage 121, 147, 149
    managing risk 147, 149, 152
    metal products 147
    steel 139, 140, 145, 150, 151, 205
        green 139, 151, 205–206
maritime *see* transportation
market-based mechanisms *see* Kyoto Protocol
Mediterranean climates 46, 48, 52, 84, 127, 195; *see also* climate, types
metal products *see* manufacturing

metals *see* mining
methane *see* agriculture, emissions; mining, emissions
Milankovitch cycles *see* forcings, orbital
minerals *see* mining
mining 13, 58, 85, 91, 96, 107–108, 110, 112, 113, 116, 119, 135, 138–144, 145, 146, 147, 148, 149, 150, 159, 186, 203, 204, 206
    climate impacts and risks 139–141, 144
    climate-resilient development pathways 142–144, 146
        adaptation pathways 144, 146
        mitigation pathways 142–144
    crude oil 108, 109, 117, 138, 139, 142, 144, 145
    decommissioning 141, 146
    emissions 142–143
        fugitive 12, 142
    employment 138
    environmental management 140, 141
        remediation 141
    footprint 138
    infrastructure 140, 144, 146
    legacy sites 141
    managing risk 139, 140
    metals and minerals
        aluminium ore (bauxite) 135, 151
        black coal 135, 138, 142–143 (*see also* energy, non-renewable)
        brown coal (*see* energy, non-renewable)
        coal 14, 72, 138, 139, 140, 142, 144, 145, 147, 150, 162
        critical minerals 135, 139, 144, 151
        iron ore 108, 135, 138, 151, 205
        lithium 135
        rare earth elements 135, 139, 144, 151
    natural gas (methane) 12, 78, 108, 114, 138–139, 142–143, 145, 150

natural gas *(cont.)*
    operations 140, 146
    relocation 146
    royalties 138
    workplace health and safety 140,
        141, 146
mitigation *see* climate
mitigation pathways *see* climate,
        mitigation
MJO *see* Madden-Julian Oscillation
monsoon 27, 33–34, 43
Montreal Protocol 11, 52
Murray-Darling Basin 41–42, 49,
        52, 91, 95, 122, 126, 134,
        201

N

National Climate Resilience and
        Adaptation Strategy
        78–79, 152, 206
National Electricity Market (NEM)
        109, 111
Nationally Determined
        Contributions *see* Paris
        Agreement
natural capital *see* land
natural forcing *see* forcings
natural gas *see* energy,
        non-renewable
natural systems *see* ecosystems
NDCs *see* Nationally Determined
        Contributions
NEM *see* National Electricity Market
net-zero emissions *see* emissions
nitrous oxide *see* greenhouse
        gases
no regrets practices *see* agriculture
non-renewable energy *see* energy
Northern Australia *see*
        Development of Northern
        Australia Agenda

O

oceans 3, 4, 10, 26, 28, 33, 35,
        49–50, 52–53, 57, 58,
        61, 86, 88, 96, 102, 129,
        137, 181
    acidity 28, 34, 53, 54, 88, 93, 98,
        159, 202–202

deoxygenation 28, 34, 54, 88, 93,
        98, 201–202
heating (warming) xi, 28, 34, 53,
        88, 93, 96, 159 (*see also*
        temperature)
salinity 28, 34, 96
orbital forcing *see* forcings

P

Pacific Decadal Oscillation 40–41
Paleo-reference periods 21–23, 37
Paris Agreement xi, 12, 32, 54, 62,
        63, 69, 71–74, 77, 115, 128,
        142, 156, 203
    nationally determined
        contributions (NDCs)
        71–72
    targets 32, 54, 71–72,
        74–75
pasture *see* agriculture
pathogens *see* biosecurity; human
        health
pathways *see* climate, adaptation;
        climate, mitigation
PDO *see* Pacific Decadal Oscillation
pests *see* biosecurity
planned adaptation *see* climate,
        adaptation
plants *see* biodiversity; biosecurity
pleistocene 21–23, 39, 181
population xi, 7, 30, 31, 62, 65,
        93, 105, 109, 120, 124,
        136, 153, 169–172, 180,
        194–197, 207
    ageing 170–171, 178, 179
    disadvantaged groups 172,
        178–179, 181, 190, 192 (*see*
        also Indigenous)
    distribution 169–172, 185, 187
    fertility rate 170
    growth 169–170
    life expectancy 170
    lower-socio-economic groups
        172, 178, 190, 192
    median age 170–171
    projections 171–172
    sex ratios 170–171
    size 169–170, 186
    urbanised 169, 185
    vulnerability 172, 190, 192, 207

precipitation 27–28, 33–34, 43,
        48–49, 52–53, 91, 93
    rainfall 33, 43, 53, 54, 93, 175,
        201
    snowfall 33, 43, 52, 92
property *see* insurance
protected areas 99, 100, 102
    Indigenous 184

R

radiation balance 9–11
radiative forcing (RF) *see* forcings
rail *see* transportation
rainfall *see* precipitation
rainforests *see* ecosystems,
        terrestrial
Ramsar sites *see* ecosystems,
        freshwater
rare earth elements *see* mining,
        metals and minerals
real estate *see* insurance
Red Centre *see* tourism
reforestation *see* ecosystems, habitat
regional *see* built environments
renewable energy *see* energy
reptiles *see* biodiversity
resilience xii, 58, 60, 61, 65, 67, 71,
        72, 74, 76, 77, 98, 103, 120,
        133, 144, 148, 155, 157,
        165, 172, 182, 189, 192,
        194, 197, 206
    ecological 88, 89, 98–99,
        100–102, 103, 206
    economic 117, 118, 133–134, 135,
        136–137, 144, 146, 156,
        157
    social 182, 189, 190, 191, 192,
        194, 206
restoration *see* habitats
RF *see* radiative forcing
rice *see* agriculture, field crops
risks *see* climate
road *see* transportation
rural *see* built environments

S

safeguard mechanism *see* clean
        energy regulator
salination *see* land

saltwater intrusion *see* land
SAM *see* Southern Annular Mode
scenarios *see* climate
seagrasses *see* ecosystems, marine
sea ice *see* cryosphere
sea level xi, 3, 5, 21–24, 26, 28, 29,
        34, 39–40, 50, 53, 54,
        60, 65, 90, 92, 93, 95, 96,
        97, 98, 101, 118, 119, 127,
        140–141, 158, 159, 161,
        172, 173, 177–178, 182,
        183, 185, 189, 190, 193,
        201, 203
sea surface temperature *see*
        temperature
sea urchins *see* biodiversity
sequestration *see* carbon
settlements *see* built environments
SDGs *see* Sustainable Development
        Goals
Shared Socio-economic Pathways
        (SSPs) 9, 30, 31–32, 37,
        62–63, 174
small-medium enterprises (SMEs)
        118, 147, 153, 154, 162, 163
SMEs *see* small-medium enterprises
snowfall *see* precipitation
Snowy 2.0 Hydro-electricity
        Scheme 116; *see also*
        energy, renewable
social systems *see* built
        environments;
        Indigenous; human
        health
socio-ecological system 61
soils 13, 84, 87, 91, 99, 123, 124, 122,
        131, 132, 180, 205
    carbon 78, 99, 123, 128, 130,
        143, 205
    degradation 88, 123, 124, 126,
        137
solar forcing *see* forcings
Southern Annular Mode 44, 48,
        51–51
species *see* biodiversity
SSPs *see* shared socio-economic
        pathways
SST *see* sea surface temperature
steel *see* manufacturing
storage *see* energy, renewable
storm surges *see* extreme events

streamflow *see* hydrology
structural adaptation *see* climate,
        adaptation
structural adjustment policies
        (SAPs) 107
sugar cane *see* agriculture, field
        crops
supply chains 111, 121, 139–141,
        145, 150, 152, 159, 185,
        202, 204–205
sustainable buildings *see*
        construction
Sustainable Development Goals
        (SDGs) 6, 57, 62, 65–68,
        75–77, 88, 102–103, 107,
        118, 120, 135–137, 147–
        148, 153, 155, 164–165,
        172, 192, 196–198, 207
    goals 75–76, 102, 120, 136–137,
        148, 155, 165, 196–197
    indicators 75–76, 102, 120,
        136–137, 148, 155, 165,
        196–197
    targets 75–76, 102, 120, 136–137,
        148, 155, 165, 196–197

T

temperature
    average 48, 53, 92, 93, 141, 201
    extreme 48, 53, 54, 141
    global mean surface air
        temperature (GSAT)
        32–33
    global mean surface
        temperature (GMST)
        21–24, 27, 201
    ocean (*see* oceans)
    sea surface temperature (SST)
        41, 43, 49, 51, 53
threats *see* climate
tourism xii, 52, 58–59, 64, 83, 84,
        86, 96, 105, 107, 110,
        113, 124, 154, 156–159,
        161–164, 165, 169, 185,
        202, 203, 204, 206
    climate impacts and risks
        157–159
    climate-resilient development
        pathways 159–164
        adaptation pathways 163–164

    mitigation pathways 162–163
    destinations 156–157, 158–159
    emissions 157, 159, 161–162, 204
    employment 154, 156
    managing risk 156–157, 163
    Red Centre 157–158, 163–164,
        169 (*see also* World
        heritage sites)
    tourism infrastructure 157–158,
        163
    travel decisions 157–158
        tourism climatic index (TCI)
        157–158
    visitor numbers 154, 158
    visitor health and well-being 157
transformational adaptation *see*
        climate, adaptation
transformative adaptation *see*
        climate, adaptation
transmission grid *see* energy
transportation 13, 14, 58, 73, 74,
        105, 107, 108–119, 120,
        124, 140–142, 149, 153,
        157, 159, 161–162, 172,
        175, 183, 186, 187, 198,
        203, 204, 206
    aviation 17, 20, 73, 110, 112, 117,
        156, 157, 161–162, 204
    climate impacts and risks
        111–112, 175
    climate-resilient development
        pathways 112–118
        adaptation pathways 117–118
        mitigation pathways 113,
        115–117
    emissions 14, 112–113, 161–162
    employment 110
    managing risk 110–111
    maritime 110, 157, 161–162
    rail 14, 109–110, 111, 112, 116,
        124, 140, 141, 156, 173,
        175, 178, 185, 204
    renewables (*see* energy)
    road 14, 31, 65, 100, 106, 109,
        111, 112, 114, 116–117,
        140, 141, 157, 173, 185, 196
    routes and networks 38, 111,
        112, 116, 118, 124, 139,
        140–141, 144
    transport infrastructure 112,
        116, 118, 186

travel decisions *see* tourism
tree crops *see* agriculture, horticulture
tropical cyclones *see* cyclones

U

United Nations 2030 Agenda for Sustainable Development *see* Sustainable Development Goals (SDGs)
United Nations Framework Convention for Climate Change (UNFCCC) 6, 9, 57, 68–69; *see also* Kyoto Protocol Paris Agreement
urban environments *see* built environments

V

value chains *see* supply chains
vertebrates *see* biodiversity; biosecurity
viticulture *see* agriculture
volcanic eruptions *see* volcanic forcing
volcanic forcing *see* forcings
vulnerability *see* climate

W

Walker Circulation *see* El Niño-Southern Oscillation (ENSO)
warming *see* temperature
water cycle *see* hydrology
water quality *see* hydrology
water security *see* hydrology
weeds *see* biosecurity
well-being *see* human health
wetlands *see* ecosystems, freshwater
wet tropics of Queensland *see* World heritage sites
wheat *see* agriculture, field crops
wine grapes *see* agriculture, viticulture
woody weeds *see* biosecurity
World heritage sites xi, 6, 58, 83, 85, 86, 89, 90, 92, 93, 95, 96, 157, 159–161, 202

Y

yields *see* agriculture